Oil and Oilseed Processing

About the IFST Advances in Food Science Book Series

The Institute of Food Science and Technology (IFST) is the leading qualifying body for food professionals in Europe and the only professional organization in the UK concerned with all aspects of food science and technology. Its qualifications are internationally recognized as a sign of proficiency and integrity in the industry. Competence, integrity, and serving the public benefit lie at the heart of the IFST philosophy. IFST values the many elements that contribute to the efficient and responsible supply, manufacture, and distribution of safe, wholesome, nutritious, and affordable foods, with due regard for the environment, animal welfare, and the rights of consumers. IFST Advances in Food Science is a series of books dedicated to the most important and popular topics in food science and technology, highlighting major developments across all sectors of the global food industry. Each volume is a detailed and in-depth edited work, featuring contributions by recognized international experts, and which focuses on new developments in the field. Taken together, the series forms a comprehensive library of the latest food science research and practice, and provides valuable insights into the food processing techniques that are essential to the understanding and development of this rapidly evolving industry. The IFST Advances series is edited by Dr. Brijesh Tiwari, who is Senior Research Officer at Teagasc Food Research Centre in Ireland.

Forthcoming titles in the IFST series

Recent Advances in Micro- and Macroalgal Processing: Food and Health Perspectives edited by Gaurav Rajauria and Yvonne V. Yuan.

Oil and Oilseed Processing

Opportunities and Challenges

Edited by

Tomás Lafarga
University of Almería, Almería
Spain

Gloria Bobo
IRTA, Lleida
Spain

Ingrid Aguiló-Aguayo
IRTA, Lleida
Spain

WILEY Blackwell

This edition first published 2021

Registered Office(s)
John Wiley & Sons, Inc., 111 River Street, Hoboken, NJ 07030, USA
John Wiley & Sons Ltd, The Atrium, Southern Gate, Chichester, West Sussex, PO19 8SQ, UK

Editorial Office
The Atrium, Southern Gate, Chichester, West Sussex, PO19 8SQ, UK
For details of our global editorial offices, customer services, and more information about Wiley products visit us at www.wiley.com.

Wiley also publishes its books in a variety of electronic formats and by print-on-demand. Some content that appears in standard print versions of this book may not be available in other formats.

Library of Congress Cataloging-in-Publication Data

Names: Lafarga, Tomás, editor. | Bobo, Gloria, editor. | Aguiló-Aguayo, Ingrid, 1981– editor.
Title: Oil and oilseed processing : opportunities and challenges / edited by Toma's Lafarga, Gloria Bobo, Ingrid Aguiló-Aguayo.
Description: Hoboken, NJ : Wiley-Blackwell, [2021] | Series: IFST advances in food science | Includes bibliographical references and index.
Identifiers: LCCN 2020049757 (print) | LCCN 2020049758 (ebook) | ISBN 9781119575276 (cloth) | ISBN 9781119575269 (adobe pdf) | ISBN 9781119575337 (epub)
Subjects: LCSH: Oilseeds. | Oilseed products.
Classification: LCC TP680 .O355 2021 (print) | LCC TP680 (ebook) | DDC 665/.2–dc23
LC record available at https://lccn.loc.gov/2020049757
LC ebook record available at https://lccn.loc.gov/2020049758

Cover Design: Wiley
Cover Image: © Amarita/Getty Images

Set in 9.5/11.5pts Times Ten by SPi Global, Pondicherry, India

10 9 8 7 6 5 4 3 2 1

Contents

Preface

Oils are essential components and the largest source of commercial oils is oilseeds, which include the seeds of annual plants. This book comprises various segments – from the evaluation of different oil sources to aspects related to quality and safety, oil and oilseed processing and applications. The production of oil from oilseeds has been steadily increasing over the past 20 years. This book provides an intense review of the oilseeds and oils in the human diet and surveys current and future trends. The book also reviews vegetable and animal-derived oils together with novel sources for oil production that are of interest for industrial applications because of their similarity to plant oils and for their economic overproduction of seed oils and nutritional value. Safety concerns regarding fungal occurrence and mycotoxin, pesticide, and heavy metals contamination in agri-food commodities is an important issue. The most common contaminants detected in oil and oilseed and current legislation on mycotoxin, pesticide, and heavy metal contamination of foods is discussed. In addition, as the maximum allowed concentration of some of these compounds is extremely low, their detection and determination is investigated, focusing on the use of chromatographic techniques and mass spectrometry. Current applications and future perspectives for the utilization of these techniques in the food and non-food industry are presented. Finally, safety concerns regarding oil and oilseed processing and waste valorization is an issue of great concern and is discussed in depth in this book.

The authors of the chapters are international leading experts in the fields covered in the book and we would like to thank all of them for their valuable contributions.

Tomás Lafarga
Gloria Bobo
Ingrid Aguiló-Aguayo
Editors

List of Contributors

Ingrid Aguiló-Aguayo, IRTA, Postharvest Pogramme, Edifici Fruitcentre, Parc Científic i Tecnològic Agroalimentari de Lleida, Lleida, Spain

Gerard Aragonès, Universitat Rovira i Virgili, Departament de Bioquímica i Biotecnologia, Nutrigenomics Research Group, Tarragona, Spain

Javier Ávila-Román, Universitat Rovira i Virgili, Departament de Bioquímica i Biotecnologia, Nutrigenomics Research Group, Tarragona, Spain

Gloria Bobo, IRTA, Postharvest Pogramme, Edifici Fruitcentre, Parc Científic i Tecnològic Agroalimentari de Lleida, Lleida, Spain

Francisca Isabel Bravo, Universitat Rovira i Virgili, Departament de Bioquímica i Biotecnologia, Nutrigenomics Research Group, Tarragona, Spain

América Chávez-Martínez, Faculty of Zootechnics and Ecology, Autonomous University of Chihuahua, Chihuahua, México

Gemma Echeverria, IRTA, Postharvest Program, Edifici Fruitcentre, Parc Científic i Tecnològic Agroalimentari de Lleida, Lleida, Catalonia, Spain

María Antonia Flores-Córdova, Faculty of Agrotechnological Sciences, Autonomous University of Chihuahua, Chihuahua, México

Jan Fritsche, Institute of Safety and Quality of Milk and Fish Products, Max Rubner-Institute, Federal Research Institute of Nutrition and Food, Kiel, Germany

Nirupama Gangopadhyay, Kusuma School of Biological Sciences, Indian Institute of Technology Delhi, New Delhi, India

Marco Garcia-Vaquero, School of Agriculture and Food Science, University College Dublin, Dublin, Ireland
TEAGASC, Food Research Centre, Dublin, Ireland

Antonia Garrido-Frenich, Department of Chemistry and Physics, Analytical Chemistry Area, Research Centre for Mediterranean Intensive Agrosystems and Agri-Food Biotechnology (CIAIMBITAL), Agrifood Campus of International Excellence ceiA3, University of Almería, Almería, Spain

Editha Giese, Institute of Safety and Quality of Milk and Fish Products, Max Rubner-Institut, Federal Research Institute of Nutrition and Food, Kiel, Germany
Department of Life Sciences, Hamburg University of Applied Sciences (HAW Hamburg), Hamburg, Germany

Jordi Giné-Bordonaba, IRTA, Postharvest Program, Edifici Fruitcentre, Parc Científic i Tecnològic Agroalimentari de Lleida, Lleida, Catalonia, Spain

Andreu Gual-Grau, Universitat Rovira i Virgili, Departament de Bioquímica i Biotecnologia, Nutrigenomics Research Group, Tarragona, Spain

Néstor Gutiérrez-Méndez, Faculty of Chemical Sciences, Autonomous University of Chihuahua, Chihuahua, México

Douglas G. Hayes, Department of Biosystems Engineering and Soil Science, University of Tennessee, Knoxville, TN, USA

Maria Hayes, Food BioSciences Department, Teagasc Food Research Centre, Ashtown, Dublin, Ireland

José L. Hidalgo-Ruiz, Department of Chemistry and Physics, Analytical Chemistry Area, Research Centre for Mediterranean Intensive Agrosystems and Agri-Food Biotechnology (CIAIMBITAL), Agrifood Campus of International Excellence ceiA3, University of Almería, Almería, Spain

Brijesh K. Tiwari, TEAGASC Food Research Centre, Dublin, Ireland

Tomás Lafarga, Department of Chemical Engineering, University of Almería, Almería, Spain

Chloe Leclerc, IRTA, Postharvest Program, Edifici Fruitcentre, Parc Científic i Tecnològic Agroalimentari de Lleida, Lleida, Catalonia, Spain

José Luis Martínez-Vidal, Department of Chemistry and Physics, Analytical Chemistry Area, Research Centre for Mediterranean Intensive Agrosystems and Agri-Food Biotechnology (CIAIMBITAL), Agrifood Campus of International Excellence ceiA3, University of Almería, Almería, Spain

Begoña Muguerza, Universitat Rovira i Virgili, Departament de Bioquímica i Biotecnologia, Nutrigenomics Research Group, Tarragona, Spain

Miquel Mulero, Universitat Rovira i Virgili, Departament de Bioquímica i Biotecnologia, Nutrigenomics Research Group, Tarragona, Spain

Iolanda Nicolau-Lapeña, Food Technology Department, University of Lleida (UDL), Lleida, Spain

Ramona Pérez-Leal, Faculty of Agrotechnological Sciences, Autonomous University of Chihuahua, Chihuahua, México

María Janeth Rodríguez-Roque, Faculty of Agrotechnological Sciences, Autonomous University of Chihuahua, Chihuahua, México

Agustí Romero, IRTA, Oliviculture, Oil Science, and Nuts, Mas de Bover, Constantí, Tarragona, Spain

Roberto Romero-González, Department of Chemistry and Physics, Analytical Chemistry Area, Research Centre for Mediterranean Intensive Agrosystems and Agri-Food Biotechnology (CIAIMBITAL), Agrifood Campus of International Excellence ceiA3, University of Almería, Almería, Spain

Nora Aideé Salas-Salazar, Faculty of Agrotechnological Sciences, Autonomous University of Chihuahua, Chihuahua, México

Rogelio Sánchez-Vega, Faculty of Zootechnics and Ecology, Autonomous University of Chihuahua, Chihuahua, México

Mayra Cristina Soto-Caballero, Faculty of Agrotechnological Sciences, Autonomous University of Chihuahua, Chihuahua, México

Manuel Suárez, Universitat Rovira i Virgili, Departament de Bioquímica i Biotecnologia, Nutrigenomics Research Group, Tarragona, Spain

Juan Manuel Tirado-Gallegos, Faculty of Zootechnics and Ecology, Autonomous University of Chihuahua, Chihuahua, México

Cristina Torres-Fuentes, Universitat Rovira i Virgili, Departament de Bioquímica i Biotecnologia, Nutrigenomics Research Group, Tarragona, Spain

1 Production and Consumption of Oils and Oilseeds

Tomás Lafarga
Department of Chemical Engineering, University of Almería, Almería, Spain

1.1 Introduction

Lipids, which are together with proteins and carbohydrates the main constituents of plant and animal cells, can be broadly defined as substances such as fats, oils, or waxes that dissolve in organic solvents but not in water. Lipids, oil-bearing nuts, and animal fats have been used by man for centuries. These were used for lighting, as cosmetics, applied to weapons or other utensils, and as foods, mainly as sources of energy or as medicines. Nowadays, the use of oils has expanded to several food applications which include cooking, frying, baking shortenings, salad dressing, and flavor carriers, among others. Each of these applications requires oils with specific physical and chemical properties (Gupta 2017). These properties will therefore determine the quality of the vegetable oil, which starts with high quality oilseeds or oil-bearing nuts or fruits. Several factors which include maturity of the oilseed, climatic conditions, harvest conditions, handling of the harvested raw material, and storage can significantly affect the quality of the end product.

Broadly, oil extraction from seeds or beans is achieved by pressing and/or extraction using an organic solvent, generally hexane. Mechanical and thermal pre-treatments can also be used to improve extraction yields (Savoire et al. 2013). Other oils such as avocado, palm, or olive oil are obtained after pressing of the whole fruit. Although some oils such as olive oil are used without further treatment than filtering, most of the currently commercialized edible oils are refined in some measure prior to commercialization (Gunstone 2011b). The term refining refers to the removal of several major and minor impurities and its main goal is to produce high quality oils with optimal properties to satisfy the different oil applications.

Oil production from oilseeds has been steadily increasing by an average of 12.3% annually over the last three decades: for the main oils – palm (palm kernel plus palm

Oil and Oilseed Processing: Opportunities and Challenges, First Edition.
Edited by Tomás Lafarga, Gloria Bobo, and Ingrid Aguiló-Aguayo.

oil), rapeseed, soybean, and sunflower seed oil – there is a production increase of 231.1%, from 45.73 Mt in 1990 to 151.42 Mt in 2017 (UN 2019). Despite recent advances in oil extraction techniques, this impressive increase in oil production could not have been achieved without an increase in the quantity of seeds harvested. Indeed, during the same time period, the production of palm fruits, rapeseeds, soybean, and sunflower seeds followed the same trend from 159.29 Mt in 1990 to 794.31 Mt in 2017, representing an 17.7% annual increase (UN 2019).

The aim of the current chapter is to give an overview of the production of the most relevant vegetable oils and oilseeds, focusing on soybeans, rapeseeds, sunflower seeds, palm fruits, and the oils derived thereof. This chapter will also discuss the production quantities and trade of other common oilseeds such as groundnuts, maize, or sesame seeds as well as novel sources for oil production that are expected to soon gain an increased industrial relevance.

1.2 Oilseeds and Oils: Production and Trade

The majority of the currently utilized vegetable oils are obtained from seeds or beans. Sources for vegetable oil extraction can be divided into three main categories: (i) those plants that are not grown for oil production, where oil can be considered as a by-product; (ii) those crops that cannot be changed on a yearly basis, generally trees; and (iii) those crops that are planted annually such as sunflower, linseed, or rape (Gunstone 2011a). Currently, from the total oil and fat production, approximately 14% is utilized as a starting material for the oleochemical industry, 79–80% is used for human food as spreads, frying oils, or salad oils, and the remaining 6% is used as animal feed, and therefore, indirectly used for human food production (Gunstone 2011a). Crushing of soybeans or oilseeds into cake and oil dominates total usage and it is expected that 90% of the world soybean production and 86% of world production of other oilseeds will be crushed in 2027 (FAO 2018). The current section will discuss current applications as well as the production and trade of the most important edible oils and oil sources.

1.2.1 Copra and Coconut Oil

Coconut oil is obtained from the fruit of the coconut palm (*Cocos nucifera* L.), a tropical plant normally utilized for edible and non-edible purposes. Two major types of oil can be obtained from coconut: (i) coconut or copra oil, which is obtained from the dried coconut flesh, also known as copra, and (ii) virgin coconut oil, obtained from fresh coconut flesh (Zuknik et al. 2016). Copra oil can be found either unrefined or refined, bleached, and deodorized (Kumar and Krishna 2015).

As a portable source of water (and food), coconuts played an important role in the ability of humans to voyage and colonize regions throughout the tropics (Gunn et al. 2011). Currently, coconut is an important crop in tropical countries where it plays an important role in diets and livelihoods (MacDonald et al. 2018). A large number of food products containing coconut oil have been launched into the market during recent years (Table 1.1).

Besides being used for frying or as an ingredient in baked goods, coconut oil is currently used for pharmaceutical and cosmetic applications, among others (Kumar and

Table 1.1 Foods manufactured using vegetable (and microalgal) oils.

Product	Brand	Company	Country of commercialization	Oil(s) used
Sour cream and onion flavored potato snack	Pringles	Pringles, USA	Argentina	Cottonseed and coconut oil
Roasted and salted peanuts	Pami	Pami, Greece	Croatia	Cottonseed and sunflower oil
Corn "tostadas"	Mamá Lycha	Productos Mamá Lycha, USA	Canada	Cottonseed, palm, and/or soybean oil
Chocolate cake with chocolate topping	Betty Crocker Mug Treats	General Mills, USA	Saudi Arabia	Cottonseed, palm, sunflower, and rapeseed oil
Tomato soup	Campbell's Soup at Hand	Campbell, Canada	Canada	Canola, corn, cottonseed, and or/soybean oil
Hazelnut cream with cocoa and breadsticks	Nut Bari	DKC Grup Gida San, Turkey	Turkey	Sunflower, cottonseed, and palm oil
Spicy vegetable tajine with bulgur, spelt, and raisins	Carrefour Veggie	Carrefour, France	Spain	Cottonseed and extra virgin olive oil
Breakfast cereals	Kellogg's Froot Loops	Kellogg Sales, USA	USA	Coconut, soybean, and cottonseed oil
Cookie covered almonds	7-Select	7-Eleven, USA	USA	Canola, cottonseed, palm kernel, and/or peanut oil
Apple pie	Royal Classic Dutch	De Specialiteitenbakkerij, the Netherlands	Switzerland	Cottonseed, rapeseed, and sunflower oil
Rice mini squares	Kellogg's Rice Krispies Treats Mini Squares	Kellogg Sales, USA	USA	Soybean and palm oil
Sweet kale salad	Eat Smart Salad Shake Ups	Apio, USA	USA	Canola, soybean, and sunflower oil

(*Continued*)

Table 1.1 (Continued)

Product	Brand	Company	Country of commercialization	Oil(s) used
Mr Kipling lemon whirls	Mr Kipling	Premier Foods, Ireland	Ireland	Rapeseed and palm oil
Cookies drizzled in chocolate	Bake Shop Bites	Cookies United, USA	Puerto Rico	Palm kernel oil, hydrogenated palm oil, and soybean oil
Special bread tapas biscuits	Quely	Quely, Spain	Spain	Sunflower and olive oil
Taralli biscuits	Continental Taralli Biscuits	Continental Taralli Biscuits, Australia	Australia	Olive oil
Quinoa salad	Pur	PUR Bio Feinkost Manufaktur – Heinz Gierze, Germany	Germany	Sunflower and olive oil
Tuna salad	The Yummy Tummy Co,	The Yummy Tummy Company, UK	UK	Olive oil
Artichoke pesto	Casa de la Torre	Converfrut, Argentina	Argentina	Olive oil
Infant formula	Enfamil Neuro Pro EnfaCare	Mead Johnson, USA	USA	Sunflower, soybean, coconut, and *Schizochytrium* sp. oil
Apple drink	Life Mix	W Pesquisa, Tecnologia e Indústria de Alimentos, Brazil	Brazil	*Schizochytrium* sp. oil
Evaporated milk	Laive Niños	Laive, Peru	Peru	*Mortierella alpine* and *Schizochytrium* sp. oil

Source: Data accessed on June 7, 2019 from MINTEL, available at http://www.mintel.com.

Krishna 2015). Over 12.30 Mha of coconut are currently planted across 89 tropical countries, mainly in Asia (Gunn et al. 2011) – Figure 1.1. Coconut plantations can now be found on practically every tropical and sub-tropical coastline worldwide, distributed mainly in coastal regions between 20°N and 20°S (Ramanatha Rao et al. 2005). Coconuts are even cultivated far from the coast in many regions with adequate rainfall and altitude (Foale 2005). Over 95% of the farmers who grow coconuts are smallholders tending less than four hectares and lack the resources to invest in technologies that would improve production yields (Batugal et al. 2005). However, production and consumption of coconut oil is expected to increase in coming years and production of virgin coconut oil shows potential for improving coconut farm incomes by five to eightfold over traditional copra production (Bawalan and Chapman 2006).

Production of coconut oil during 2014 was estimated as 3.10 Mt (UN 2019). During the period 2018/2019, coconut oil production was reported to be 3.63 Mt (USDA 2019). Approximately 2.70 Mt of coconut oil were produced in Asia, representing approximately

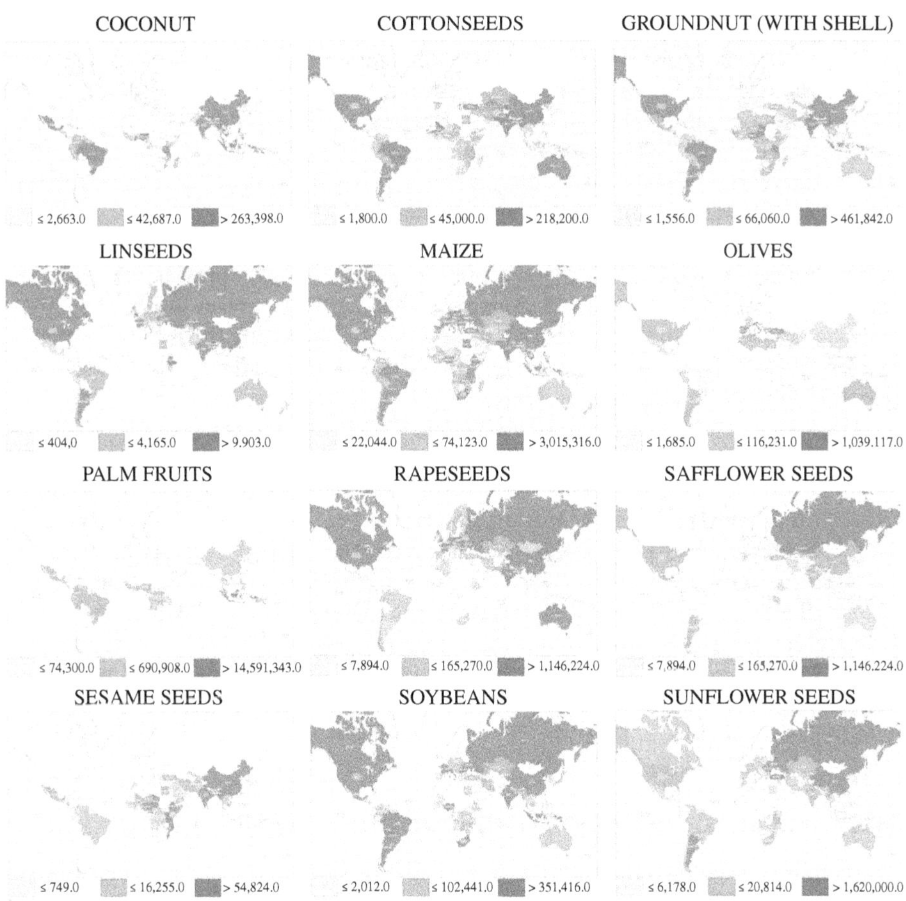

Figure 1.1 World production of palm fruit, olives, and most economically important oilseeds. Values represent tonnes produced during 2017. *Source:* Data accessed on April 8, 2019 from FAOSTAT, the Food and Agriculture Organization Corporate Statistical Database, available at http://www.fao.org/faostat.

88.4% of the total production. Largest coconut oil producers are the Philippines, Indonesia, and India with 1.10, 0.91, and 0.39 Mt, respectively. The Philippines and Indonesia are also the biggest coconut exporters with approximately 0.75 and 0.60 Mt of coconut oil exported during 2016, respectively. Moreover, the US and the Netherlands are the major coconut oil importers followed by Malaysia and Germany with approximately 0.51, 0.34, 0.15, and 0.15 Mt of coconut oil imported during 2017. Global copra production is expected to increase by 2%, mainly due to expanding coconut plantations in Vietnam and improving yields in the Philippines (USDA 2017).

1.2.2 Cottonseeds and Cottonseed Oil

Cotton, which is obtained from the cotton plants of the genus *Gossypium* in the family *Malvaceae*, is a profit crop for over 20 M farmers in developing countries, mainly cultivated to meet the basic requirements for cotton fabrics (Saxena et al. 2011). Cottonseed is a valuable by-product of the cotton plant: for every 1 kg of cotton fiber, 1.65 kg of cottonseeds are produced (Saxena et al. 2011). The cottonseed is used to produce oil, mainly for human consumption, and oilseed cake for animal feed. Cottonseed oil fits within the previously mentioned group of plants that are not grown for oil production, and where oil can be considered as a by-product. However, besides being a "by-product," cottonseed oil dominated the US vegetable oil market for almost 100 years (O'Brien et al. 2005). Table 1.1 lists some products containing cottonseed oil that were recently launched into the market.

Cottonseeds were the fifth most produced commodity within the "processed crops" classification of FAOSTAT, surpassed only by barley, sugar, molasses, and palm oil. World production of cottonseeds was measured as 46.98 Mt in 2014 (UN 2019). Although cottonseed production decreased significantly during the period 2015/2016 to 37.76 Mt, it is now slowly recovering and reached 43.39 Mt produced during the period 2018/2019 (USDA 2019a). Major cottonseed producers were China, India, and the US with a production of 12.32, 12.30, and 4.64 Mt. Cottonseed oil production is largely concentrated in India, the US, Pakistan, and China (FAO 2018). In the EU, cotton is produced only in three Member States on around 300 000 ha, mainly in Greece (approximately 80% of European production), Spain (20%, mainly grown in the region of Andalucía), and Bulgaria that produces cotton on less than 1 ha. Globally, major cottonseed exporters are the US, Greece, Australia, and Brazil with average exports of 182.12, 176.42, 166.96, and 95.45 kt/year. Despite being one of the most produced oilseeds, cottonseed oil production during the period 2018/2019 was reported to be 5.17 Mt, significantly lower than the amount of soybean, rapeseed, palm, or sunflower seed oil, which were 57.07, 27.78, 73.49, and 19.49 Mt during the same period (USDA 2019a).

1.2.3 Groundnuts and Groundnut Oil

Groundnut or peanut (*Arachys hypogaea* L.) is one of the most important oilseed crops in the world. Groundnut consumption as well as commercial products containing groundnuts vary in large proportions as peanuts have been developed into a large number of products like roasted peanuts, peanut butter, peanut paste, peanut flour, peanut "milk," or peanut cheese analogues (Arya et al. 2016). In Western countries, groundnuts are mainly used for direct consumption as salted and roasted nuts. In the

US, groundnuts are also utilized for making peanut butter and confectionary. In turn, in China and India groundnuts are primarily crushed for oil production (Pandey et al. 2014).

The cultivation of peanuts, which was originated in South America, has expanded and groundnuts are currently being produced in over 100 countries of Asia, Africa, and the Americas (Pandey et al. 2014). Over 70% of the groundnut growing area comes under arid and semi-arid regions (Shasidhar et al. 2017). Figure 1.1 shows the annual production of groundnuts per country during 2017.

World production during 2017 totaled approximately 47.09 Mt, which represents an 18% increase from the 39.9 Mt produced in 2013 (Torres et al. 2014). World area harvested with groundnuts has also increased over the last decades and is currently over 27.94 Mha. This trend is likely to continue to grow. Major groundnut producers are China, India, and the US with 17.09, 9.18, and 3.28 Mt produced in 2017. The US is one of the biggest groundnut exporters, with 0.36 Mt exported in 2016. Argentina, which was the seventh major producer during 2017, was the third major exporter during that same year with 0.29 Mt of groundnuts exported. Argentina, the US, Sudan, Senegal, and Brazil account for over 71% of total world exports (Torres et al. 2014). Other countries like Vietnam, India, and several African countries periodically enter the world market depending on market demands and their crops quality (Torres et al. 2014). Indeed, in 2016, India was the major exporter with 0.61 Mt of groundnuts exported. Moreover, major groundnut importers in 2017 were the Netherlands, China, Indonesia, and Mexico with 0.32, 0.19, 0.14, and 0.13 Mt respectively. Peanut imports into the EU, Canada, and Japan account for approximately 78% of the world's imports (Torres et al. 2014). Approximately 5.03 Mt of groundnut oil were produced in 2014, mainly in Asia (69.4%), Africa (23.8%), and the Americas (4.5%). The top five peanut oil producers are China, India, Nigeria, Myanmar, and Sudan with an approximate production of 1.87, 1.25, 0.27, 0.21, 0.15 Mt respectively.

1.2.4 Linseed

Flax (*Linum usitatissimum* L.) belonging to the family *Lineaceae* is a blue flowering herb that produces small flat seeds, which vary from golden yellow to reddish brown and have a nutty flavor (Kajla et al. 2015). Flaxseeds, also known as linseeds, are an economically important oilseed crop used for industrial, food, and feed purposes. Generally the term flaxseed refers to the seed grown for fiber production (linen), while linseed refers to the oilseed grown for industrial and food uses. In the current chapter, both terms will be used indistinctively.

Linseeds are currently being cultivated in more than 50 countries, predominantly in the Northern hemisphere as shown in Figure 1.1. The harvested area in 2017 amounted to 2.78 Mha representing a 14.04% annual increase during the last decade. For example, in the EU, the area harvested increased from 52 000 ha in 2008 to 70 000 ha in 2019, representing an average annual increase of 13.5% (European Commission 2019). World production of linseed increased from 1.98 Mt in 2000 to around 2.79 Mt in 2017. At the beginning of the century, Canada was the world's largest producer and exporter of flax, and represented about 40% of the world's production and 75% of the global flax trade (Oomah 2001). Currently, Canada produces 0.51 Mt of linseed per year and ranks third in the list of top producers, surpassed by Kazakhstan and Russia, which produced 0.68 and 0.61 Mt, respectively. Currently, most of the linseed currently produced is grown in Asia followed by Europe and the Americas, with production shares of 44.7, 28.6, and 22.7%, respectively. Canada is still the biggest linseed exporter, followed by Russia, Kazakhstan, Belgium, and the US. Exports of these countries during 2017 were approximately 0.62, 0.61, 0.27, 0.18, 0.03 Mt, respectively. Major linseed importers were Belgium, China, Turkey, and Germany with 0.58, 0.47, 0.19, and 0.15 Mt of seeds imported during 2017.

Belgium, which imports large amounts of linseeds is one of the world's most important linseed oil producers. In 2014, 0.12 Mt of linseed oil were produced in Belgium, only surpassed by China with an annual production of 0.19 Mt third, fourth, and fifth major linseed oil producers were the US, Turkey, and Germany. Top five linseed oil exporters during 2016 were Belgium, Germany, Turkey, Kazakhstan, and Gambia. In turn, Gambia was the major linseed oil importer followed by China, Germany, the Netherlands, and the UK.

1.2.5 Maize

Maize (*Zea mays* L.), also known as corn, is a cereal grain from the family *Gramineae*, which includes plants such as wheat, rice, or sugarcane. Cereal grains are key-importance crops as they provide humans with more nourishment than any other food class and nearly 50% of the total caloric requirement (Ranum et al. 2014). Maize was first domesticated by indigenous peoples in the mid- to low-land regions of what today is known as Mexico, around 9000 years before present (van Heerwaarden et al. 2011). Today, maize is a staple food in many parts of the world, with the total production of maize surpassing that of other staples including wheat (UN 2019). Table 1.1 lists some products recently launched into the market manufactured using maize oil. Production of maize showed an important increase in both area harvested and total production over the last decade. Currently, world area harvested is approximately 197.15 Mha, showing an important increase from 2007 when approximately 158.67 Mha of maize were harvested. World production of maize was 1134.74 Mt in 2017. Maize ranked fifth in the most produced commodities during 2017 according to data accessed from FAOSTAT, only surpassed by "total cereals," rice, sugar cane, and "total coarse grain." Approximately 50.9% of the maize produced during 2017 was cultivated in the Americas, followed by Asia and Europe with production shares of 31.9 and 9.7%, respectively. Largest producers were the US, China, Brazil, Argentina, and India with production quantities of 370.96, 259.07, 97.72, 49.47, and 28.72 Mt in 2017. Recently, the USDA predicted lower maize imports and production for 2018/2019. Larger maize exporters during 2016 were the US, Argentina, Brazil, and Ukraine with 55.99, 24.50, 21.87, and 11.01 Mt, respectively. Increased corn exports in Argentina and Ukraine are expected, partially offset by reductions for South Africa and Mexico (USDA 2019b). Japan, Mexico, the Republic of Korea, and Vietnam were the larger importers during this period with approximately 15.34, 14.10, 9.78, and 8.05 Mt, respectively (UN 2019). As a crop, maize is highly productive and versatile, and responded successfully to investments in research, breeding, and promotion. However, the current corn production system can be considered inefficient, at least at feeding people. Only a small fraction of the maize grown over the world is consumed directly by humans. Most of the maize currently harvested today is being indirectly used for human consumption as animal feed or as raw material for the production of maize-derived products such as corn starch or corn syrup, or used for other industrial purposes such as the production of ethanol. An important part of maize production, approximately 40%, is being used as a raw material for the production of ethanol used as a motor fuel (Ranum et al. 2014). The utilization of maize for producing biofuels has a strong impact on production and trade. Indeed, total corn use was predicted to decrease by 165 M bushels (approximately 4.19 Mt) mainly caused by a 40 M bushel decrease (approximately 1.01 Mt) in the amount of corn used for ethanol production (USDA 2019b).

Only a small amount of the maize currently cultivated is used for oil production. Total maize oil production during 2014 was 3.18 Mt, led by the US, China, Turkey, and Brazil with oil production quantities of 1.82, 0.26, 0.19, and 0.09 Mt. Major oil exporters during 2016 were the US, Turkey, and Spain with approximately 0.47, 0.05, and 0.04 Mt of oil exported. Spain was also one of the largest importers, with 0.09 Mt, only surpassed by Libya, who imported 0.10 Mt of maize oil during 2016. Overall, although human consumption of maize has remained steady over the last years, its global production has slightly but steadily increased mainly caused by an increase in its use for animal feed and for biofuel production.

1.2.6 Olive Oil

The olive tree (*Olea europea* L.) is one of the oldest known cultivated trees in the world and olive oil has become one of the most widely accepted and used oils in culinary applications. The amount of food products manufactured using olive oil increases every year. Some of these are listed in Table 1.1. Olive oil is a staple food for people living in Mediterranean countries but its use is now expanding to other parts of the world. As can be seen in Figure 1.1, the olive tree is currently being cultivated in several countries including Spain, Italy, Greece, Tunisia, Portugal, Turkey, Morocco, Argentina, Australia, and the US, among others. The world production of olives was approximately 20.87 Mt in 2017. Approximately 62% of the total world production of olives was carried out in Europe, with Spain, Greece, and Italy the biggest producers with 6.54, 2.72, and 2.57 Mt/year in 2017, respectively. The EU is the leading producer, consumer, and exporter of olive oil: indeed, in the last five marketing years the EU produced, consumed, and exported 67, 55, and 67% of the worlds' olive oil. The biggest non-European producers were Morocco and Egypt with an annual production of 1.03 and 0.92 Mt in 2017 (UN 2019). Major exporters of olive fruit in 2016 were Portugal, Spain, Mexico, Greece, and Jordan: 24.69, 12.81, 10.72, 4.47, and 3.39 kt, respectively. Portugal was also the most important importer of olives, with approximately 19.25 kt imported during 2016, most of these from Spain (EU 2019). Spain is the biggest producer of olive oil followed by Italy and Greece with a production of 1.73, 0.29, and 0.21 Mt of virgin olive oil during 2014, respectively. These countries were also the most important exporters with approximately 920.82, 354.29, and 161.00 kt exported during 2016. Major non-European olive oil producers are Tunisia and Morocco, followed by Syria with an annual production of approximately 179.70, 137.40, and 100.64 kt, respectively. Tunisia ranks fifth just behind Portugal among the major olive oil exporters with 107.95 and 117.83 kt of oil exported in 2016. During 2018/2019, from the total 124.06 kt of olive oil exported from the EU to non-EU countries, 41.80, 12.80, 9.66, and 5.74 kt were imported by the US, China, Japan, and Canada, respectively (Ferraris 2019). These values show a 22.6, 0.5, 18.4, and 34.4% increase when compared to the period 2017/2018.

1.2.7 Palm and Palm Kernel Oil

The oil palm fruit, which is obtained from the palm tree (*Elaeis guineensis* L.), is a drupe formed in tight bunches. Palm fruit is mainly used for the extraction of palm oil driven by a lower price, and the fact that one hectare of oil palm plantation can produce up to 10 more times oil than other leading oilseed crops (Mba et al. 2015). Palm oil can be divided into crude palm oil, obtained from the fruits mesocarp, and palm kernel oil. Both are key important in world trade. Indeed, palm oil was the fourth most produced commodity within the "processed crops" classification of FAOSTAT, surpassed only by beer of barley, sugar, and molasses. Palm oil production has boomed over the last decades caused by an increasing use as frying oil, as an ingredient in foods and other products such as detergents, and biodiesel production (Pirker et al. 2016). As shown in Figure 1.2, total world production of palm oil increased from 45.78 Mt in the year 2000 to 57.33 Mt in 2014. In 2019, the total world production of palm oil has been calculated as 70.90 Mt, representing a 16.16% annual increase during the last decade.

Malaysia and Indonesia have been the center of this impressive increase. As an example, planted areas increased from 3.37 and 2.01 Mha in 2000 to 5.11 and 9.28 Mha in 2017, which represent an averaged 8.90 and 27.09% annual increase during the period 2000–2017 in Malaysia and Indonesia, respectively. The USDA recently forecasted the Indonesian oil palm harvested area for 2018/2019 as 11.3 Mha. In 2016, Vijay et al. (2016) assessed the impact of oil palm on deforestation and biodiversity

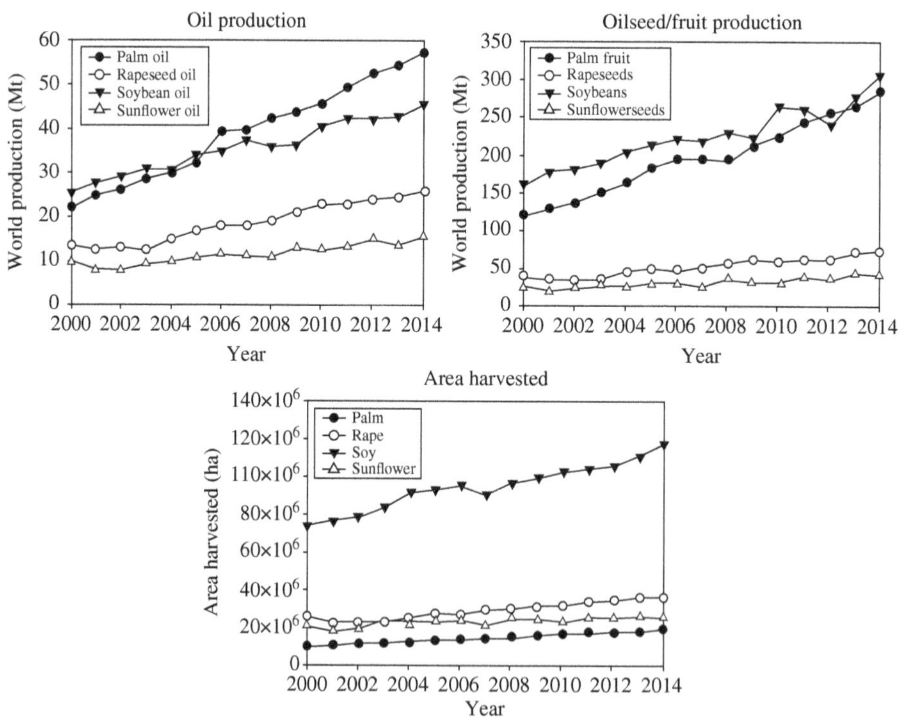

Figure 1.2 Area harvested and production of palm, rapeseed, soybean, and sunflower seed oil. *Source:* Data accessed on April 8, 2019 from FAOSTAT, available at http://www.fao.org/faostat.

loss and reported that 45% of sampled oil palm plantations evaluated in Southern Asia came from areas that were forests in 1989 (less than three decades). Similar conclusions were obtained by Henders et al. (2015), who observed that over 50% of Malaysian and Indonesian oil palm plantations in 2005 were on land that was forest in the 1990s. This has led not only to obvious negative environmental issues, but also to closer scrutiny from consumers. However, as global demand for food increases and land availability in those countries becomes scarcer, governments of developing and emerging countries such as Brazil, Peru, and some African countries are increasingly promoting palm tree cultivation (Pirker et al. 2016). Nowadays, the palm tree is cultivated in several countries with adequate temperature, sunlight, and rainfall (Figure 1.1). However, major palm oil producers continue to be Indonesia, Malaysia, and Thailand with an annual productions in 2017 of approximately 29.27, 19.67, and 1.85 Mt. As shown in Figure 1.3, Malaysia and Indonesia are also the largest palm oil exporters with approximately 22.76 and 13.81 Mt exported during 2017.

Major palm oil importers in 2017 were India, China, and Pakistan with 8.26, 4.48, and 2.60 Mt respectively.

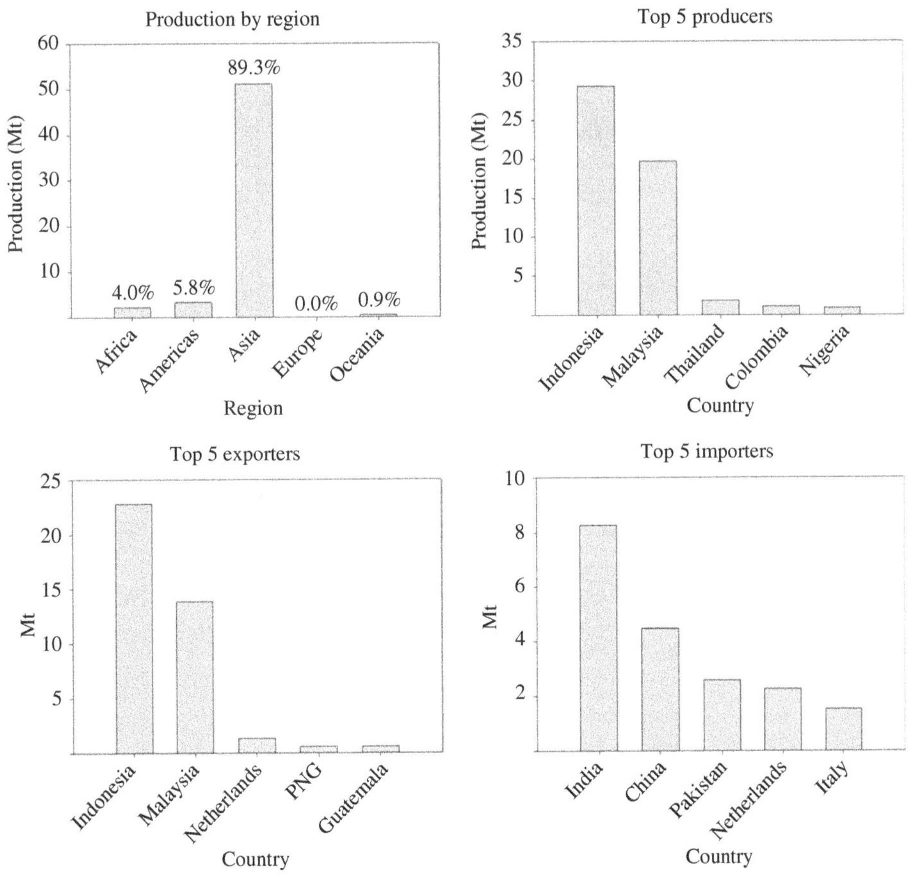

Figure 1.3 Palm oil: Production and trade. Abbreviation: PNG, Papua New Guinea. *Source:* Data accessed on April 8, 2019 from FAOSTAT, available at http://www.fao.org/faostat.

1.2.8 Rapeseed and Canola Oil

The genus *Brassica*, which is the most important within the *Brassicaceae* family, includes the six interrelated species *Brassica nigra*, *Brassica oleracea*, *Brassica rapa*, *Brassica carinata*, *Brassica juncea*, and *Brassica napus* (Cartea et al. 2016). Both rapeseed and canola belong to the Brassicaceae family. In some countries, the term rapeseed oil refers to the oils for industrial use, while canola oil is used to refer to the edible cooking oil. Canola is a subgroup within the rapeseeds and include those Brassica species having seeds containing oil with under 2% erucic acid and low levels of glucosinolates. As an estimate, Daun (2015) recently suggested that, to identify the amount of rapeseed crops that are canola quality, a good estimate can be obtained by assuming that all production in North America, Australia, Scandinavia, and Western Europe is canola quality, while only 10 and 75% of the production in China and Eastern Europe are canola. In the current chapter, all *Brassica* seeds utilized for oil production will be referred to as rapeseeds. Currently, rapeseeds rank second in world production of oilseeds (third if considering palm oil). According to data accessed from the USDA, world production of rapeseed during 2018/2019 was 70.37 Mt (USDA 2019a). Production of rapeseed in the EU increased from 19.02 Mt in 2009 to 19.80 Mt in 2018, with a historical maximum production of 24.26 Mt in 2014 (European Commission 2019). Rapeseed has a long history of use in Europe, Scandinavia, China, and India and has been more recently introduced into Canada and Australia (Daun 2015). Rapeseeds are adapted to the cool extremes of the temperature zones: although the optimum growth temperature is around 10 °C, minimum temperatures for growth have been reported to be near 0 °C (Daun 2015). Major rapeseed growing areas, shown in Figure 1.1, include temperate areas of Europe, Asia, and to a lesser extent, North America. Indeed, 41.5 and 39.3% of the total rapeseed oil production was carried out in Europe and Asia, respectively (Figure 1.4). Rapeseed oil was ninth most produced commodity within the "processed crops" classification of FAOSTAT, and the third most produced oil only surpassed by palm oil and soybean oil, which ranked fourth and sixth, respectively.

In the EU-28 the area harvested increased from 6.17 Mha in 2008 to 6.96 Mha in 2018 (European Commission 2019). As shown in Figure 1.4, biggest producers in 2017 were China, Germany, and Canada (UN 2019). Canada is currently the major rapeseed oil exporter. Indeed, according to data obtained from FAOSTAT, 2.93 Mt of rapeseed oil were exported during 2017 (UN 2019). Although there will continue to be a market for rapeseed oil, mainly for industrial processes because of its high heat stability and for specialty foods, the world trend is to increase the production of canola-quality seeds (Daun 2015). Global rapeseed production is forecast to rise 3.9 Mt on the strength of a 9% increase in are harvested and an increased production in Canada, Ukraine, India, the EU, and the US (USDA 2017).

1.2.9 Sesame Seeds and Sesame Oil

Sesame (*Sesamum indicum* L.) is considered as one of the world's major oilseeds and one of the most ancient crops cultivate by humans (Wan et al. 2015). The origin of sesame has been a matter of discussion for more than a century, but it is now widely accepted that domestication occurred on the Indian subcontinent (Bedigian 2010a). Sesame seeds were a major oilseed in the ancient world mainly because of their stability,

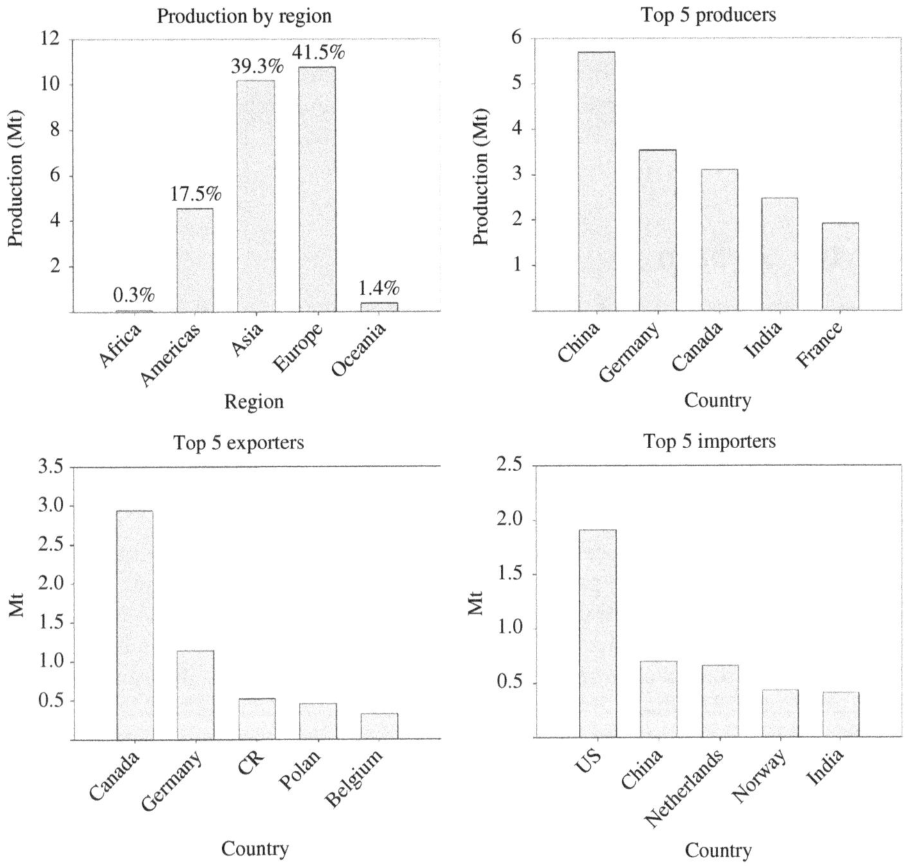

Figure 1.4 Rapeseed oil: Production and trade. Abbreviations: CR, Czech Republic; US, United States of America. *Source:* Data accessed on April 8, 2019 from FAOSTAT, available at http://www.fao.org/faostat.

resistance to drought, and their high oil content. Currently, sesame is grown in varied environments from semiarid tropics and subtropics to temperate regions. World area harvested with sesame in 2017 was calculated as 9.98 Mha, most of it in African and Asian countries (Figure 1.1). During the last decades, world production of sesame seeds has increased steadily Total world production of sesame seed in 2018 was estimated to be 5.53 Mt, most of it being cultivated in Asia (39.7%) and Africa (56.9%). Top five sesame seed producers during 2018 were the United Republic of Tanzania, Myanmar, India, Nigeria, and Sudan, with an estimated production of 805.7, 764.3, 751.0, 550.0, and 550.0 kt respectively. On average, 70% of the worlds' sesame seeds are consumed within producer countries, leaving an approximately 30% excess for global trade (Bedigian 2010b). Major sesame seed exporters during 2016 were Ethiopia, India, Sudan, Nigeria, and Burkina Faso with approximately 382.0, 325.9, 258.54, 172.8, and 159.8 kt respectively. Approximately 65% of the sesame seeds harvested are dedicated for oil production while only 35% are consumed as a food ingredient (Wan et al. 2015). Global sesame oil production was estimated to be 1.63 Mt in 2014. Major sesame oil producers in 2014 were the United Republic of Tanzania, Myanmar, China, India, and Japan with an annual production of 544.3, 318.0, 234.0, 108.0, and 45.44 kt respectively. The top five sesame oil exporters during 2016 were Mexico, Ghana, India, and Japan. In turn, major sesame oil importers were the US, the UK, Australia, Taiwan, and Vietnam with approximately 16.39, 3.52, 3.24, 3.18, and 3.08 kt imported during 2016 respectively.

1.2.10 Soybean

The soybean (*Glycine max* L.) or soya bean is a legume native to East Asia, widely cultivated for its edible bean. It is thought that domestication of soybean started in China between 7000 and 6600 BCE, in Japan between 5000 and 3000 BCE, and in Korea in approximately 1000 BCE (Lee et al. 2011). Currently, cultivation of soybean is successful in climates with hot summers, with optimum growing conditions in mean temperatures of 20–30 °C. Soybean is a globally important crop used for the production of soybean meal and soybean oil, which is used in a large variety of processed foods, some of them listed in Table 1.1.

While soybeans are produced around the world, production and exports are dominated by the America's. The production of soybeans will continue to grow strongly in Latin America, with Argentina and Paraguay producing 66 Mt and 12 Mt by 2027 (FAO 2018). Soybean production is also expected to increase in Ukraine, Russian Federation, and several countries in Africa (FAO 2018). In the EU, the area harvested increased from 0.27 Mha in 2008 to 0.98 Mha in 2018, representing an average annual increase of 36.3% (European Commission 2019). These increased production will result in an estimated 360.6 Mt of soybean produced during 2018/2019, led by the US, Brazil, and Argentina with a forecasted production of 123.7, 117.0, and 55.0 Mt (USDA 2019b). However, soybean production in the US is estimated to decrease by approximately 1.52 Mt for the period 2018/2019 (USDA 2019b) and by 2027, it is expected that Brazil will overtake North America as the world's largest exporter of soybean with a 41.8% rise in its share in the global soybean exports (FAO 2018). The 2018/2019 global soybean outlook of the USDA includes a 1.7 Mt reduction in soybean exports from Brazil, Uruguay, and Paraguay, which are partially offset by the above mentioned higher exports for Argentina. Indeed, the highly developed crushing industry in Argentina, together with a small domestic market, and a lower export tax on processed commodities than on unprocessed commodities favors the export of soybean oil and meal from Argentina.

Soybeans are not only used as a protein source but also for producing soybean oil. Soybean oil is the sixth most produced commodity within the "processed crops" classification of FAOSTAT, surpassed by beer of barley, sugar, molasses, palm oil, and cottonseed. Figure 1.5 shows the world production of soybean oil during 2014.

World production of soybean was calculated as 45.7 Mt, the major producers being China, the US, Brazil, and Argentina with 11.7, 9.7, 7.4, and 7.1 Mt, respectively (UN 2019). Soybean oil trade is expected to further expand mainly due to India's growing demand (USDA 2017).

1.2.11 Sunflower

The common name sunflower refers to the popular annual species *Helianthus annuus* L., grown as food crops for humans, cattle, and poultry or as ornamental plants. The oil content of sunflower seeds is around 50% (w/w) (Salas et al. 2015). Sunflower

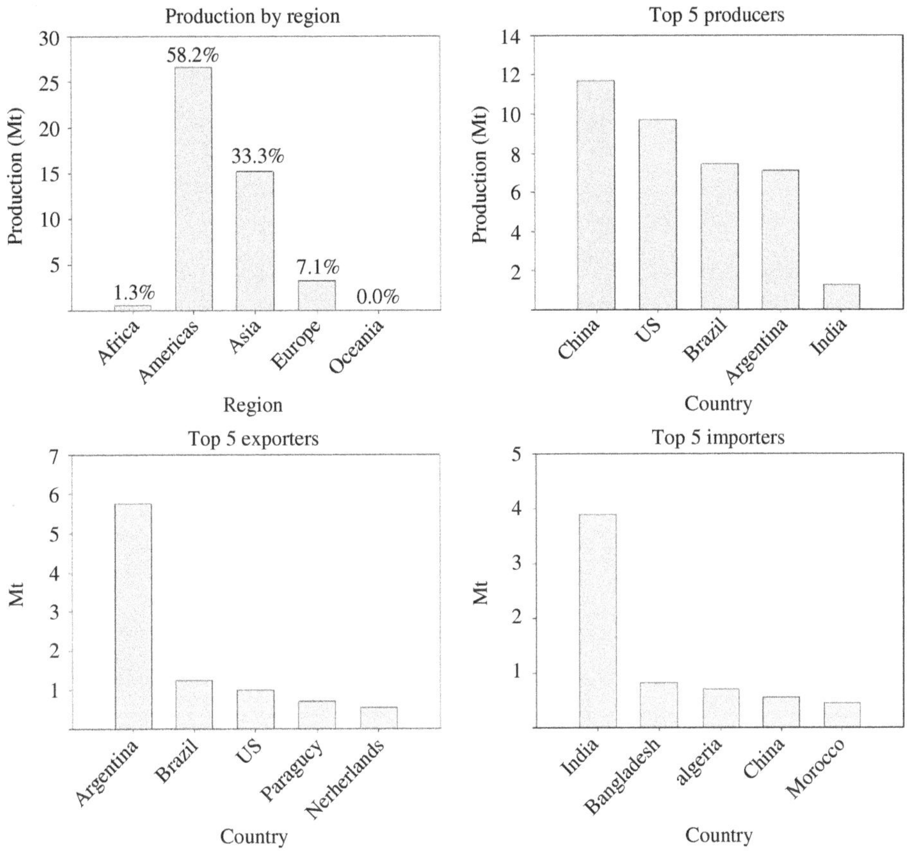

Figure 1.5 Soybean oil: Production and trade. Abbreviation: US, United States of America. *Source:* Data accessed on April 8, 2019 from FAOSTAT, available at http://www.fao.org/faostat.

seeds are crushed for oil recovery via either hard pressing or prepress solvent extraction. The former relies on exerting high pressure on the seeds to separate the oil from the solid fraction obtaining recovery yields close to 25% (Williams 2005). In turn, the latter relies on exerting medium pressure followed by solvent extraction and allows recovering approximately 40% of the seed's oil (Le Chef and Kemper 2015). Crude sunflower oil contains several impurities which are removed by degumming, to remove phosphatides, chemical neutralization to remove free fatty acids and residual phosphatides, and bleaching, which aims at removing colored compounds such as chlorophylls and carotenes. Other processes which include dewaxing and deodorization generally by stripping under vacuum with steam at elevated temperatures are also applied (Dijkstra 2015).

Sunflower grows better at temperate (20–25 °C) and dry climates with high irradiation and deep soils (Salas et al. 2015). Argentina was the biggest sunflower seed and oil producer until the late twentieth century, when it was surpassed by Ukraine and Russia after important economic and technological reforms (Feoll and Ingaramo 2015). Currently, over 75% of the sunflower oil commercialized worldwide is made in Europe (Figure 1.6). The total area dedicated to cultivate sunflower seeds increased from

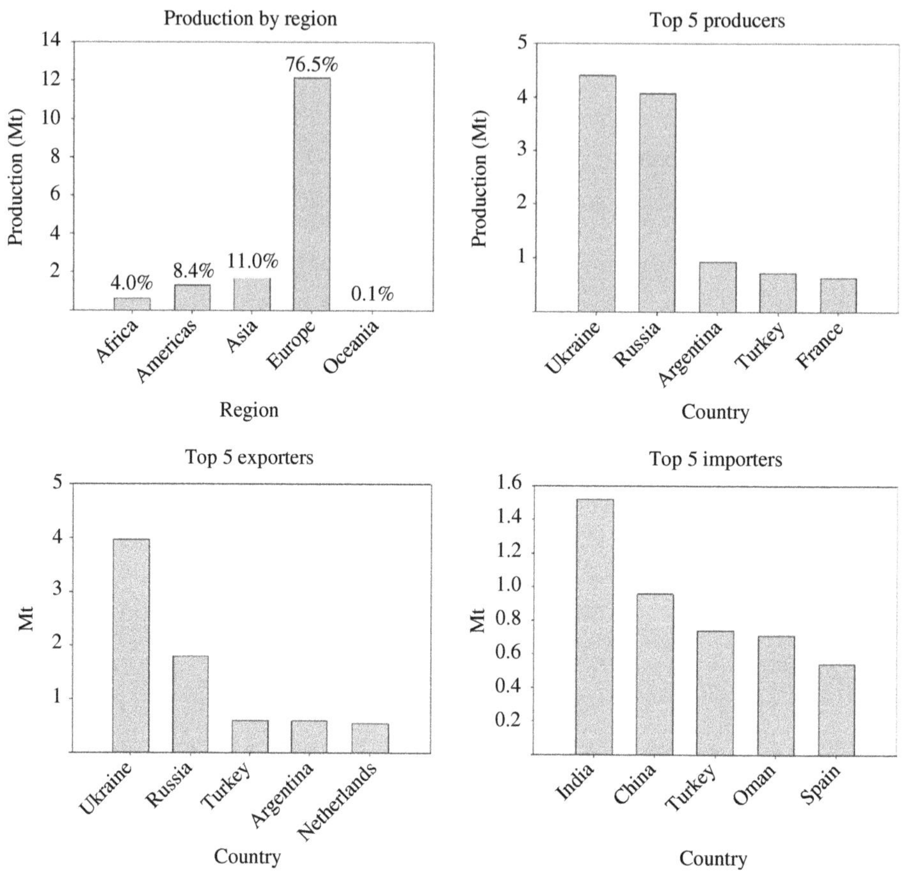

Figure 1.6 Sunflower seed oil: Production and trade. *Source:* Data accessed on April 8, 2019 from FAOSTAT, available at http://www.fao.org/faostat.

3.79 Mha in 2008 to 4.17 Mha in 2018, representing an average 11.3% annual increase (European Commission 2019).

Despite being a Native American plant, traditionally, the main regions of sunflower oil consumption are the EU and former soviet countries. Ukraine and Russia are currently the largest producers, followed by Argentina. In 2017, Ukraine, Russia, and Argentina produced 12.23, 10.48, and 3.54 Mt, respectively. Expected production gains for Ukraine (to 15.5 in 2018/2019 from 13.7 Mt in 2017/2018) will be caused by improved sunflower yields, as Ukraine's crop area is expected to stay unchanged in 2018/2019 at 6.4 Mha (USDA 2018). In the EU-28, sunflower seed production increased from 7.21 Mt in 2008 to 10.17 Mt in 2018 (European Commission 2019). World sunflower seed production was 51.49 Mt during 2018/2019 (USDA 2019a). Russia and especially Ukraine are not only the biggest producers but also the biggest sunflower oil exporters. In 2017, Ukraine exported 3.97 Mt of sunflower oil. According to the International Sunflower Oil Association, in 2016, sunflower oil was exported to more than 110 countries in the world. Top importers in 2017 are shown in Figure 1.6 and are led by India and China, which imported approximately 1.52 and 0.96 Mt of sunflower oil in 2017, respectively.

1.3 Novel Sources for Oil Production

Several unconventional raw materials are emerging as promising alternatives to conventional oilseeds for the production of edible oil for use in food, feed, and other industrial applications. These include, but are not limited to insects, microorganisms such as bacteria or microalgae, and seaweed. Insects with potential for being used as sources of edible oils were revised recently (Mariod et al. 2017). Most commonly consumed insects are beetles (*Coleoptera*, 31%), caterpillars (*Lepidoptera*, 18%), bees, wasps, and ants (14%), and grasshoppers, locusts, and crickets (*Orthoptera*, 13%) (FAO 2013). Their lipid content varies significantly among species and depends on what they have fed on, sex, environmental factors, or stage of development – overall, fat content varies between 7 and 77% on a dry weight basis (Dobermann et al. 2017). Insect-derived lipids are generally liquid at room temperature and are therefore called "insect oils." These are rich in unsaturated and essential fatty acids and have been suggested as potential ingredients for use in mayonnaises, vinaigrettes, as frying oils, and as food grade lubricants (Sosa and Fogliano 2017). Although insect consumption is not new for people living in some African and South East Asian countries, where at times up to 50% of dietary protein come from insects (Dobermann et al. 2017), it is still unfamiliar and has been associated with unappealing and negative sensory attributes in Western countries (Baker et al. 2016). In Thailand, for example, locusts (*Patanga succincta*) are such a popular snack that farmers grow crops specifically to feed them (Dobermann et al. 2017). Information on production and trade of edible insects is scarce. Zimbabwe exports caterpillars to Botswana, South Africa, Zambia, and the Democratic Republic of the Congo. Moreover, a case study conducted in the Central African Republic reported that the main importers of caterpillars were Chad, Nigeria, and Sudan, while the Central African Republic also exports caterpillars to African communities in Europe (FAO 2013). Thailand imports approximately 700 t of edible insects from Lao People's Democratic Republic and Cambodia (Yen et al. 2013).

Moreover, microalgae have been suggested as one of the most promising resources for generating oil for biodiesel production. Despite the huge efforts and significant amount of money invested over the last two decades the large-scale technology needed to produce biodiesel from microalgae is still in its early stages of development (Islam et al. 2017; Lafarga 2020). Recently, Garrido-Cardenas et al. (2018) reviewed the recent worldwide trends in the area of microalgae biotechnology and reported "lipids" as one of the most relevant keywords that appear in over 1000 scientific publications studied. Microalgae-derived oil for human consumption shows potential for further development as, for example, docosahexaenoic acid- and eicosapentaenoic acid-rich oil obtained from the microalgae *Schizochytrium* sp. has been accepted as a novel food in the EU according to Regulation (EU) 2015/2283 (EC 2015) and microalgae are one of the top trends in the food industry (Lafarga 2019). Currently, world production of microalgal biomass is estimated to be over 20 kt per year and increases an average of 10% per year (Acién-Fernández et al. 2018). This value is likely to increase as studies have suggested a potential of producing 50 Mt of dried microalgal biomass in Europe alone (Skarka 2012). Macroalgae were also suggested as potential sources for oil production (Wang et al. 2013), although their lipid content is lower when compared to that of microalgae and are especially rich in carbohydrates (Lafarga et al. 2020). Overall, the production quantities and market value of these novel sources for oil production are very low. However, their huge potential suggests that microalgae- or insect-derived oils will play an important role in the food industry of the future.

1.4 Summary

While discussing the trade in oils and oilseeds, in 2011, Gunstone (2011a) divided countries or regions into four main categories: (i) countries like Argentina, Canada, or Australia which have a small population but produce large amounts of oilseeds and oils, dominating world trade; (ii) countries like Brazil, US, or Indonesia which produce large amounts of oilseeds and oils, and despite the need to feed their own population can still export large quantities; (iii) countries like China, India, and other countries in Asia which despite local production still need to import oilseeds and oil; and (iv) regions like the EU which are essentially traders and produce, consume, import, and export these commodities. Vegetable oil has one of the highest trade shares of production of all agricultural commodities – around 41%. Vegetable oil production will also continue to increase in coming years. Oil production depends on both the crush of oilseeds and the production of tropical oil plants, namely oil palm. Asia, or more explicitly Indonesia and Malaysia are the biggest palm oil exporters (70 and 80% of the oil produced in Indonesia and Malaysia is currently exported) and will continue to dominate oil exports in coming years (FAO 2018). However, environmental policies from the major palm oil importers together with the mainstreaming of global sustainable agricultural norms, are expected to slow the expansion of the oil pam area in Indonesia and Malaysia (FAO 2018).

Acknowledgments

The author thanks the Spanish Ministry of Science, Innovation, and Universities (IJC2018-035287-I) and the BBVA Foundation.

References

Acién-Fernández, F.G., Fernández-Sevilla, J.M., and Molina-Grima, E. (2018). Contribución de las microalgas al desarrollo de la bioeconomía. In: *Bioeconomía y desarrollo sostenible* (eds. A. Aguilar, D. Ramón and F.J. Egea), 309–332. Almería, Spain: Cajamar Caja Rural.

Arya, S.S., Salve, A.R., and Chauhan, S. (2016). Peanuts as functional food: a review. *Journal of Food Science and Technology* 53 (1): 31–41.

Baker, M.A., Shin, J.T., and Kim, Y.W. (2016). An exploration and investigation of edible insect consumption: the impacts of image and description on risk perceptions and purchase intent. *Psychology and Marketing* 33 (2): 94–112.

Batugal, P., Rao, V.R., and Oliver, J. (2005). *Coconut Genetic Resources*. Selongor, Malaysia: International Plant Genetic Resources Institute – Regional Office for Asia, the Pacific and Oceania.

Bawalan, D.D. and Chapman K.R. (2006). Virgin coconut oil: Production manual for micro-and village-scale processing. FAO Regional Office for Asia and the Pacific, Bangkok, Thailand.

Bedigian, D. (2010a). Introduction: history of the cultivation and use of sesame. In: *Sesame: The Genus Sesamum* (ed. D. Bedigian), 1–32. Boca Raton, FL: CRC Press.

Bedigian, D. (2010b). Current market trends: critical issues and economic importance of sesame. In: *Sesame: The Genus Sesamum* (ed. D. Bedigian), 423–490. Boca Raton, FL: CRC Press.

Cartea, M.E., Lema, M., and Francisco, M. (2016). Basic information on vegetables Brassica crops. In: *Genetics, Genomics, and Breeding of Vegetables Brassicas* (eds. J. Sadowski and C. Kole), 1–33. Enfield: CRC Press.

Daun, J.K. (2015). Origin, distribution, and production. In: *Canola: Chemistry, Production, Processing, and Utilization* (eds. J.K. Daun, N.A. Michael Eskin and D. Hickling), 1–27. Champaign, IL: AOAC Press.

Dijkstra, A.J. (2015). Oil refining. In: *Sunflower: Chemistry, Production, Processing, and Utilization* (eds. E. Martinez-Force, N.T. Dunford and J.J. Salas), 227–258. Champaign, IL: AOAC Press.

Dobermann, D., Swift, J., and Field, L. (2017). Opportunities and hurdles of edible insects for food and feed. *Nutrition Bulletin* 42 (4): 293–308.

EC (2015). Commission implementing decision (EU) 2015/546 of 31 March 2015 authorising an extension of use of DHA and EPA-rich oil from the micro-algae Schizochytrium sp. as a novel food ingredient under Regulation (EC) No 258/97 of the European Parliament and of the Council. *Official Journal of the European Union* 90: 11–13.

EU (2019). EUROSTAT – EU Statistical Office.

European Commission (2019). *EU Crops Market Observatory – oilseeds and protein crops*.

FAO (2013). *Edible Insects: Future Prospects for Food and Feed Security*. Rome, Italy: FAO.

FAO (2018). Oilseeds and oilseed products. In: *OECD-FAO Agricultural Outlook 2018–2027*, 127–138. OECD Publishing.

Feoll, C.E. and Ingaramo, J. (2015). South America perspectives on sunflower production and processing. In: *Sunflower: Chemistry, Production, Processing, and Utilization* (eds. E. Martinez-Force, N.T. Dunford and J.J. Salas), 517–553. Champaign, IL: AOAC Press.

Ferraris, S. (2019). Market situation in the olive oil and table olives sector. Committee for the Common Organisation of the Agricultural Markets: European Comission.

Foale, M. (2005). An introduction to the coconut palm. In: *Coconut Genetic Resources* (eds. P. Batugal, V.R. Rao and J. Oliver), 1–12. Selangor, Malaysia: International Plant Genetic Resources Institute – Regional Office for Asia, the Pacific and Oceania.

Garrido-Cardenas, J.A., Manzano-Agugliaro, F., Acién-Fernández, F.G. et al. (2018). Microalgae research worldwide. *Algal Research* 35: 50–60.

Gunn, B.F., Baudouin, L., and Olsen, K.M. (2011). Independent origins of cultivated coconut *(Cocos nucifera L.)* in the old world tropics. *PLoS One* 6 (6): e21143.

Gunstone, F. (ed.) (2011a). Production and trade of vegetable oils. In: *Vegetable Oils in Food Technology: Composition, Properties and Uses*, 1–24. Pondicherry, India: Wiley.

Gunstone, F. (2011b). *Vegetable Oils in Food Technology: Composition, Properties and Uses*. Pondicherry, India: Wiley.

Gupta, M. (2017). *Practical Guide to Vegetable Oil Processing*. San Diego, CA: Elsevier.

van Heerwaarden, J., Doebley, J., Briggs, W.H. et al. (2011). Genetic signals of origin, spread, and introgression in a large sample of maize landraces. *Proceedings of the National Academy of Sciences of the United States of America* 108 (3): 1088–1092.

Henders, S., Persson, U.M., and Kastner, T. (2015). Trading forests: land-use change and carbon emissions embodied in production and exports of forest-risk commodities. *Environmental Research Letters* 10 (12): e125012.

Islam, M.A., Heimann, K., and Brown, R.J. (2017). Microalgae biodiesel: current status and future needs for engine performance and emissions. *Renewable and Sustainable Energy Reviews* 79: 1160–1170.

Kajla, P., Sharma, A., and Sood, D.R. (2015). Flaxseed—a potential functional food source. *Journal of Food Science and Technology* 52 (4): 1857–1871.

Kumar, P.P. and Krishna, A.G. (2015). Physicochemical characteristics of commercial coconut oils produced in India. *Grasas y Aceites* 66 (1): e 062.

Lafarga, T. (2019). Effect of microalgal biomass incorporation into foods: nutritional and sensorial attributes of the end products. *Algal Research* 41: 101566.

Lafarga, T. (2020). Cultured microalgae and compounds derived thereof for food applications: strain selection and cultivation, drying, and processing strategies. *Food Reviews International* 36: 559–586.

Lafarga, T., Acién-Fernández, F.G., and García-Vaquero, M. (2020). Bioactive peptides and carbohydrates from seaweed for food applications: natural occurrence, isolation, purification, and identification. *Algal Research* 48: 101909.

Le Chef, E. and Kemper, T. (2015). Sunflower seed preparation and oil extraction. In: *Sunflower: Chemistry, Production, Processing, and Utilization* (eds. E. Martinez-Force, N.T. Dunford and J.J. Salas), 187–226. Champaign, IL: AOAC Press.

Lee, G.A., Crawford, G.W., Liu, L. et al. (2011). Archaeological soybean (Glycine max) in East Asia: does size matter? *PLoS One* 6 (11): e26720.

MacDonald, I., Oghale, O.U., Sheena, O.E. et al. (2018). Physicochemical properties, antioxidant activity and phyto-nutritional composition of cold and hot pressed coconut oils. *GSC Biological and Pharmaceutical Sciences* 5 (1): 56–66.

Mariod, A.A., Mirghani, M.E.S., and Hussein, I. (eds.) (2017). *Unconventional Oilseeds and Oil Sources*. London: Academic Press.

Mba, O.I., Dumont, M.J., and Ngadi, M. (2015). Palm oil: processing, characterization and utilization in the food industry – a review. *Food Bioscience* 10 (1): 26–41.

O'Brien, R., Jones, L.A., King, C.C. et al. (2005). Cottonseed oil. In: *Bailey's Industrial Oil and Fat Products* (ed. F. Shahidi), 136–159. Hoboken, NJ: Wiley.

Oomah, B.D. (2001). Flaxseed as a functional food source. *Journal of the Science of Food and Agriculture* 81 (9): 889–894.

Pandey, M.K., Wang, M.L., Qiao, L. et al. (2014). Identification of QTLs associated with oil content and mapping FAD2 genes and their relative contribution to oil quality in peanut (*Arachis hypogaea L.*). *BMC Genetics* 15 (1): ID133.

Pirker, J., Mosnier, A., Fraxner, F. et al. (2016). What are the limits to oil palm expansion? *Global Environmental Change* 40 (1): 73–81.

Ramanatha Rao, V., Hodgkin, T., and Bourdeix, R. (2005). Locating coconut genetic diversity. In: *Coconut Genetic Resources* (eds. P. Batugal, V.R. Rao and J. Oliver), 13–74. Selangor, Malaysia: International Plant Genetic Resources Institute – Regional Office for Asia, the Pacific and Oceania.

Ranum, P., Peña-Rosas, J.P., and Garcia-Casal, M.N. (2014). Global maize production, utilization, and consumption. *Annals of the New York; Academy of Sciences* 1312 (1): 105–112.

Salas, J.J., Bootello, M.A., and Garcés, R. (2015). Food uses of sunflower oil. In: *Sunflower: Chemistry, Production, Processing, and Utilization* (eds. E. Martinez-Force, N.T. Dunford and J.J. Salas), 441–464. Champaign, IL: AOAC Press.

Savoire, R., Lanoisellé, J.L., and Vorobiev, E. (2013). Mechanical continuous oil expression from oilseeds: a review. *Food and Bioprocess Technology* 6 (1): 1–16.

Saxena, D.K., Sharma, S.K., and Sambi, S.S. (2011). Comparative extraction of cottonseed oil by n-hexane and ethanol. *ARPN Journal of Engineering and Applied Sciences* 6 (1): 84–89.

Shasidhar, Y., Vishwakarma, M.K., Pandey, M.K. et al. (2017). Molecular mapping of oil content and fatty acids using dense genetic maps in groundnut (*Arachis hypogaea L.*). *Frontiers in Plant Science* 8 (1): 794.

Skarka, J. (2012). Microalgae biomass potential in Europe. *Technikfolgenabschätzung – Theorie und Praxis* 21 (1): 72–79.

Sosa, D.A.T. and Fogliano, V. (2017). Potential of insect-derived ingredients for food applications. In: *Insect Physiology and Ecology* (ed. V.D.C. Shields), 215–231. Rijeka, Croatia: IntechOpen.

Torres, A.M., Barros, G.G., Palacios, S.A. et al. (2014). Review on pre- and post-harvest management of peanuts to minimize aflatoxin contamination. *Food Research International* 62 (1): 11–19.

UN FAOSTAT – FAO statistical database (United Nations). (2019).
USDA (2017). *Oilseeds: World Markets and Trade*. Washington, DC: USDA Foreign Agricultural Service.
USDA (2018). *Oil Crops Outlook*. Washington, DC: USDA.
USDA (2019a). *Oilseeds: World Markets and Trade*. Washington, DC: USDA Foreign Agricultural Service.
USDA (2019b). *World Agricultural Supply and Demand Estimates*. Washington, DC: USDA.
Vijay, V., Pimm, S.L., Jenkins, C.N. et al. (2016). The impacts of oil palm on recent deforestation and biodiversity loss. *PLoS One* 11 (7): e0159668.
Wan, Y., Li, H., Fu, G. et al. (2015). The relationship of antioxidant components and antioxidant activity of sesame seed oil. *Journal of the Science of Food and Agriculture* 95 (13): 2571–2578.
Wang, S., Wang, Q., Jiang, X. et al. (2013). Compositional analysis of bio-oil derived from pyrolysis of seaweed. *Energey Conversion and Management* 68: 273–280.
Williams, M.A. (2005). Recovery of oils and fats from oilseeds and fatty materials. In: *Bailey's Industrial Oil and Fat Products* (ed. F. Shahidi), 101–135. Hoboken, NJ: Wiley.
Yen, A.L., Hanboonsong, Y., and van Huis, A. (2013). The role of edible insects in human recreation and tourism. In: *The Management of Insects in Recreation and Tourism* (ed. R.H. Lemelin), 169–185. New York;: Cambridge University Press.
Zuknik, M.H., Norulaini, N.A., Wan Nursyazreen Dalila, W.S. et al. (2016). Solubility of virgin coconut oil in supercritical carbon dioxide. *Journal of Food Engineering* 168: 240–244.

2
Conventional Oils and Oilseeds: Composition and Nutritional Importance

Gloria Bobo[1], Iolanda Nicolau-Lapeña[2] and Ingrid Aguiló-Aguayo[1]
[1] *IRTA, Postharvest Programme, Edifici Fruitcentre, Parc Científic i Tecnològic Agroalimentari de Lleida, Lleida, Spain*
[2] *Food Technology Department, University of Lleida (UDL), Lleida, Spain*

2.1 Introduction

According to the FAO Commodity Snapshot (OECD-FAO 2017) global soybean production increased in the United States and Brazil in 2016 registering record crops. Soybean is one of the most important duel-purpose crops with a different use in food, animal feed, biochemistry, and industrial purposes (Pratap et al. 2016). Other important conventional oilseed crops include rapeseed, canola, sunflower, safflower, peanut, cottonseed, linseed, and sesame. The predictions for 2026 are an overall increase of 90% of world soybean production and 86% of world production of other oilseeds due to direct food consumption (OECD-FAO 2017). Crushing oilseeds into meal (cake) and oil are the main usage. However, the raw materials act as a renewable source of energy and their role in power generation make them a powerful material for oleochemical industries (Jankowski and Budzyński 2003). Therefore, the biofuel industry is one of the applications of these conventional oilseed crops as an alternative to petrochemicals leading the biochemical industries to produce genetically modified oilseed crops to produce high amounts of biofuel (Moser et al. 2013; Waseem et al. 2017).

Focusing on the nutritional properties of oilseed crops, their high amounts of proteins present in the meal fraction obtained after oil removal is a high-value product for animal feed (Mailer 2016). On the other hand, the increasing demand for healthy vegetable oils has increased the production of oilseed oils use as a commodity due to the presence of polyunsaturated fatty acids and their positive contribution to health due to their antioxidant properties (Intelligence 2006). Polyunsaturated fatty acids in vegetable oils are of importance in human diets. The major polyunsaturated fatty acids include omega-3, omega-6 and linoleic acid (Balbino 2017). However, the fatty acid composition will strongly depend on the cultivar, climate, and agronomic

Oil and Oilseed Processing: Opportunities and Challenges, First Edition.
Edited by Tomás Lafarga, Gloria Bobo, and Ingrid Aguiló-Aguayo.

conditions as well as the maturity of the oilseeds (Panda 2010). In this context, these oilseed oils represent an alternative to palm oil, which is under discussion due to the high level of saturated fatty acids. High oleic sunflower and high oleic rapeseed oil are considering good solutions (Hinrichsen 2016). However, palm oil is still the most worldwide consumed vegetable oil with 74.62 million metric tons (MT) followed by soybean oil with 56.84 MT and sunflower seed oil with 19.06 MT as recorded in Shahbandeh (2020).

2.2 Oilseeds

2.2.1 Description of Oilseeds

Oilseeds have three basic parts, the seed coat, the endosperm, and the cotyledon. The seed coat is marked with a seed scar and its main function is protecting the embryo from infections. The cotyledons work as food reserve structures. For this reason, in the oilseeds cotyledons there are discrete cellular organelles called proteins and lipid bodies (aleurone grains and spherosomes and oleosomes, respectively) throughout the germ cells (Rosenthal et al. 1996). The lipid particles consist of a core of triacylglycerols and or steryl esters. A phospholipid monolayer surrounds the core with only a small quantity of proteins (oleosins) that help to stabilize the spherosomes of oilseeds during desiccation in the seed maturation (Zweytick et al. 2000).

During the last decades the consumption of animal fats has been decreasing, and, on the other hand, the consumption of vegetable oils has increased. The most important vegetable oils can be classified as four major oils (palm, soybean, rapeseed, and sunflower), two lauric oils (coconut and palm kernel) and the remaining oils (cottonseed, groundnut, and olive). Another kind of classification is distinguishing between oils and fats obtained from trees as coconut, palm, and olive, or those from annual seeds or beans as soybean, rapeseed, groundnuts, and cottonseeds. Some of them, for example, cottonseed and peanuts (groundnut seeds), have woody hulls and shells, which have to be removed before processing (Gunstone 2011).

Edible oils and fats are essential nutrients. Vegetable oils usually have high concentration of polyunsaturated fatty acids (PUFAs). There are some exceptions as coconut (copra), palm, and palm kernel oils which are mainly saturated. Olive and canola oils have less PUFAs and more monounsaturated fatty acids (MUFAs) (Dupont 2003).

The nutritional and also the industrial value of a vegetable oil depend on its fatty acid composition. There is a particular importance in PUFAs because of animal cells cannot synthesize linoleic acid (C18:2 n-6) and α-linolenic acid (C18:3 n-3), which have essential metabolic functions. For this reason they are considered as dietary essential fatty acids (Balbino 2017). In general, oils and fats supply calories and help with the absorption of the fat-soluble vitamins A, D, E, and K. They also work as functional ingredient because help to improve the sensory characteristics of food products (Salunkhe et al. 1992). Also the proportions of fatty acids determine the physical form of the fat, which mean that a higher amount of PUFAs and MUFAs content in oil the more liquid it will be at room temperature. When the content of saturated fatty acids is high they are usually more solid at room temperature; sometimes they are referred as "solid fats" (e.g. coconut).

Linoleic acid (C18:2 n-6) and α-linolenic acid (C18:3 n-3) are called essential because the human body can produce two whole families of polyunsaturated fatty acids from them as a precursors. For example, from *n-6* family (linoleic acid) it can obtain arachidonic acid (ARA) (20:4 n-6) and docosapentaenoic acid (22:5 n-6), among others. In the case of n-3 family (α-linolenic acid) it can be produced eicosapentaenoic acid (20:5n-3) and docosahexaenoic acid (22:6 n-3, DHA). Some of these molecules have great importance in human metabolisms. For example ARA is abundant in the brain, muscles, and liver. It has been explained that it is not an essential fatty acid, however if there is a deficiency in linoleic acid, it becomes essential (Smith et al. 2011). This acid is involved in muscle growth and brain. It is present in the brain in similar quantities as DHA both of them are involved in the neurological health. Furthermore, there are studies like Birch et al. (2000) about supplementation of infant formula with DHA+ARA where an increase on the Mental Development Index (MDI) was observed. Also at cognitive and motor subscales of the MDI there was a significant developmental age advantage for supplementation with DHA or DHA+ARA (Birch et al. 2000).

2.2.2 Physicochemical Properties of Oilseeds Oils

Relative density, viscosity, refractive index, smoke point, flash point, fire point, melting point, and crystallization as well as thermal behavior are some of the chemical properties of importance in oilseeds oils to consider for food applications. The crystallization or polymorphisms properties greatly affects the textural and functional properties of oils. Soybean oil tends to form β-crystals in contrast to the hydrogenated form that can be crystallized in β', form achieving desired functional properties in the final product (Gunstone 2011). Temperature affects the density of oilseed oils leading to a density decrease with an increment of temperature. Predictions by mathematical model to estimate changes in the viscosity of oils have been developed helping to predict the physical changes (Rodenbush et al. 1999). Table 2.1 shows some of the key physical characteristics of some of the key oilseed oils comparing to a standard olive oil. Refractive index could help to predict oil viscosity and in turn key physical properties (Gunstone 2011). In the case of flash point (Table 2.1), temperatures of the different oilseed oils around 300 °C differ depending at which temperature the volatiles are produced in amounts that ignite but do not appear as a flame (Gunstone 2011). Peroxide values indicate the level of primary oxidation, which could increase when the oils are exposed to light. The food industry frequently adds sources of antioxidants to enhance oil protection from oxidation changes. Carelli et al. (2005) reported the effectiveness of adding α- and δ-tocopherol, citric acid, ascorbic acid, and ascorbyl palmitate in sunflower oil observing that ascorbic acid improved the rancimat oxidative stability of the oil. The measure of iodine value is linked to the degree of unsaturation of fat and oil. Therefore, a low iodine value is related to a high level of saturated and MUFA as observed in palm-based or coconut oils (Table 2.1).

2.2.3 Nutritional Properties

Oilseeds are a source of fiber and minerals such as phosphorus, iron, and magnesium. Generally, they are a source of vitamin E (an antioxidant), folate, and niacin. Also contain phytoestrogens, which is a group of substances including isoflavones (McKevith 2005). Soybean and sunflower seeds are a rich source of lipids with triacylglycerols as the main

Table 2.1 Some physical characteristics of conventional oilseed oils.

Parameters	Palm kernel oil	Palm oil	Soybean oil	Rapeseed oil	Sunflower oil	Groundnut oil	Olive oil	Coconut oil
Specific gravity value (relative to pure water)	0.904	0.9020	0.9150–0.9280	0.9123	1.521517	0.9155	0.910–0.915	0.91
Refractive index (t = 40 °C)	14 490 – 14 520	14 530 – 14 560	14 660 – 14 700	1465–1.469	14 610 – 14 680	1.47	14 670 – 14 710 (at 20 °C)	14 480 – 14 500
Saponification value (mg KOH/g)	280.5	200.05	188–195	172.29	177.06	191.5	189.30	257.5
Acid value (mg KOH/g)	2.7	0.84	≤0.5	1.78	0.22	9.0	0.84	5.5
Peroxide value (mEq/kg)	14.3	7.98	≤10	0.63	9.99	9.99	7.98	–
Iodine value (mg KOH/g)	15.86	602.7	75–94	1198.9	94.35	9.4	831	8.5
Viscosity, mPA s (30 °C)	43	45 (at 45 °C)	45	72–82 (at 20 °C)	47	45	55	39
Flash point, °C	267	314	317–324	317	316	315	300–330	300/288 /270

Source: Chadwick 1988; Bailey and Shahidi 2005; Firestone 2006; Tan et al. 2009; Azlan et al. 2010; Neagu et al. 2013; Amira et al. 2014; Roiaini et al. 2015; Alajtal et al. 2018; Godswill et al. 2018; Xu et al. 2018; Beszterda and Nogala-Kałucka 2019.

constituents. Proteins and phospolipds and bioactive compounds such as vitamin E are also present (White et al. 2006). This is of particular importance in the field of pharmacology, food, and personal care (Sukhotu et al. 2016). Table 2.2 shows the main composition table of the main oilseed oils (USDA 2020). As observed, sunflower oil is a potential source of vitamin E and soybean oil of vitamin K. Regarding the PUFA content, sunflower, soybean, flaxseed, and poppy seed oil present the highest values, whereas coconut has a considerably high saturated fatty acid content (Table 2.2).

Table 2.3 presents the main fatty acids in oilseed oils. Linoleic acid, oleic, and α-linolenic acids are the primarily unsaturated fatty acids accounting for 88.83% of total lipids. Palmatic acid is the major saturated fatty acid (10.42%). As reported in Ramos et al. (2009), linoleic acid (54.0%), oleic (23.9%), palmitic acid (11.3%), and α-linolenic acids (6.1%) are the major fatty acids present in soybean oil. Canola oil, also known in Europe as rapeseed oil, has low concentrations of saturated lipids and, in line with soybean, the protein content is high. In particular, percentages of 23 and 40% have been reported in rapeseed and soybean respectively (Bauer and Kostik 2014).

2.2.4 Bioactive Properties

Phenolic, tocopherol, and sterol profiles of oilseed oils are the main important bioactive compounds reported. Palm oil is one of the potential sources of tocotrienol, part of Vitamin E, and in turn, palm oil is a great source of this vitamin helping as antioxidant protector against the oxidation of lipophilic molecules and linked to skin protection when consuming in diet (Enig et al. 1983).

Tocols are synthesized only by plants; however they are very important dietary nutrients for humans and animals (Dunford and Dunford 2004). Tocopherols are present in oilseeds, leaves, and other green parts of higher plants. In the extracted seed oil the main compounds with antioxidant activity are tocopherols and also tocotrienols, which are compounds with vitamin E activity. These compounds protect the oil from oxidative deterioration (Soldo et al. 2019). Furthermore, they prevent metabolic, proliferative, inflammatory, and oxidative damage (Galli and Azzi 2010).

Sunflower, safflower, and cottonseed oils are the richest source of vitamin E (Table 2.2). Sunflower has a high content of α-tocopherol. Groundnut and cottonseed present balanced levels of α- and γ-tocopherol. Olive oil presents a similar tocochromanol profile to sunflower, however its α-tocopherol content is lower (Velasco and Ruiz-Méndez 2015). The content of tocopherol in soybean at pH 6.8 is β-, γ-tocopherol 0.27 ± 0.02, δ-tocopherol 0.47 ± 0.00 and α-tocopherol 0.02 ± 0.00 mg/g neutral lipid of oil bodies (Chen et al. 2014).

Some phytosterols are important components, they are part of the steroids group, and participate in the chemical defense of the plants, as for example, triterpenoids or campesterol which is one of the most important phytosterols and has a similar structure to cholesterol. Because of this, dietary phytosterols reduce intestinal absorption of cholesterol and reduce its levels (Plat and Mensink 2005). Technologically, phytosterols increase thermal stability of oils with high levels of polyunsaturated fatty acids (Winkler and Warner 2008).

These compounds can be present in the oil in free form or esterified with glucosides, ferulic acid, or fatty acids. Ferulic acid-esterified sterol is known as oryzanol and has been reported to have diverse health benefits, including hypolipidemic and antioxidant effects or stimulation of growth and hypothalamus activity

Table 2.2 Composition table from Food Data Central form USDA (values per 100 g of oil).

OILS	Energy (kcal)	Total lipid (g)	FA total saturated (g)	FA total MU[a] (g)	FA total PU[b] (g)	Vitamin E[c] (mg)	Vitamin K[d] (μg)
Sunflower oil	884	100	10.3	19.5	65.7	41.1	5.4
Canola Oil	884	100	7.4	63.3	28.1	17.5	71.3
Soybean oil	857	100	15.7	22.7	57.7	8.2	183.9
Groundnut oil	884	100	16.9	46.2	32	15.69	0.7
Cottonseed oil	884	100	25.9	17.8	51.5	35.3	24.7
Flaxseed oil	884	100	9	18.4	67.8	0.47	9.3
Poppyseed oil	884	100	13.5	19.7	62.4	11.4	---
Sesame oil	884	100	14.2	39.7	41.7	1.4	13.6
Safflower oil	884	100	7.5	74.8	12.2	34.1	7.1
Olive oil	884	100	13.8	72.9	10.5	14.35	60.2
Coconut oil	892	99.1	82.5	6.3	1.7	0.1	0.6

[a] monounsaturated
[b] polyunsaturated
[c] α-tocopherol
[d] phylloquinone.

Table 2.3 Composition of main fatty acids from oilseed oils.

Oilseed oil	Main Fatty Acids	Quantity (%)	References
Soybean oil	Palmitic acid C16:0	10.42 ± 0.07	Zhong et al. (2015)
	Oleic acid C18:1	22.85 ± 0.04	
	Linoleic acid C18:2	53.62 ± 0.05	
	α-linolenic acid C18:3 (n-3)	7.33 ± 0.07	
Sunflower oil	Palmitic acid C16:0	53.8 ± 1.28	Ye et al. (2019)
	Oleic acid C18:1	27.8 ± 0.01	
	Linoleic acid C18:2	60.6 ± 0.50	
	α-linolenic acid C18:3 (n-3)	0.12 ± 0.01	
Olive oil	Palmitic acid C16:0	12.0 ± 0.74	Skiada et al. (2020)
	Oleic acid C18:1	76.7 ± 1.96	
	Linoleic acid C18:2	6.1 ± 1.60	
	α-linolenic acid C18:3 (n-3)	0.7 ± 0.07	
Sesame oil	Palmitic acid C16:0	8.3 ± 0.01	Soldo et al. (2019)
	Oleic acid C18:1	39.8 ± 0.05	
	Linoleic acid C18:2	45.0 ± 0.12	
	α-linolenic acid C18:3 (n-3)	0.4 ± 0.00	
Palm oil	Palmitic acid C16:0	37.1 ± 0.93	Ye et al. (2019)
	Oleic acid C18:1	45.9 ± 0.88	
	Linoleic acid C18:2	10.4 ± 0.10	
	α-linolenic acid C18:3 (n-3)	0.12 ± 0.00	
Rapeseed oil	Palmitic acid C16:0	3.6 ± 0.06	Ye et al. (2019)
	Oleic acid C18:1	76.4 ± 0.12	
	Linoleic acid C18:2	12.2 ± 0.01	
	α-linolenic acid C18:3 (n-3)	2.30 ± 0.02	
Flaxseed oil	Palmitic acid C16:0	5.1 ± 0.02	Ye et al. (2019)
	Oleic acid C18:1	21.0 ± 0.43	
	Linoleic acid C18:2	23.4 ± 0.29	
	α-linolenic acid C18:3 (n-3)	45.03 ± 0.95	

(Dunford 2004). β-sitosterol is the phytosterol predominant in the profile of vegetable oils. Brassinosteroids, which are plant hormones essential for normal growth and development (Bishop and Yokota 2001) are present in high concentration in rapeseed oil. Soybean decreases its content of β-sitosterol with pH changes, at pH 6.8 0.82 ± 0.10 mg/g neutral lipid of oil bodies, at pH 9.5 with 0.73 ± 0.05 mg/g, and at pH 11.0 with 0.62 ± 0.04 mg/g (Chen et al. 2014).

One of the compounds most biologically active in vegetable oils is squalene, which is a terpenoid hydrocarbon that in plants is synthesized as a biochemical intermediate of the phytosterol biosynthetic pathway. Squalene is a high-value compound with some applications in the pharmaceutical and cosmetics industries (Dunford 2004). Virgin olive oil and crude sunflower are rich sources of this compound (Grompone 2005).

Other common compounds in vegetable oils are pigments as carotenoids and chlorophylls. There are two types of carotenoids, xanthophyll and carotenes, where the differences is the presence of oxygen in the molecule or not, respectively. In sunflower oil xanthophyll and lutein are the majority pigments. In sunflower oil β-carotene is added to increase its oxidation stability because of a synergistic effect with tocopherols (Yanishlieva et al. 2001).

Phospholipids are the major constituents of biological membranes. They have great antioxidant effects due to their activity as metal scavengers, their synergistic action with tocopherols, and their capacity to decompose hydroperoxides (Smouse 1995; Carelli et al. 1997). To maintain the fluidity of membranes, phospholipids have a high degree of fatty acid unsaturation, which make them susceptible to oxidation. Crude sunflower has high levels of phospholipids as phosphatidylcholines, phosphatidylethanolamines, phosphatidylinositols, and phosphatidic acids (Gupta et al. 2002).

2.2.5 Antinutritional Factors

"Antinutritional factors are present in different food substances in varying amounts, depending on the kind of food, mode of its propagation, chemicals used in growing the crop as well as those chemicals used in storage and preservation of the food substances" (Inuwa et al. 2011). Oxalates, phytates, tannins, trypsin inhibitors, and hemagglutinins are examples of antinutritional compounds present in vegetable oils. For example, palm and soybean oil have oxalate content of 4.95 and 1.16 g/kg, respectively, and the lethal dose established for oxalates is 2–5 g/kg. The content of alkaloids in palm oil is 0.16/100 mg and in soybean oils is 16.3/100 mg, where the lethal dose for alkaloids is 20/100 mg (Inuwa et al. 2011; Chatepa et al. 2019).

Trypsin inhibitor can bind with trypsin to inhibit the enzyme. The phytic acid or inositol could form insoluble compounds in the form of stable phytates inducing the suppression of living bodies. This compound is a strong chelating agent affecting to the bio-availability of some health-related minerals including zinc, calcium, magnesium, phosphorus, and iron. It has been demonstrated in some studies conducting in rats fed with soy products that the bio-availability of zinc was low. Interactions of diets rich in oilseed crops are still under investigation as many factors could be of influence, such as food processing conditions or digestibility of the foods (Wang 2016).

Tannins not only could provide a bitter taste but also can react with trypsin and amylase or their substrates by the hydroxide radicals and reduce the utilization rate of protein or carbohydrate. They could also interfere with the absorption of some minerals such as iron ions and form insoluble complexes on the intestinal mucosal surface damaging the intestinal wall. Similarly, the combination of tannin with metal ions including calcium, iron, and zinc could form complexes with vitamin B12 and negatively affect their absorption (Wang 2016).

2.3 Factors Affecting Oil Yield

Oil yield depends on a number of interrelated factors. The potential yield is defined as a yield of a cultivar when grown in environments to which it is adapted; with non-limiting nutrients and water; and with pests, diseases, weeds, lodging, and other stresses being effectively controlled (Evans and Fischer 1999). There is a gap between the potential yield and the actual yield of a crop. Once the type, variety, and location are determined, climatic conditions and agronomic practices have an impact on the amount of oil that can be obtained. The potential yield is primarily determined by the type of crop, as each has its characteristic fruit or seed density, and those have an average amount of percentage oil in them. Most oilseeds typically accumulate 20–50% of the dry weight of their seed tissue as storage oil, usually as triacylglycerol. However, there are some oilseed species, such as candlenut or sesame, which contain as much as 60–76% oil in their seeds (Walker 2008). Among the cultivated oil sources around the world, Palm fruit has the highest yield, followed by soybeans, olives, rapeseed, and sunflower, whereas poppy seed and sesame have the lowest yields (FAOSTAT 2018).

According to the definition, the potential yield is calculated in the absence of stresses, but there exist some abiotic factors – drought, extreme temperatures, and salinity – or biotic factors – diseases – that severely affect the growth and yield of the crops. Climatic periods of drought may affect crops around the world, and according to the studies in climatic changes, such extreme periods might affect crop yield in the near future (Müller et al. 2010). Drought conditions change physiological traits like stomatal conductance, turgor potential, relative water content, osmotic adjustment, crop temperature stability, and leaf temperature (Pasban 2009). As inflorescences, leaves, and roots respond differently to drought conditions, stress tolerance may depend on the type of crop (Azzeme et al. 2016). Olive trees are well adapted to drought conditions, and sunflower is usually characterized as a low-medium sensitivity crop to drought, whereas rapeseed's yield decreases markedly with it (Balbino 2017). Nevertheless, in some crops like palm trees, resistance can be induced by the addition of adequate doses of boron and silicon.

It is well known that extreme temperatures may affect oil content, and also oil composition. In sunflower, high temperature during the development of the seed leads to stress on the biosynthesis of fatty acids, and not only is the yield decreased, but also the percentage of linoleic acid is decreased (Harris et al. 1978). For olive crops, the beginning of the oil accumulation period is the most sensitive to temperature increases. (García-Inza et al. 2014) found that olive oil content decreased 1.1% for each 1 °C increase. Moreover, even a tropical crop such as coconut its resistance to extremely high temperatures depends on the humidity of the zone, being less productive in a dry zone than it is in a wet one (Pathmeswaran et al. 2018).

Salt stress alters the morphological, physiological, and biochemical performance of crops. It occurs as result of limited water uptake, which is caused by a reduction in hydrolysis and translocation of the nutrient reserves, thus reducing seedling vigor (Parveen et al. 2016). For instance, soybean grain yield decreased by 20% when salinity was 4.0 dS/m and by 56% at 6.7 dS/m. There are some methods, which can prevent salt stress, such as the addition of glutathione, which takes part in the detoxification system of the plant. When applied at 50 mM at the reproductive stage of soybean, it improved plant growth, yield, number of pods, and number of seeds (Akram et al. 2017).

Crops can be classified according to their lifecycle: annual or biannual, such as sunflower, groundnut, or rapeseed, or perennial, such as olive, coconut, or palm (Sharma et al. 2012). There are structural differences in several ways. In annual crops, new seeds for each growing season might be advantageous, while the potential yield for perennial ones is fixed for each planting cycle (Woittiez et al. 2017). The perennial crops have a yield cycle which depends on the age of the tree, and varies among the species. Four yield phases have been described: (i) the immature or building phase, before harvestable production, (ii) the young mature phase, when yield increases linearly along the years, (iii) the mature phase, when yield remains stable, and (iv) the phase of yield decline (Díez et al. 2016). In palm oil tree, the mature phase is reached at the eighth year after plantation and stands for eight years more (Goh et al. 2003). In coconut oil tree, full production is reached between the 15th and 20th years after plantation, and lasts until the plant's 80th year (Broschat and Crane 2017). On the other hand, as calculated by Vega et al. (2001), seed number per plant in annual crops can be related with their growth rate. Seed number varied from 0 to 890 in soybean, which was directly related with the growth rate of the plant. In sunflower it varied from 0 to 4096 in sunflower, and from 0 to 1348 in maize, which both have a curvilinear relationship with the growth rate.

Variety is an important factor that affects the average oil yield of a crop, which will be later affected by stresses. The oil yield has been one of the main objectives in genetics and breeding studies in oil crops all over the years (Knowles 1983; Murphy 2014). In soybeans, oil content between varieties may range from 0.9 to 11.9% (Kim et al. 2013). Regarding palm, the variety determines the thickness of the endocarp, whose development depends on the major effect shell gene. The thickness of the endocarp is highly correlated with the oil content, which may vary by 30% between varieties (Barcelos et al. 2015). Apart from the oil content of the seed, the amount of formed seeds per area unit also contributes to the yield of a crop. For instance, it is clear for sunflower that variety affects this trait (Khan et al. 2007), but some authors report that this is not a significant factor affecting sesame yield (Mesera and Mitiku 2015). However, breeders not only look for maximum yields, but also the stability of the cultivars across a range of production environments, including the cases where several stresses occur. Sometimes, maximum and stable yields do not occur in the same genotypes, soybean cultivars being an example (Gurmu et al. 2009).

Radiation also plays a role in potential yields, as it is the driving force for photosynthesis, which will allow converting carbon dioxide into organic compounds. Regions with eight sunshine hours/day would have 60% higher potential yield than regions with only three hours/day (Van Kraalingen et al. 1989). In palm-oil, for instance, productivity is constrained if radiation is less than 5.5 hours a day. Above this, one additional hour a day yields an additional 15–20 kg bunch dry matter/palm/year compared with productivity under cloudy conditions (Paramananthan 2003). Light incidence has also a direct relationship with the cell number and cell expansion of olives, with the ones at the top of the canopy being the ones that yield the most oil (Reale et al. 2019). So, in general, under no-stressed environmental conditions, the amount of dry matter produced by a crop is linearly related to the amount of solar radiation, specifically photo synthetically active radiation, intercepted by the crop (Campillo et al. 2012).

The ability of a plant to intercept the incident radiation depends on the leaf area available. To maximize it, most production strategies relay on achieving an optimal distribution of the plants in a way that canopies can capture sufficient solar energy (Hemming et al. 2007). In rapeseed crops, plant density has the greatest effect on yield. Seedling rate establishes the competition for nutrients and water within the canopy, and evenly distribution makes yield more stable because there are fewer losses in seed growth (Diepenbrock 2000). In olive orchards, density between 200 and 550 tree/ha is translated into a higher fraction of intercepted radiation per area, which leads to higher productivity (Gucci et al. 2012). Optimal density also depends on the type of soil. For palm oil, a maximum income can be obtained from planting 148 palms/ha on mineral soil, whereas on peat soil income is still on the increase when density is at 200 palms/ha (Latif et al. 2003). Whereas a high plant density may benefit by covering all the areas with leafy tissue to intercept light, a balance must be found, as it may also increase the competition between plants for water or nutrients.

Density of the crop also has a great impact on the uptake of nutrients and nitrogen from the plant. Nitrogen uptake can be improved by using N-efficient management strategies like choice of variety, form and timing of N-application adapted to site conditions (Rathke et al. 2006). The canola oil content increased, parallel to the crop yield increase in canola seeds when treated with the adequate proportion of nitrogen and phosphorus 90/60 kg/ha, but when a higher ratio was applied, oil content was inversely proportional to crop yield (Cheema et al. 2001). The amount of N application must be taken into account, as an over-fertilization can be also negative. For instance in olive trees, an excess of N may lead to polyphenol decrease in oil, which in turn, provokes the decrease in quality and stability to oxidation (Fernández-Escobar et al. 2006).

Finally, harvest date has to be properly selected in order to optimize the oil content on the fruit or seed. Different parameters are related to harvest, including number of seeds, moisture of the seeds, weight of 1000 seeds, and oil content, which will determine the final oil yield of production. In sunflower, the highest oil content was recorded on late harvesting dates (Mikliĉ et al. 2008), and in olive, it is important to harvest the fruits at full maturation, so they can have the maximum oil content per piece (Lavee and Wodner 2004). In perennial crops, the harvesting frequency also has an impact, with a yield increase of 5–20% when reducing length of harvesting round from 14 to 10 days in palm tree (Woittiez et al. 2017).

2.4 Overview of Oilseed Processing and Current Applications

On the whole, current applications of oilseeds' oils for food, fats, and oils have been long been used anciently and in many cultures. Most of the oilseeds oils are used for food applications with purposes such as shortenings, texture, flavor, or flavor-base since most flavors are soluble in oil. For instance, palm oil is well known as cholesterol-free oil used in cooking, margarines, spreads, confectionary items, ice cream, emulsifiers, and vanaspati (Pande et al. 2012). Sustainable alternatives to palm oil are nowadays on demand in a wide range of food products. Their high level of saturated fatty acids in comparison to other oils such as rapeseed or sunflower oil is still a health issue connected to the cultivation of palm oil (Hinrichsen 2016). A fully hydrogenated fat could potentially be a replacement for palm oil in food applications. For instance, blending liquid rape, sunflower, or soybean oils or coconut oil in combination with fully hydrogenated oils are some options currently in use. As reported by Hinrichsen (2016) recommendations of high level of oleic acid for frying applications such as high oleic sunflower and high oleic rapeseed oil are good alternatives to palm oil. However, other oilseed oils with high flash point are commonly used for frying including canola, safflower, groundnut, or cottonseed oils. On the other hand, the use of partial hydrogenated oils is always under discussion for their content in transfats and the associated health risk (Canada 2016). In addition, the food industry frequently used synthetic antioxidants such as butylated hydro butylated hydroxytoluene, butylated hydroquinone and tert-butylhydroquinone to inhibit deterioration caused by fats. However, according to Balasubramaniam et al. (2020), finger millet polyphenols delayed the oxidation of both refined and unrefined peanut oils confirming the advantage of providing food safety while maintain quality. Soybean oil is one of the world-consumed cooking oils offering a neutral flavor and favorable fat profiles used widely in the food industry. Emulsifiers in the food industry are essential for sauces, chocolate, ice cream, and mayonnaise formulations. Commercial lecithin comes from soybean oil and the improvement of food-grade emulsifiers in the properties and stability is of great interest. Sunflower lecithin extracted by a cold-press process is an alternative option. Another highly consumed oil is the rapeseed oil, which can be used in salad dressings, mayonnaise, and other emulsified products including biscuits and canning (Perrier et al. 2017).

Regarding animal feed, the remaining part of the seed after oil extraction is called meal fraction and is an important source of high-protein in animal feed products. Soybean meal is accepted worldwide as the most important protein ingredient in animal diets as a potential source of protein and essential amino acids while making a significant contribution to the dietary energy requirement compared with other oilseed meals. The chicken meat industry is the major user, accounting for 50% of all soybean meal livestock, while the pig industry consumes around 35% (Willis 2003). Also, oilseeds are frequently used as an energy source in diets of lactating dairy cows as ruminal microorganisms highly degraded the protein from whole oilseeds and oilseed meals (Bernard 2016).

Another industrial application is the use of oilseeds in sustainable food packaging. Epoxidized soybean oil has been used to reduce biopolymers hydrophilicity with its low cost, biodegradable and environmentally friendly characteristics, and its renewable origin (Willis 2003; Ciannamea et al. 2018). A partial replacement of glycerol by 20% of epoxidized soybean oil increased the heat seal strength in synthetic polymers as reported by Ciannamea et al. (2018). Other applications are thought highly of in the chemical industry. For instance, cold-pressed rapeseed meal also has techno-functional characteristics like emulsification and foaming which are highly valuable for application in adhesives, detergents, paints, varnishes, and are biodegradable (Chew 2020). Linked to industrial applications, flaxseed oil, which is rich in the unsaturated fatty acid linolenic is a drying oil used in protective coatings including paints or varnishes. Other commercial applications are cosmetics, in which oilseed oils suit cosmetic formulations due to their high fatty-acid composition. Production of lubricants, soaps, and personal care products are some examples of the topical treatments in which oilseeds oils are the main ingredients as base oils (Zimba et al. 2005; Chivandi et al. 2008).

Biotechnology has helped to improve the fatty acid composition of oilseed oils in food, feed, and bio-based industrial materials. For instance, the production of oilseeds of fish oil-type omega-3 polyunsaturated fatty acids resulted in an improvement of oil quality for direct human consumption with health benefits (Wu et al. 2005; Truksa et al. 2006). Metabolic engineering through the synthesis and metabolism of novel fatty acids with hydroxy and epoxy residues has proved valuable for industrial applications including lubricants, plasticizers, and nylon precursor (Burgal et al. 2008; Li et al. 2010).

Rapeseed oil and soybeans are the main oilseed crops for the production of biodiesel in Europe and the United States, respectively. However, other emerging oilseed crops such as mustard or hazelnuts are being explored as promising biodiesel feedstocks (Farm Energy 2019). In the case of rapeseed oil, the production of biodiesel is made through the rapeseed oil methyl ester with a limitation value of contamination lower than 25 mg/kg according to the "Quality standard for rapeseed oil as a fuel" (Fetzer et al. 2018; USDA 2019).

Acknowledgments

Ingrid Aguiló-Aguayo thanks the National Programme for the Promotion of Talent and Its Employability of the "Ministerio de Economía, Industria y Competitividad" of the Spanish Government and the European Social Fund for the Postdoctoral Senior Grant "Ramon y Cajal" (RYC-2016-19949). This work was supported by the CERCA Programme of Generalitat de Catalunya.

References

Akram, S., Siddiqui, M.N., Hussain, B.N. et al. (2017). Exogenous glutathione modulates salinity tolerance of soybean [Glycine max (L.) Merrill] at reproductive stage. *Journal of Plant Growth Regulation* 36 (4): 877–888.

Alajtal, A.I., Sherami, F.E., and Elbagermi, M.A. (2018). Acid, peroxide, ester and saponification values for some vegetable oils before and after frying. *AASCIT Journal of Materials* 4 (2): 43–47.

Amira, P.O., Babalola, O., and Oyediran, A. (2014). Physicochemical properties of palm kernel oil. *Current Research Journal of Biological Sciences* 6 (5): 205–207.

Azlan, A., Prasad, K.N., Khoo, H.E. et al. (2010). Comparison of fatty acids, vitamin E and physicochemical properties of Canarium odontophyllum Miq.(dabai), olive and palm oils. *Journal of Food Composition and Analysis* 23 (8): 772–776.

Azzeme, A.M., Abdullah, S.N.A., Aziz, M.A. et al. (2016). Oil palm leaves and roots differ in physiological response, antioxidant enzyme activities and expression of stress-responsive genes upon exposure to drought stress. *Acta Physiologiae Plantarum* 38 (2): 52.

Bailey, A.E. and Shahidi, F. (2005). *Bailey's Industrial Oil & Fats Products*. Chichester: Wiley.

Balasubramaniam, V.G., Sathvika, S., Ayyappan, P. et al. (2020). Improved oxidative stability of peanut oil through addition of finger millet (Eleusine coracana) seed coat polyphenols. *Journal of Food Process Engineering* 43 (3): e13194.

Balbino, S. (2017). Vegetable oil yield and composition influenced by environmental stress factors. In: *Oilseed Crops: Yield and Adaptations Under Environmental Stress* (ed. P. Ahmad), 80–101. Chichester: John Wiley & Sons.

Barcelos, E., Rios, S.d.A., Cunha, R.N. et al. (2015). Oil palm natural diversity and the potential for yield improvement. *Frontiers in Plant Science* 6: 190.

Bauer, B. and Kostik, V. (2014). The content of saturated, monounsaturated and polyunsaturated fatty acids in the seeds of different canola varieties. In: *Proceedings of the Eighth Conference on Medicinal and Aromatic Plants of Southeast European Countries* (eds. Z.D. Stevanović and A. Ibraliu). Belgrade: Academy of Science of Albania, Agricultural University of Tirana and Association for Medicinal and Aromatic Plants of Southeast European Countries (AMAPSEEC).

Bernard, J.K. (2016). Oilseed and oilseed meals. In: *Reference Module in Food Science* (ed. G.W. Smithers). Amsterdam: Elsevier.

Beszterda, M. and Nogala-Kałucka, M. (2019). Current research developments on the processing and improvement of the nutritional quality of rapeseed (Brassica napus L.). *European Journal of Lipid Science and Technology* 121 (5): 1800045.

Birch, E.E., Garfield, S., Hoffman, D.R. et al. (2000). A randomized controlled trial of early dietary supply of long-chain polyunsaturated fatty acids and mental development in term infants. *Developmental Medicine and Child Neurology* 42 (3): 174–181.

Bishop, G.J. and Yokota, T. (2001). Plants steroid hormones, brassinosteroids: current highlights of molecular aspects on their synthesis/metabolism, transport, perception and response. *Plant and Cell Physiology* 42 (2): 114–120.

Broschat, T.K. and Crane, J.H. (2017). *The Coconut Palm in Florida*. Gainseville, FL: Institute of Food and Agricultural Sciences (IFAS), HS40.

Burgal, J., Shockey, J., Lu, C. et al. (2008). Metabolic engineering of hydroxy fatty acid production in plants: RcDGAT2 drives dramatic increases in ricinoleate levels in seed oil. *Plant Biotechnology Journal* 6 (8): 819–831.

Campillo, C., Fortes, R., Prieto, M.D.H. et al. (2012). Solar radiation effect on crop production. *Solar Radiation* 1: 494.

Canada, H. (2016). Consultation on the health implications of alternatives to trans fatty acids: summary of responses from experts. https://web.archive.org/web/20070225021532/

http:/www.hc-sc.gc.ca/fn-an/nutrition/gras-trans-fats/tf-ge/tf-gt_app9iii_e.html (accessed March 12, 2020).

Carelli, A.A., Brevedan, M.I., and Crapiste, G.H. (1997). Quantitative determination of phospholipids in sunflower oil. *Journal of the American Oil Chemists' Society* 74 (5): 511–514.

Carelli, A.A., Franco, I.C., and Crapiste, G.H. (2005). Effectiveness of added natural antioxidants in sunflower oil. *Grasas y Aceites* 56 (4): 303–310.

Chadwick, S.S. (1988). Ullmann's encyclopedia of industrial chemistry. *Reference Services Review* 16 (4): 31–34.

Chatepa, L.E.C., Uluko, H., and Masamba, K. (2019). Comparison of oil quality extracted from selected conventional and non conventional sources of vegetable oil from Malawi. *African Journal of Biotechnology* 18 (8): 171–180.

Cheema, M., Malik, M., Hussain, A. et al. (2001). Effects of time and rate of nitrogen and phosphorus application on the growth and the seed and oil yields of canola (Brassica napus L.). *Journal of Agronomy and Crop Science* 186 (2): 103–110.

Chen, Y., Cao, Y., Zhao, L. et al. (2014). Macronutrients and micronutrients of soybean oil bodies extracted at different pH. *Journal of Food Science* 79 (7): C1285–C1291.

Chew, S.C. (2020). Cold-pressed rapeseed (Brassica napus) oil: chemistry and functionality. *Food Research International* 131: 108997.

Chivandi, E., Davidson, B.C., and Erlwanger, K.H. (2008). A comparison of the lipid and fatty acid profiles from the kernels of the fruit (nuts) of Ximenia caffra and Ricinodendron rautanenii from Zimbabwe. *Industrial Crops and Products* 27 (1): 29–32.

Ciannamea, E.M., Castillo, L.A., Barbosa, S.E. et al. (2018). Barrier properties and mechanical strength of bio-renewable, heat-sealable films based on gelatin, glycerol and soybean oil for sustainable food packaging. *Reactive and Functional Polymers* 125: 29–36.

Diepenbrock, W. (2000). Yield analysis of winter oilseed rape (*Brassica napus* L.): a review. *Field Crops Research* 67 (1): 35–49.

Díez, C.M., Moral, J., Cabello, D. et al. (2016). Cultivar and tree density as key factors in the long-term performance of super high-density olive orchards. *Frontiers in Plant Science* 7: 1226.

Dunford, N.T. (2004). Oil-and oilseed-based bioactive compounds and their health effects. In: *Nutritionally Enhanced Edible Oil and Oilseed Processing* (eds. N.T. Dunford and H.B. Dunford), 1–24. Champaign, IL: AOCS Press.

Dunford, N.T. and Dunford, H.B. (2004). *Nutritionally Enhanced Edible Oil and Oilseed Processing*. Champain, IL: AOCS Press.

Dupont, J. (2003). Vegetable oils: dietary importance. In: *Encyclopedia of Food Sciences and Nutrition* (eds. B. Caballero, L.C. Trugo and P.M. Finglas), 5921–5925. San Diego, CA: Academic Press.

Enig, M., Pallansch, L., Sampugna, J. et al. (1983). Fatty acid composition of the fat in selected food items with emphasis on trans components1. *Journal of the American Oil Chemists' Society* 60 (10): 1788–1795.

Evans, L. and Fischer, R. (1999). Yield potential: its definition, measurement, and significance. *Crop Science* 39 (6): 1544–1551.

Farm Energy (2019). Oilseed crops for biodiesel production. https://farm-energy.extension.org/oilseed-crops-for-biodiesel-production, (accessed September 9, 2020).

FAOSTAT (2018). FAO Statistics Division http://www.fao.org/faostat/en/#data/QC (accessed March 15, 2019).

Fernández-Escobar, R., Beltrán, G., Sánchez-Zamora, M. et al. (2006). Olive oil quality decreases with nitrogen over-fertilization. *HortScience* 41 (1): 215–219.

Fetzer, A., Herfellner, T., Stäbler, A. et al. (2018). Influence of process conditions during aqueous protein extraction upon yield from pre-pressed and cold-pressed rapeseed press cake. *Industrial Crops and Products* 112: 236–246.

Firestone, D. (2006). *Physical and Chemical Characteristics of Oils, Fats and Waxes*. Urbana, IL: AOCS Press.

Galli, F. and Azzi, A. (2010). Present trends in vitamin E research. *BioFactors* 36 (1): 33–42.

García-Inza, G.P., Castro, D.N., Hall, A.J. et al. (2014). Responses to temperature of fruit dry weight, oil concentration, and oil fatty acid composition in olive (*Olea europaea* L. var.'Arauco'). *European Journal of Agronomy* 54: 107–115.

Godswill, A.C., Amagwula, I.O., Victory, I.S. et al. (2018). Effects of repeated deep frying on refractive index and peroxide value of selected vegetable oils. *International Journal of Advanced Academic Research* 4 (4): 116–119.

Goh, K.-J., Härdter, R., and Fairhurst, T. (2003). Fertilizing for maximum return. In: *Oil Palm: Management for Large and Sustainable Yields* (eds. T.F. Fairhurst and R. Härdter), 279–306. Penang, Malaysia: Interntional Plant Nutrition Institute.

Grompone, M.A. (2005). Sunflower oil. In: *Bailey's Industrial Oil and Fat Products*, 6e (ed. F. Shahidi), 655–730. Hoboken, NJ: Wiley.

Gucci, R., Fereres, E., and Goldhamer, D. (2012). Olive. In: *Yield Response to Water of Fruit Trees and Vines: Guidelines*, vol. 66 (eds. E. Fereres, D.A. Goldhamer and V.O. Sadras), 246–497. FAO.

Gunstone, F.D. (2011). *Vegetable Oils in Food Technology: Composition, Properties and Uses*. Chichester: Wiley.

Gupta, M., Prasad, A., Ram, M. et al. (2002). Effect of the vesicular–arbuscular mycorrhizal (VAM). fungus Glomus fasciculatum on the essential oil yield related characters and nutrient acquisition in the crops of different cultivars of menthol mint (Mentha arvensis). under field conditions. *Bioresource Technology* 81 (1): 77–79.

Gurmu, F., Mohammed, H., and Alemaw, G. (2009). Genotype x environment interactions and stability of soybean for grain yield and nutrition quality. *African Crop Science Journal* 17 (2): 87–99.

Harris, H.C., McWilliam, J., and Mason, W. (1978). Influence of temperature on oil content and composition of sunflower seed. *Australian Journal of Agricultural Research* 29 (6): 1203–1212.

Hemming, S., Dueck, T., Janse, J. et al. (2007). The effect of diffuse light on crops. In: *Proceedings of the International Symposium on High Technology for Greenhouse System Management: Greensys* (ed. S. De Pascale), 801. Leuven: International Society for Horticultural Science.

Hinrichsen, N. (2016). Commercially available alternatives to palm oil. *Lipid Technology* 28 (3–4): 65–67.

Intelligence, C.M. (2006). *CBI Trends: Vegetables Oils in Europe*. CBI Ministry of Foreing Affairs.

Inuwa, H.M., Aina, V.O., Gabi, B. et al. (2011). Comparative determination of antinutritional factors in groundnut oil and palm oil. *Advance Journal of Food Science and Technology* 3 (4): 275–279.

Jankowski, K. and Budzyński, W. (2003). Energy potential of oilseed crops. *Electronic Journal of Polish Agricultural Universities* 6 (2).

Khan, H., Muhammad, S., Shah, R. et al. (2007). Genetic analysis of yield and some yield components in sunflower. *Sarhad Journal of Agriculture* 23 (4): 985.

Kim, W.W., Rho, H.S., Hong, Y.D. et al. (2013). Determination and comparison of seed oil triacylglycerol composition of various soybeans (Glycine max (L.)) using 1H-NMR spectroscopy. *Molecules* 18 (11): 14448–14454.

Knowles, P. (1983). Genetics and breeding of oilseed crops. *Economic Botany* 37 (4): 423–433.

Latif, J., Mohd, N., Tayeb, D. et al. (2003). Economics of higher planting density in oil palm plantations. *Oil Palm Industry Economic Journal* 3 (2): 32–39.

Lavee, S. and Wodner, M. (2004). The effect of yield, harvest time and fruit size on the oil content in fruits of irrigated olive trees (Olea europaea), cvs. Barnea and Manzanillo. *Scientia Horticulturae* 99 (3–4): 267–277.

Li, R., Yu, K., Hatanaka, T. et al. (2010). Vernonia DGATs increase accumulation of epoxy fatty acids in oil. *Plant Biotechnology Journal* 8 (2): 184–195.

Mailer, R.J. (2016). Oilseeds: Overview. In: *Encyclopedia of Food Grains* (eds. C.W. Wrigley, H. Corke, K. Seetharaman, et al.). Oxford: Academic Press.

McKevith, B. (2005). Nutritional aspects of oilseeds. *Nutrition Bulletin* 30 (1): 13–26.

Mesera, T. and Mitiku, M. (2015). Performance evaluation of sesame (*Sesamum indicum* L.) varieties in lowland area of South Omo Zone, SNNPR, Ethiopia. *International Journal of Agricultural Science* 7 (2): 434–439.

Miklič, V., Jocić, S., Miladinović, D. et al. (2008). Changes in seed oil content of sunflower (Helianthus annus L.), as affected by harvesting date. In: *Proceedings of the 17th International Sunflower Conference, Cordoba, Spain* (ed. L. Velasco), 281–285. Seville: Consejería de Agricultura y Pesca.

Moser, D., Eckerstorfer, M., Pascher, K. et al. (2013). Potential of genetically modified oilseed rape for biofuels in Austria: land use patterns and coexistence constraints could decrease domestic feedstock production. *Biomass and Bioenergy* 50: 35–44.

Müller, T., Lüttschwager, D., and Lentzsch, P. (2010). Recovery from drought stress at the shooting stage in oilseed rape (Brassica napus). *Journal of Agronomy and Crop Science* 196 (2): 81–89.

Murphy, D.J. (2014). Using modern plant breeding to improve the nutritional and technological qualities of oil crops. *OCL* 21 (6): D607.

Neagu, A.-A., Niţa, I., Botez, E. et al. (2013). A physico-chemical study for some edible oils properties. *Analele Universitatii Ovidius Constanta-Seria Chimie* 24 (2): 121–126.

OECD-FAO (2017). *Commodity Snapshots*. FAO.

Panda, H. (2010). *Fruits, Vegetables, Corn and Oilseeds Processing Handbook*. Delhi: Asia Pacific Business Press.

Pande, G., Akoh, C.C., and Lai, O.-M. (2012). Food uses of palm oil and its components. In: *Palm Oil: Production, Processing, Characterization, and Uses* (eds. O.-M. Lai, C.-P. Tan and C. Akoh), 561–586. Urbana, IL: Elsevier.

Paramananthan, S. (2003). Land selection for oil palm. In: *Oil Palm – Managament for Large and Sustainable Yields* (eds. R. Härdter and T.H. Fairhurst), 27–58. Singapore: Potash & Phosphate Institute/Potash & Phosphate Institute of Canada (PPI/PPIC) and International Potash Institute (IPI).

Parveen, A.-u.-H.M., Akhtar, J., and Basra, S.M. (2016). Interactive effect of salinity and potassium on growth, biochemical parameters, protein and oil quality of soybean genotypes. *Pakistan Journal of Agricultural Sciences* 53 (1): 69–78.

Pasban, E.B. (2009). Evaluation of physiological indices, yield and its components as screening techniques for water deficit tolerance in oilseed rape cultivars. *Journal of Agricultural Science and Technology* 11 (4): 413–422.

Pathmeswaran, C., Lokupitiya, E., Waidyarathne, K. et al. (2018). Impact of extreme weather events on coconut productivity in three climatic zones of Sri Lanka. *European Journal of Agronomy* 96: 47–53.

Perrier, A., Delsart, C., Boussetta, N. et al. (2017). Effect of ultrasound and green solvents addition on the oil extraction efficiency from rapeseed flakes. *Ultrasonics Sonochemistry* 39: 58–65.

Plat, J. and Mensink, R.P. (2005). Plant stanol and sterol esters in the control of blood cholesterol levels: mechanism and safety aspects. *The American Journal of Cardiology* 96 (1): 15–22.

Pratap, A., Gupta, S.K., Kumar, J. et al. (2016). Soybean. In: *Breeding Oilseed Crops for Sustainable Production* (ed. S.K. Gupta), 293–315. San Diego: Academic Press.

Ramos, M.J., Fernández, C.M., Casas, A. et al. (2009). Influence of fatty acid composition of raw materials on biodiesel properties. *Bioresource Technology* 100 (1): 261–268.

Rathke, G.-W., Behrens, T., and Diepenbrock, W. (2006). Integrated nitrogen management strategies to improve seed yield, oil content and nitrogen efficiency of winter oilseed rape (*Brassica napus* L.): a review. *Agriculture, Ecosystems & Environment* 117 (2–3): 80–108.

Reale, L., Nasini, L., Cerri, M. et al. (2019). The influence of light on olive (Olea europaea L.) fruit development is cultivar dependent. *Frontiers in Plant Science* 10: 385.

Rodenbush, C., Hsieh, F., and Viswanath, D. (1999). Density and viscosity of vegetable oils. *Journal of the American Oil Chemists' Society* 76 (12): 1415–1419.

Roiaini, M., Ardiannie, T., and Norhayati, H. (2015). Physicochemical properties of canola oil, olive oil and palm olein blends. *International Food Research Journal* 22 (3): 1227–1233.

Rosenthal, A., Pyle, D., and Niranjan, K. (1996). Aqueous and enzymatic processes for edible oil extraction. *Enzyme and Microbial Technology* 19 (6): 402–420.

Salunkhe, D.K., Adsule, R., Chavan, J. et al. (1992). *World Oilseeds*. Berlin: Springer Science & Business Media.

Shahbandeh, M. (2020). Vegetable oils: global consumption by oil type 2013/14 to 2019/2020. https://www.statista.com/statistics/263937/vegetable-oils-global-consumption/#statisticContainer (accessed September 28, 2020).

Sharma, M., Gupta, S., and Mondal, A. (2012). Production and trade of major world oil crops. In: *Technological Innovations in Major World Oil Crops*, vol. 1 (ed. S.K. Gupta), 1–15. New York: Springer.

Skiada, V., Tsarouhas, P., and Varzakas, T. (2020). Comparison and discrimination of two major monocultivar extra virgin olive oils in the southern region of Peloponnese, according to specific compositional/traceability markers. *Food* 9 (2): 155.

Smith, G.I., Atherton, P., Reeds, D.N. et al. (2011). Omega-3 polyunsaturated fatty acids augment the muscle protein anabolic response to hyperinsulinaemia–hyperaminoacidaemia in healthy young and middle-aged men and women. *Clinical Science* 121 (6): 267–278.

Smouse, T.H. (1995). Factors affecting oil quality and stability. In: *Methods to Assess Quality and Stability of Oils and Fat-Containing Foods* (eds. K. Warner and N.A.M. Eskin), 17–36. Urbana, IL: AOCS Press.

Soldo, B., Andelic, I., Nikolov, N. et al. (2019). The effect of selected herb extracts on oxidative stability of vegetable oils. *Croatica Chemica Acta* 92 (3): 331–336.

Sukhotu, R., Guo, S., Xing, J. et al. (2016). Changes in physiochemical properties and stability of peanut oil body emulsions by applying gum arabic. *LWT- Food Science and Technology* 68: 432–438.

Tan, C.H., Ghazali, H.M., Kuntom, A. et al. (2009). Extraction and physicochemical properties of low free fatty acid crude palm oil. *Food Chemistry* 113 (2): 645–650.

Truksa, M., Wu, G., Vrinten, P. et al. (2006). Metabolic engineering of plants to produce very long-chain polyunsaturated fatty acids. *Transgenic Research* 15 (2): 131–137.

USDA (2019). EU-28 Oilseeds and Products Annual. http://www.fas.usda.gov/data/eu-28-oilseeds-and-products-annual-3 (accessed March 8, 2020).

USDA (2020). Food Data Central. https://fdc.nal.usda.gov (accessed March 8, 2020).

Van Kraalingen, D., Breure, C.J., and Spitters, C. (1989). Simulation of oil palm growth and yield. *Agricultural and Forest Meteorology* 46 (3): 227–244.

Vega, C.R., Andrade, F.H., Sadras, V.O. et al. (2001). Seed number as a function of growth. A comparative study in soybean, sunflower, and maize. *Crop Science* 41 (3): 748–754.

Velasco, L. and Ruiz-Méndez, M.V. (2015). Sunflower oil minor constituents. In: *Sunflower: Chemistry, Production, Processing, and Utilization* (eds. E. Martínez-Force, N. Durnford and J. Salas), 297–329. Urbana, IL: Elsevier.

Walker, J. (2008). Book review: People, plants and genes. The story of crops and humanity by D.J. Murphy. *Experimental Agriculture* 44 (4): 571.

Wang, Q. (2016). *Peanuts: Processing Technology and Product Development*. Amsterdam: Academic Press.

Waseem, S., Imadi, S.R., Gul, A. et al. (2017). Oilseed crops: present scenario and future prospects. In: *Oilseed Crops: Yield and Adaptations Under Environmental Stress* (ed. P. Ahmad), 1–18. Chichester: John Wiley & Sons.

White, P.M., Wolf, D.C., Thoma, G.J. et al. (2006). Phytoremediation of alkylated polycyclic aromatic hydrocarbons in a crude oil-contaminated soil. *Water, Air, and Soil Pollution* 169 (1–4): 207–220.

Willis, S. (ed.) The use of soybean meal and full fat soybean meal by the animal feed industry. Preseented at the 12th Australian soybean conference Soy Australia, Bundaberg. 2003.

Winkler, J.K. and Warner, K. (2008). The effect of phytosterol concentration on oxidative stability and thermal polymerization of heated oils. *European Journal of Lipid Science and Technology* 110 (5): 455–464.

Woittiez, L.S., van Wijk, M.T., Slingerland, M. et al. (2017). Yield gaps in oil palm: a quantitative review of contributing factors. *European Journal of Agronomy* 83: 57–77.

Wu, G., Truksa, M., Datla, N. et al. (2005). Stepwise engineering to produce high yields of very long-chain polyunsaturated fatty acids in plants. *Nature Biotechnology* 23 (8): 1013–1017.

Xu, B., Wei, B., Ren, X. et al. (2018). Dielectric pretreatment of rapeseed 1: influence on the drying characteristics of the seeds and Physico-chemical properties of cold-pressed oil. *Food and Bioprocess Technology* 11 (6): 1236–1247.

Yanishlieva, N., Raneva, V., and Marinova, E. (2001). β-carotene in sunflower oil oxidation. *Grasas y Aceites* 52 (1): 10–16.

Ye, Z., Li, R., Cao, C. et al. (2019). Fatty acid profiles of typical dietary lipids after gastrointestinal digestion and absorbtion: a combination study between in-vitro and in-vivo. *Food Chemistry* 280: 34–44.

Zhong, Y., Wang, Z., and Zhao, Y. (2015). Impact of radio frequency, microwaving, and high hydrostatic pressure at elevated temperature on the nutritional and antinutritional components in black soybeans. *Journal of Food Science* 80 (12): C2732–C2739.

Zimba, N., Wren, S., and Stucki, A. (2005). Three major tree nut oils of southern central Africa: their uses and future as commercial base oils. *International Journal of Aromatherapy* 15 (4): 177–182.

Zweytick, D., Athenstaedt, K., and Daum, G. (2000). Intracellular lipid particles of eukaryotic cells. *Biochimica et Biophysica Acta (BBA) - Reviews on Biomembranes* 1469 (2): 101–120.

3 Novel Sources for Oil Production

Marco Garcia-Vaquero[1,2] and Brijesh K. Tiwari[2]
[1] *School of Agriculture and Food Science, University College Dublin, Dublin, Ireland*
[2] *TEAGASC, Food Research Centre, Dublin, Ireland*

3.1 Introduction

Fat is an essential macronutrient in human diet as a source of energy, accounting for approximately 25 or 30% of the total energy of a well-balanced diet as determined by the *Codex Alimentarius* (Bialek et al. 2017). Moreover, dietary fat is also a source of other natural antioxidants such as tocopherols, carotenoids, and phenolic compounds with proven health benefits (Bialek et al. 2017; Mariod et al. 2017).

The biological relevance of dietary fat is related mainly to the type and proportion of fatty acids: monounsaturated fatty acids (MUFAs) and polyunsaturated fatty acids (PUFAs) (Bialek et al. 2017; Xue et al. 2018). PUFAs have important health and nutritional benefits, providing protection against cardiovascular diseases, rheumatoid arthritis, and some types of cancer (Maurer et al. 2012). PUFAs – omega-6 (n-6) and omega-3 (n-3) – are essential in the diet at a ratio of 5:1 (n-6:n-3) according to the European Nutritional Society (Dawczynski et al. 2007). European diets regularly provide a ratio n-6:n-3 ratio of 15–17:1, suggesting a deficit in n-3 and an excess of n-6 (Dawczynski et al. 2007; Holdt and Kraan 2011). Fish and fish oil, have been traditionally used as source of n-3 PUFA, with levels of n-6 ranging from 0.9 to 12 g/100 g oil and higher levels of n-3, ranging between 11.9 to 35.3 g/100 g oil (Rubio-Rodríguez et al. 2010). However, the production and processing of fish oil is costly, requiring the removal of pigments and volatiles responsible for undesirable sensorial properties (odor and flavor) and other contaminants such as dioxins (Maurer et al. 2012). Moreover, there is currently a concern regarding the sustainability and environmental impacts of fisheries and aquaculture (Vannuccini et al. 2019).

Meeting the future demands of food and quality nutrients for the growing population of the World, while reducing the environmental impact of food production, is a major agricultural and global challenge (Chaudhary et al. 2018). The Food and Agriculture

Oil and Oilseed Processing: Opportunities and Challenges, First Edition.
Edited by Tomás Lafarga, Gloria Bobo, and Ingrid Aguiló-Aguayo.

Organization (FAO) identified five main principles to support a sustainable agriculture and food production which include (i) improvement of resource efficiency; (ii) preservation, protection, and enhancement of natural resources; (iii) protection of rural livelihoods, equity, and social well-being; (iv) development of resilient people, communities, and ecosystems; and (v) responsible and effective governance (FAO 2014).

Novel sources of oil and other health-promoting nutrients such as algae – microalgae and macroalgae – (Garcia-Vaquero et al. 2019a, 2020; Hildebrand et al. 2020), insects (Guil-Guerrero et al. 2018; Elhassan et al. 2019; Mba et al. 2019), and alternative seeds from plants and trees (Mariod et al. 2017) are currently being explored as sustainable sources of high-quality nutrients in substitution of traditional food ingredients. Moreover, no oil from any single source has been found to be suitable for all purposes due to the variable fatty acid composition of the lipids extracted from multiple sources (Mariod et al. 2017). Thus, the fatty acid profile of novel sources of oil is currently being evaluated as they may influence the nutritional quality and the future application of these oils in the food industry. For example, oils containing a high proportion of unsaturated fatty acids may experience important chemical changes such as oxidation during storage and thermal treatments, increasing the generation of off-flavors and toxic compounds (Bialek et al. 2017).

This chapter summarizes the main novel sources of dietary oil, mainly algae (microalgae and macroalgae) and insects, but also other non-conventional herbs and seeds. The fatty acid profile, nutritional quality, and uses of these novel oils will be discussed together with the challenges and sustainability of the industrial exploitation of these unconventional natural resources.

3.2 Algae

Algae, including microalgae and macroalgae, are an extremely diverse group of photosynthetic organisms, with varied physiological, morphological, and genetic traits enabling the production of multiple biologically active compounds (Garcia-Vaquero et al. 2017, 2018, 2019; Ramesh Kumar et al. 2019). Algae are able to grow in aquatic environments and can also be cultivated in waste water, avoiding the competition for arable land and other natural resources with traditional crops (Xue et al. 2018), being a promising source of quality feed and food ingredients (García-Vaquero 2019; Garcia-Vaquero et al. 2019a, b).

3.2.1 Microalgae

Microalgae are a promising source of bioactive compounds including proteins and peptides, carbohydrates, lipids, and pigments such as carotenoids (Ventura et al. 2018; Lafarga et al. 2019a; Hildebrand et al. 2020). Microalgae are one of the top trends in the food industry (Lafarga 2019) and have been used to formulate a wide variety of novel enriched foods (Fradique et al. 2013; Lafarga et al. 2019b, c). The biochemical composition of microalgae depends on the culture conditions (i.e. macro- and micronutrients in the culture media, light, temperature, etc.) that influence the growth of these microorganisms and the production of compounds for the survival of the cells (da Silva Vaz et al. 2016; Chew et al. 2017; Lafarga 2020). Thus, understanding the biosynthetic pathways for the production of the metabolites of interest in microalgae is a promising strategy to target the production and increase the yields of multiple bioactive compounds (Lafarga et al. 2019a). Microalgae are able to accumulate high amounts of lipids, accounting for up to 30–50% of the total composition of the cells, depending on the cultivation conditions (Chew et al. 2017). Thereby, stress conditions such as nitrogen starvation, high temperature, pH shifts, and high salt concentrations are promising strategies to increase the production of lipids in several microalgal species (Kwak et al. 2016). Moreover, the combination of stresses has also shown an effect on the yield of lipids produced by microalgae. For example, the combination of two cultivation stresses can enhance the production of lipids, with increases of up to 106% in oil yields when compared to single stress conditions; while the application of more than three stresses had a negative influence on the yields of these compounds (Kwak et al. 2016). Furthermore, the composition of microalgae can also be manipulated by controlling the cultivation conditions. Mathimani et al. (2018) obtained high amounts of C16:0, C16:1, C18:1 and low abundance of C18:3 fatty acids in *Chlorella vulgaris* BDUG 91771 cells cultivated in media containing low urea and superphosphate when compared to cells produced using a control culture media.

PUFAs or long fatty acids containing more than one double bond in their long carbon chain structure, represent between 25 and 60% of the total lipids of microalgae depending on the species (da Silva Vaz et al. 2016). There are two classes of PUFAs: n-6 synthesized from linoleic acid (LA) and n-3 from linolenic acid with important health beneficial properties in humans (Ramesh Kumar et al. 2019). Microalgae are a promising source of n-6, such as LA (18:2, n-6) and arachidonic acid (AA, 20:4, n-6); and also n-3 PUFAs including eicosapentaenoic acid (EPA, 20:5, n-3) and docosahexaenoic acid (DHA, 22:6, n-3) as seen in detail in Table 3.1.

Among all PUFAs, the consumption of n-3, mainly EPA and DHA, has shown promising health benefits. The Food and Drug Administration (FDA) recognized the benefits of these compounds to reduce the risk of coronary heart disease (Adarme-Vega et al. 2014). Moreover, DHA has also been linked with key roles in maintaining membrane fluidity in the brain and retina, reducing inflammation, and protecting against injury (Saini and Keum 2018). Thus, the culture of microalgae for the production of PUFAs has recently become more cost-effective than the the use of this biomass for the production of biofuels (Abdo et al. 2015).

In order to fully utilize microalgal PUFAs and other bioactive molecules, these compounds should be extracted by developing selective processes that minimize the co-extraction of other compounds, while generating minimum waste and consuming low energy during the processing of the biomass (Lafarga et al. 2019a). The downstream processing of microalgae encompasses cultivation and harvesting processes, cell disruption techniques and extraction, purification and storage of lipids and other high-value compounds (Gong and Bassi 2016; Chew et al. 2017; Lafarga et al. 2019a) as represented in Figure 3.1.

The extraction of lipophilic compounds from microalgae requires the break-down of the complex, dynamic, and rich in carbohydrates algal cell-walls (da Silva Vaz et al. 2016; Kim et al. 2016; Lafarga 2020). As seen in Figure 3.1, there are several mechanical and non-mechanical processes to disrupt microalgal cells and other methods including chemical and biological processes.

The techniques currently used to recover lipids include conventional solvent-based methods such as Soxhlet, but also innovative approaches such as ultrasound-assisted extraction, microwave-assisted extraction, pressurized liquid extraction, and supercritical fluid extraction (Chew et al. 2017). The use of conventional technologies requires high energy input and organic solvents such as hexane, to recover the lipids from the disrupted biomass (Chew et al. 2017; Ramesh Kumar et al. 2019). Innovative extraction

Table 3.1 Fatty acid composition of different microalgal species. *Source:* Compiled by Xue et al. (2018) and reproduced with permission from John Wiley & Sons. The dashes in the table indicate the fatty acid was non-detected and "nd" stands for no determined.

Fatty acid	*Palova gyran*	*Isocrysis galbana*	*Prorocentrum minimum*	*Phaeodactylum tricornutum*	*Cocconeis scutellum*	*Seminavis gracilenta*	*Nannochloropsis oculata*	*Chlorella vulgaris*	*Spirulina platensis*	*Oocystis* sp.
C14:00	15.3	9.5	5.9	6.9	28.6	4.3	3.2	0.5	–	0.8
C16:00	9.6	5.2	7.2	16.9	11	10	24.1	27.6	44.2	14.4
C18:00	4.5	8.7	5.7	0.3	6.5	14.4	2.1	1	–	–
C24:00:00	–	–	–	nd	4.7	5.6	nd	nd	1.1	0.4
Other SFA	–	0.4	3.1	5.9	8.5	3.4	1.1	1.8	–	–
Total SFA	29.4	23.8	21.9	30	59.3	37.6	30.4	30.8	45.3	15.6
C15:01	nd	nd	nd	3.5	nd	nd	nd	4	nd	nd
C16:01	12.4	10.7	9.2	19.2	2.8	3.9	26.6	2.1	6.4	5.7
C18:01	–	0.2	4.6	1.6	13.7	13.2	5.8	1.8	0.9	14.8
Other MUFA	–	0.8	2	3.2	3	–	0.7	1.9	5.2	–
Total MUFA	12.4	11.7	15.8	27.5	19.5	17.1	33.1	9.9	12.5	20.5
C16:4 n-3	nd	nd	nd	0.2	nd	nd	nd	–	nd	nd
C18:2 n-6	0.8	0.4	0.9	3.1	3.8	4.8	5.5	14.7	18.7	24.3
C18:3 n-3	–	9.5	1.3	1.1	–	1.3	nd	40.3	nd	30.8
C18:3 n-6	2	6.4	0.2	0.4	2.2	1.7	nd	4	23.4	nd
C18:4 n-3	nd	nd	nd	nd	nd	nd	nd	nd	nd	2.7
C20:4 n-6	3.7	3.4	13.8	1	2.5	15.8	9.7	0.4	nd	1.9
C20:5 n-3	34.2	10.3	14	34.1	10.1	15.7	21.3	–	nd	4.2
C22:6 n-3	16.4	33.6	28.1	1	2.6	0.2	nd	–	nd	–
Other PUFA	1.2	0.8	3.9	1.4	–	5.8	–	–	–	–
Total PUFA	58.2	64.5	62.3	42.5	21.2	45.3	36.4	59.4	42.2	63.9
References	Suh et al. (2015)	Suh et al. (2015)	Suh et al. (2015)	Matos et al. (2016)	Chen (2012)	Chen (2012)	Wei and Huang (2017)	Matos et al. (2016)	Ramadan et al. (2008)	Patil et al. (2007)

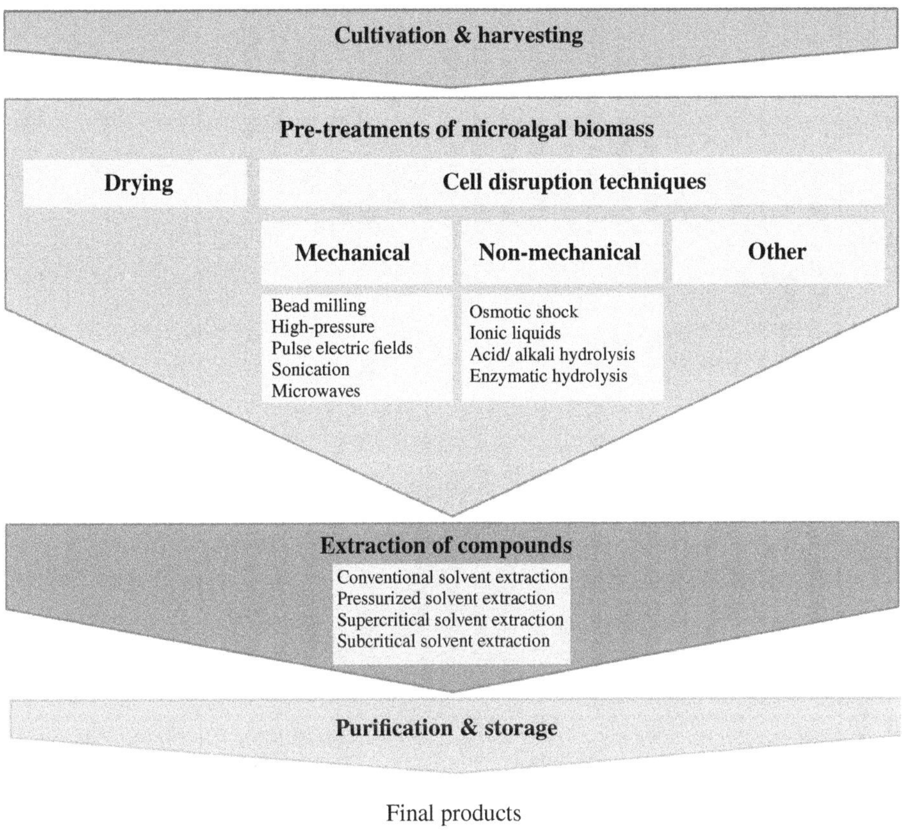

Figure 3.1 Scheme modified from Lafarga et al. (2019a) representing the downstream processing stages of microalgae to obtain lipophilic compounds.

technologies aim to reduce some of the major drawbacks of traditional techniques by developing greener and more sustainable extraction processes. Thereby, supercritical fluids such as CO_2 offer high selectivity when extracting non-polar lipids, using CO_2 that is safe and easy to recycle (Hernández et al. 2014). The main methods, solvents, and conditions used to recover lipids from microalgae are summarized in Table 3.2.

Several strategies have been suggested in the scientific literature to increase the microalgal production of EPA and DHA, such as regulating the cultivation conditions to increase the cellular metabolism and synthesis of both compounds and to follow genetic engineering approaches to increase the synthesis of both lipids (Ramesh Kumar et al. 2019). Moreover, novel trends in algal processing include the exploitation of the microalgal biomass following a biorefinery approach. The biorefinery concept aims to obtain biofuels, energy, and multiple high-value products that can be used as functional foods and nutraceuticals, through the transformation of the biomass using multiple technological processes (Garcia-Vaquero et al. 2019b; Hayes et al. 2019) as represented schematically in Figure 3.2.

Following the approach described in Figure 3.2, the main high-value compounds extracted from the microalgal biomass (oil, carbohydrates, protein, and minerals) could be used as chemicals, fuels, feed, and biogas, while the final residues of the biomass could be transformed into heat and power (Chew et al. 2017).

Table 3.2 Summary of conventional and innovative technologies and extraction protocols used in the literature to recover lipids from microalgae. *Source:* Modified from Mubarak et al. (2015) and reproduced with permission from Elsevier.

Extraction technology	Solvents	Extraction conditions	Microalgae sp.	Yields of lipids (%)	Source
Soxhlet	n-hexane	40 °C, 0.1 MPa, 18 h	*Shizochytrium limacinum*	45	Tang et al. (2011)
	Dichloromethane		*Nannochloropsis oculata*	9	Liau et al. (2010)
	n-Hexane		*Nannochloropsis oculata*	5.79	Liau et al. (2010)
	Ethanol		*Nannochloropsis oculata*	40.9	Liau et al. (2010)
	n-Hexane		*Pavlova* sp.	45.2	Cheng et al. (2011)
	Petroleum ether		*Nannochloropsis oculata*	8.2	Converti et al. (2009)
	Ethanol		*Synechocystis PCC 6803*	48	Sheng et al. (2011)
Solvent extraction	Hexane/ethanol	200 °C, 0.1 MPa, 2 h	*Synechocystis PCC 6803*	52	Sheng et al. (2011)
	Hexane/isopropanol		*Synechocystis PCC 6803*	36	Sheng et al. (2011)
	Chloroform/methanol		*Synechocystis PCC 6803*	50	Sheng et al. (2011)
	Chloroform/methanol/water		*Synechocystis PCC 6803*	42	Sheng et al. (2011)
Supercritical extraction	CO_2 and ethanol	40 °C, 35 MPa, 30 min	*Shizochytrium limacinum*	33.9	Tang et al. 2011; Sheng et al. 2011
	CO_2		*Pavlova* sp.	34	Cheng et al. (2011)
Pressurized fluid extraction	n-Hexane	60 °C, 10–12 MPa, 10 min	*Nannochloropsis oculata*	6.1	Pieber et al. (2012)
	N-Hexane:propan-2-ol (2 : 1)		*Nannochloropsis oculata*	20	Pieber et al. (2012)
	Ethanol 96 vol.%		*Nannochloropsis oculata*	36	Pieber et al. (2012)
Ultrasound-assisted extraction	Petroleum ether	40 kHz, 1 h	*Nannochloropsis oculata*	3.3	Converti et al. (2009)

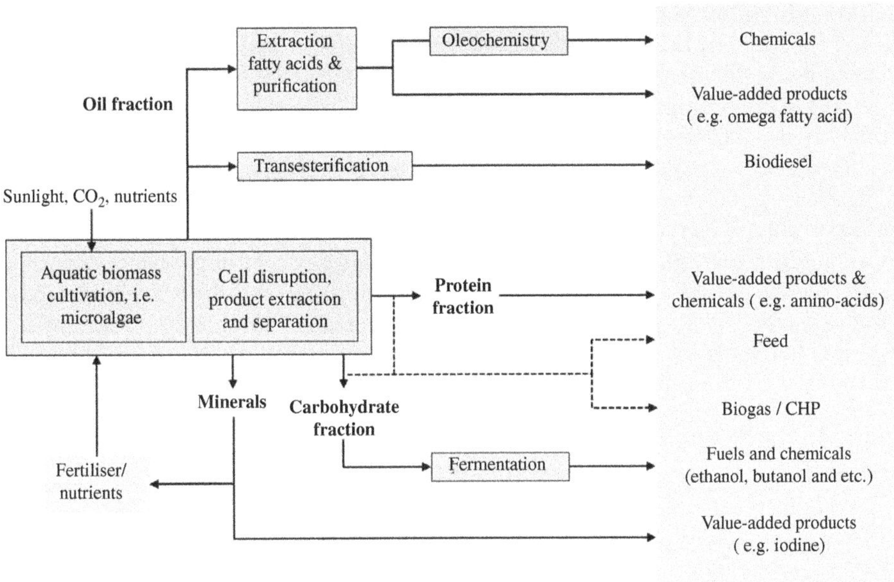

Figure 3.2 Processes and uses of microalgae to obtain biofuels and high-value compounds following a biorefinery approach. *Source:* Image reproduced from Chew et al. (2017) with permission from Elsevier.

Recent advances in the cultivation, extraction, and processing of microalgae have helped to increase the cost-effectiveness of algal biomass for the production of PUFAs (Abdo et al. 2015). Moreover, the recent recommendation to include DHA in infant formula has recently opened a new market for microalgal oil with an estimated value of 10$ billion/year globally (Ward and Singh 2005; Ventura et al. 2018). Currently, the main company producing microalgal DHA is Martek (Columbia, USA), although new companies are also producing enriched DHA oils from microalgae *Schizochytrium* sp., *Crypthecodinium cohnii,* and *Ulkenia* sp. as mentioned in detail by Ventura et al. (2018).

3.2.2 Macroalgae

Similarly to microalgae, seaweed or macroalgae are able to adapt to the marine environment by producing unique molecules with promising biological properties (Garcia-Vaquero and Hayes 2016; Garcia-Vaquero et al. 2017, 2019a; Lafarga et al. 2020). The amount and chemical structure of the macroalgal molecules will depend on the type of seaweed (class and species), the stage of development of the biomass, and the parts of the macroalgae sampled (Imbs et al. 2010; Skriptsova et al. 2011).

Overall, macroalgae are described as a biomass rich in carbohydrates (up to 60%), with medium-high protein content, and low amounts of lipids (ranging from 1 to 3%) (Kraan 2013; Dominguez and Loret 2019). Macroalgal polysaccharides (Garcia-Vaquero et al. 2018, 2019b, 2020; Lafarga et al. 2020), proteins and peptides (Garcia-Vaquero et al. 2017; García-Vaquero 2019; Lafarga et al. 2020) and lipids (da Costa et al. 2019) have been explored due to their health benefits to develop functional foods or nutraceuticals.

Although the total amount of lipids in macroalgae is relatively low (Sánchez-Machado et al. 2004; Kendel et al. 2015), the fatty acid profile of macroalgae is extremely variable depending on the macroalgal class and species, and the composition in PUFAs makes macroalgae interesting sources of lipids for nutraceuticals or functional foods as fish oils replacements (Mišurcová et al. 2011; Gosch et al. 2012). Thereby, Gosch et al. (2012) observed a huge variation in the fatty acid profile between green, red, and brown macroalgae. Green macroalgae were rich in n-3 PUFAs, while red and brown macroalgae species evaluated by the same authors had high amounts of saturated fatty acids (Gosch et al. 2012). Contrary to these authors, Sánchez-Machado et al. (Sánchez-Machado et al. 2004) reported high levels of unsaturated fatty acids in brown seaweeds and a predominance of saturated fatty acids in red macroalgae. Moreover, these authors reported that both macroalgal groups were balanced sources of n-3 and n-6 fatty acids for their use as food.

The fatty acid profile and PUFAs of different macroalgal species is a topic that is still being explored in the recent scientific literature (Sánchez-Machado et al. 2004; Mišurcová et al. 2011; Gosch et al. 2012; Kendel et al. 2015). The fatty acid content of selected macroalgae is summarized in Table 3.3.

The lipid content of dried macroalgae *Himanthalia elongata*, *Laminaria ochroleuca*, *Undaria pinnatifida*, *Palmaria* sp. and *Porphyra* sp. ranged from 0.70 ± 0.09 to 1.80 ± 0.14 g per 100 g dried seaweed (Sánchez-Machado et al. 2004). As seen in Table 3.3 the most abundant fatty acids in the species evaluated by Sánchez-Machado et al. (2004) were the C16:0, C18:1 (n-9), C20:4 (n-6), and C20:5 (n-3).

Kendel et al. (2015) observed huge differences in composition between the green seaweed *Ulva armoricana* and the red macroalgae *Solieria chordalis*. The total content of lipids of both species was low (2.6–3%), although PUFAs represented 29% and 15% of the total lipids in *U. armoricana* and *S. chordalis*, respectively. The main PUFAs identified by these authors in *U. armoricana* were C16:4 (n-3), C18:4 (n-3), C18:2 (n-3), C18:2 (n-6), and C22:6 (n-3), while in the case of *S. chordalis* the main compounds were C20:4 (n-6) and C20:5 (n-3). Recently, Patra et al. (2015) obtained essential oils from *Laminaria japonica* containing 21 volatile compounds – tetradeconoic acid (51.75%), hexadecanoic acid (16.57%), (9Z,12Z)-9,12-octadecadienoic acid (12.09%), and (9Z)-hexadec-9-enoic acid (9.25%) – with antibacterial properties against the foodborne pathogens *Staphylococcus aureus* and *Bacillus cereus*. Moreover, essential oils extracted from *U. pinnatifida* containing tetradecanoic acid (31.32%) and hexadecanoic acid (22.39%) and other minor constituents had strong antioxidant properties (Patra et al. 2017).

3.3 Insects

Approximately 2000 species of insects are currently harvested for human consumption (Nowak et al. 2016; Wilkinson et al. 2018), with about two billion people eating insects on a regular basis (Van Huis 2013). The demand for these food products is particularly high in developing countries in Asia, Africa, and South America (Wilkinson et al. 2018). The nutritional value of insects varies depending on the species, stage of development of the animal (i.e. larva or adult), and the diet received by these animals (Wilkinson et al. 2018). Overall, insects are high in protein and lipids with variable amounts of other compounds including vitamins (riboflavin, pantothenic acid, biotin, and folic acid) and minerals (i.e. copper, iron, selenium, and zinc) (Rumpold and Schlüter 2013; Nowak et al. 2016; Wilkinson et al. 2018).

Table 3.3 Relative fatty acid contents (%) of selected macroalgae analyzed by Sánchez-Machado et al. (2004). Table reproduced with permission from Elsevier.

Fatty acids	**Dried macroalgae**				
	Himanthalia elongata	***Laminaria ochroleuca***	***Undaria pinnatifida***	***Palmaria sp.***	***Porphyra sp.***
C14:0	5.85 ± 0.35	4.97 ± 0.20	3.17 ± 0.31	13.76 ± 0.61	0.53 ± 0.21
C16:0	32.53 ± 1.61	28.51 ± 1.87	16.51 ± 1.35	45.44 ± 1.84	63.19 ± 1.93
C16:1 n-7	2.79 ± 0.25	5.62 ± 0.71	3.70 ± 0.88	5.26 ± 0.63	6.22 ± 0.70
C16:3 n-4	4.38 ± 1.33	0.87 ± 0.10	2.31 ± 1.94	1.20 ± 0.16	1.56 ± 0.51
C18:0	0.68 ± 0.15	0.34 ± 0.14	0.69 ± 0.08	1.28 ± 0.12	1.23 ± 0.10
C18:1 n-9	19.96 ± 2.01	13.62 ± 1.24	6.79 ± 0.90	3.13 ± 0.47	6.70 ± 1.16
C18:1 n-7	–	–	–	2.08 ± 0.33	1.29 ± 0.68
C18:2 n-6	4.39 ± 0.34	6.79 ± 0.61	6.23 ± 0.32	0.69 ± 0.13	1.17 ± 0.13
C18:3 n-3	8.79 ± 0.71	5.15 ± 0.71	11.97 ± 1.75	0.59 ± 0.26	0.23 ± 0.16
C18:4 n-3	3.53 ± 0.56	10.77 ± 1.85	22.60 ± 2.48	0.74 ± 0.47	0.24 ± 0.35
C20:1 n-9	–	–	–	0.20 ± 0.10	4.70 ± 0.26
C20:4 n-6	10.69 ± 1.30	14.20 ± 0.66	15.87 ± 1.68	1.45 ± 0.31	6.80 ± 1.18
C20:4 n-3	0.88 ± 1.80	0.54 ± 0.90	0.70 ± 0.14	0.14 ± 0.03	0.07 ± 0.02
C20:5 n-3	5.50 ± 1.78	8.62 ± 0.56	9.43 ± 0.69	24.05 ± 2.59	6.03 ± 0.95
Saturated FAs	39.06 ± 2.11	33.82 ± 2.21	20.39 ± 1.73	60.48 ± 2.58	64.95 ± 2.24
Monounsaturated	22.75 ± 2.26	19.23 ± 1.99	10.50 ± 1.78	10.67 ± 1.55	18.91 ± 2.81
PUFAs	38.16 ± 7.84	46.94 ± 4.58	69.11 ± 9.01	28.86 ± 3.94	16.10 ± 3.31
PUFAs n-6	15.08 ± 1.64	20.99 ± 1.27	22.10 ± 2.00	2.14 ± 0.45	7.97 ± 1.31
PUFAs n-3	18.70 ± 4.84	25.08 ± 3.21	44.70 ± 5.05	25.52 ± 3.34	7.20 ± 1.48
Ratio n-6 : n-3	0.81	0.83	0.49	0.13	1.21

Insects have a positive effect in the environment (i.e. waste degradation and plant pollination) and when farmed, insects consume fewer natural resources (i.e. requires less water, feed, and space per kilogram of protein), contribute less to the emission of greenhouse gases and ammonia pollution, and have a higher feed efficiency when compared to other livestock (Nowak et al. 2016). Moreover, an increased consumption of insects is likely to have a positive impact in the livelihood, and social and economic conditions in developing countries, allowing the creation of farms with low technical and capital inputs that can benefit the economic development of poor communities (Durst et al. 2010).

Due to the benefits of insects and insect farming, the FAO advocated to increase the consumption of insects to supply food to a global human population that is estimated to reach nine billion by 2050 (FAO 2019). Despite these recommendations, major attitudinal barriers to the consumption of insects still persist in Western countries (Megido et al. 2016; Nowak et al. 2016). Western society still perceives insects as dirty, unhygienic, carriers of diseases, and unsafe to eat, facts that contribute to the feelings of neophobia and disgust toward the consumption of insects (Wilkinson et al. 2018). Food perception is influenced by psychological factors (i.e. culture, personal values, and preferences) and other factors related to the food products such as sensory attributes (e.g. appearance, taste, flavor) and other perceivable benefits such as nutritional qualities and price (Siegrist 2008). Recent research on consumer preferences mentioned that the integration of insects into the Western food culture will mainly involve the use of minced or powdered insects for the formulation of prepared or ready-to-eat foods, before the incorporation of whole insects regularly in the diet (Megido et al. 2016).

The International Network of Food Data Systems (INFOODS) in collaboration with the FAO included recently 448 food entries of 235 insect species belonging to 18 different orders (Nowak et al. 2016). The number of insect food entries in the Food Composition Database for Biodiversity (version 2.1) is represented in Figure 3.3.

The majority of the edible insects currently consumed as food are collected in the wild, and thus, little information is known on their nutritional value. As previously mentioned, insects are generally high in protein and lipids with variable amounts of other nutrients (Rumpold and Schlüter 2013; Nowak et al. 2016; Wilkinson et al. 2018). Rumpold and Schlüter (2013) reviewed the scientific literature on the composition of multiple insects from the orders Blattodea (cockroaches), Coleoptera (beetles and grubs), Diptera (flies), Hemiptera (true bugs), Hymenoptera (ants, bees and wasps), Isoptera (termites), Lepidoptera (butterflies and moths), Odonata (dragonflies and damselflies), and Orthoptera (grasshoppers, crickets, and locusts). A graphic summarizing the protein, lipids, fiber, nitrogen-free extract, and ash in the aforementioned insect orders is represented in Figure 3.4.

As seen in Figure 3.4, lipids are the second most relevant nutrient in the composition of edible insects. The percentage of fat in bugs (Hemiptera), termites (Isoptera), cockroaches (Blattodea) and caterpillars (Lepidoptera) represent 30.26, 32.74, 29.90 and 27.66%, respectively, of the total insect composition. From all the insects analyzed in the literature by Rumpold and Schlüter (2013), larval stages of *Paranerita triangularis* (Lepidoptera) and *Rhynchophorus phoenicis* (Coleoptera) had the highest fat contents, with levels of 70% or even higher (Ramos-Elorduy et al. 1997; Elorduy et al. 1998; Omotoso and Adedire 2007).

Moreover, the fatty acid composition of insects is also really variable. Saturated fatty acids represent approximately 30.83% and 41.97% of the total fatty acids in the orders Hymenoptera and Isoptera, respectively. The two main fatty acids normally described

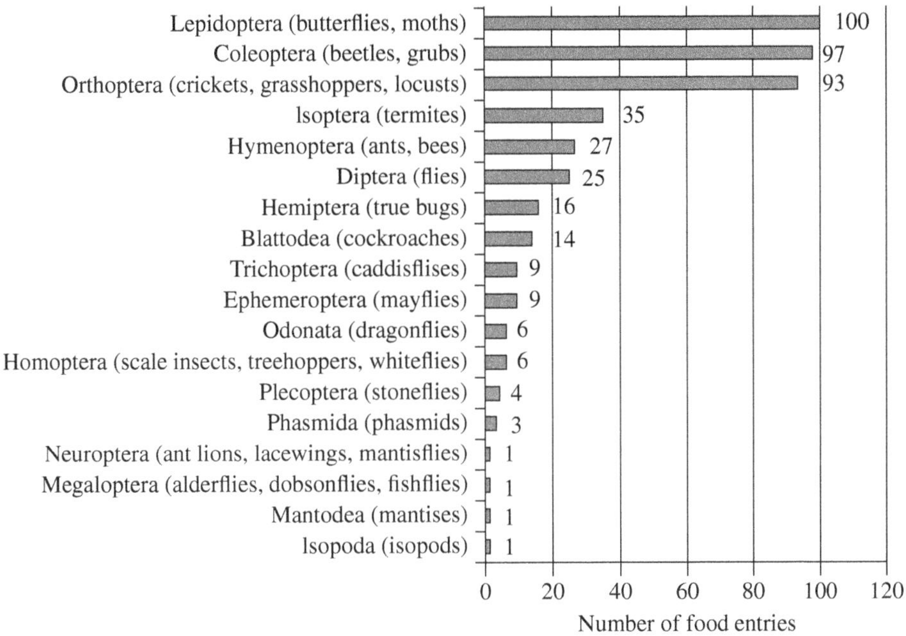

Figure 3.3 Scheme from Nowak et al. (2016) representing the food entries included in the FAO/INFOODS Food Composition Database for Biodiversity (version 2.1). *Source:* The figure is reproduced with permission from Elsevier.

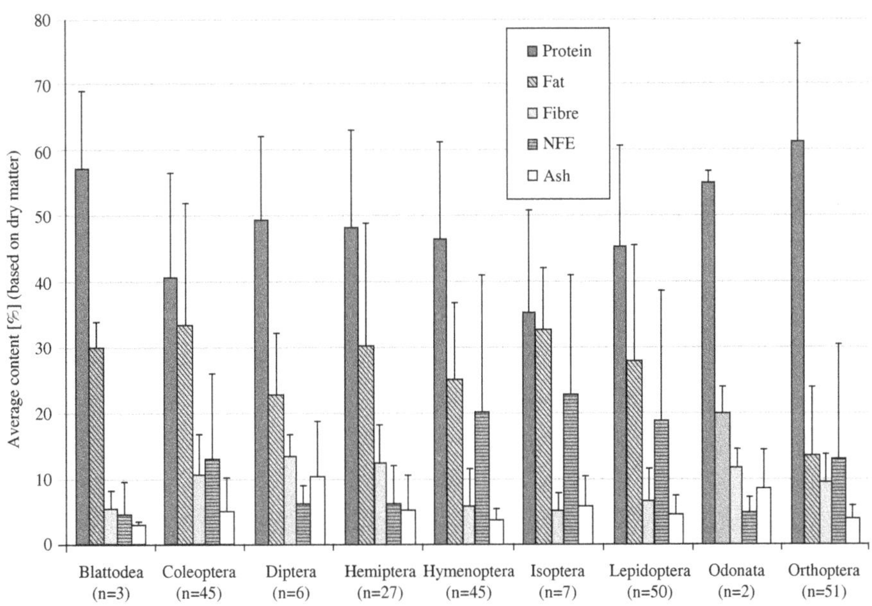

Figure 3.4 Bar chart from Rumpold and Schlüter (2013) representing the average nutrient contents (% on dry matter basis) of edible insects from multiple orders. The abbreviations in the figure are as follow: NFE (nitrogen-free extract) and n (number of insect samples). *Source:* Image reproduced with permission from John Wiley & Sons.

in insects were the palmitic acid (C16:0) and stearic acid (C18:0) (Rumpold and Schlüter 2013). The orders Orthoptera and Lepidoptera are high in PUFAs, mainly palmitoleic acid (C16:1, n-7) and oleic acid (C18:1, n-9) (Rumpold and Schlüter 2013). In general, the fatty acid profile of insects is comparable to those of beef and pork, although insects have higher amounts of PUFAs (Rumpold and Schlüter 2013).

However, the fatty acid composition is extremely variable depending on the species and place of collection. Musundire et al. (2016) identified 10 fatty acid methyl esters from stink bug (*Encosternum delegorguei*) samples including unsaturated (methyl oleate, methyl (Z)-9-hexadecenoate, methyl linoleate, methyl (Z)-9-octadecenoate, methyl octadecanoate, methyl (Z,Z)-9,12-octadecadienoate, and methyl myristoleate) and saturated fatty acids (methyl tetradecanoate, methyl hexadecanoate, and methyl eicosanoate) (Musundire et al. 2016). Oranut et al. (2010) determined the total lipid contents, lipid classes, and fatty acid composition of black (*Polyrhachis vicina* Roger) and red ants (*Oecophylla smaragdina* Fabricius). These authors noticed total lipid contents ranging from 6.3 to 15.2%, with triacylglycerol the major lipid component, representing between 37.4 and 43.4% of the total lipids in black ants and accounting for approximately 80% of the lipids in red ants. Moreover, the authors also mentioned the high abundance of unsaturated fatty acids in ant oil (68.1–77.1%), with oleic acid the most abundant fatty acid, followed by palmitic and LAs (Oranut et al. 2010). (Bophimai and Siri 2010) evaluated the fatty acid profile of six species of dung beetles (*Onthophagus seniculus*, *Onthophagus mouhoti*, *Onitis* spp., *Copris nevinsoni*, *Liatongus rhadamitus*, and *Heliocopris bucephalus*). Palmitic acid (C16:0) was the major fatty acid constituent in all species, while oleic acid (C18:1, n-9) and LA (C18:2, n-6) were the most abundant MUFA and PUFA, respectively (Bophimai and Siri 2010). Moreover, the authors also described low ratios n-6:n-3, of about 2:1 and 5:1 in some species, making these beetles specially suitable to meet the dietary recommendations of n-3 PUFAs in the human diet (Bophimai and Siri 2010). Similarly, Mariod et al. (Mariod et al. 2011) examined the fatty acid composition of two bugs, melon bug (*Aspongubus viduatus* F.) and sorghum bug (*Agonoscelis pubescens*). The average fat percentages of both species was higher than 50% on dry matter basis and in both cases the most predominant fatty acids were oleic, followed by palmitic, linoleic, and linolenic acids (Mariod et al. 2011).

Moreover, the extraction conditions used to obtain oil from insects can also affect the composition and nutritional quality of the oil. Tzompa-Sosa et al. (2014) investigated the influence of different extraction methods on the quantity and quality of oil obtained from *Tenebrio molitor*, *Alphitobius diaperinus*, *Acheta domesticus*, and *Blaptica dubia*. The authors appreciated that the aqueous extraction achieved low yields of lipids, containing also low n-6:n-3 ratio and less polar lipids compared to the Soxhlet and Folch extraction procedures.

3.4 Unconventional Plants and Seeds

Other underutilized vegetable sources including herbs, trees, and seeds have also been explored for the production of oil. Overall, the percentage of lipids in these alternative plants and seeds is lower than the one obtained from traditional crops, although the fatty acid composition of the extracted oils could be potentially suitable and beneficial for human nutrition (Mariod et al. 2017).

Seeds from *Iponema* genus (*I. indica*, *I. nil*, *I. pestigridis* and *I. quamoclit*) had low oil yields compared to other oleaginous crops, with fat contents ranging from 7.84 to 14.71% of the total composition of the seeds. Moreover, linoleic and oleic acids were the two major fatty acid constituents of the extracted oil (Taufeeque et al. 2015). Ruiz del Castillo et al. (2002) analyzed the fatty acid composition of 36 blackcurrant genotypes. These authors determined that blackcurrant seeds, currently a by-product from juice production, had total fatty acid contents ranging from 14 to 23% of the seed. The fatty acid composition of these oils revealed variable amounts of γ-linolenic acid with average values ranging from 11 to 19% of the total amount of fatty acids, although a few genotypes had values around 22–24% (Ruiz del Castillo et al. 2002).

Further information on the lipids and fatty acid composition of multiple alternative vegetable sources including seeds and whole plants, annual herbs, and trees together with the main uses of these compounds has been reviewed in detail by Mariod et al. (2017).

3.5 Opportunities, Challenges, and Future Prospects

Algae (macroalgae and microalgae) and insects are currently underutilized resources to obtain lipids and other bioactive compounds with promising applications in the food, cosmetic, and pharmaceutical industries. Moreover, the cultivation or farming of algae and insects has huge environmental, economic, and societal beneficial impacts in comparison with traditional livestock. Thereby, the cultivation of algae reduces the competition for arable land and the use of pesticides in comparison with terrestrial crops (Xue et al. 2018) and insects consume fewer natural resources and contribute less to the emission of greenhouse gases when compared to conventional livestock (Nowak et al. 2016).

The composition of alternative sources of oil is still under investigation and variable results have been obtained depending on the species, but also on the environmental factors affecting the development of these organisms. Thereby, microalgae are able to accumulate high amounts of lipids, accounting for up to 30–50% of their total weight depending on the cultivation conditions (Chew et al. 2017). Moreover, several species of algae and insects have also shown a fatty acid profile rich in n-3 and n-6 PUFAs, such as EPA and DHA, with relevant biological functions in maintaining human health. In the case of microalgae, the cultivation conditions can also be adjusted using multiple stresses to induce the metabolic pathways that increase the production of PUFAs and other lipophilic compounds such as carotenoids (Lafarga et al. 2019a). The scientific developments in genomics, transcriptomics, and proteomics of algae allowed to discover genes affecting key metabolic pathways offering huge opportunities to obtain high levels of molecules through metabolic or transcriptional engineering (Lafarga et al. 2019a).

Moreover, the exploitation of natural resources following a biorefinery approach and using multiple innovative technologies to increase the sequential extraction of compounds offers huge opportunities for a multi-purpose exploitation of the biomass (Garcia-Vaquero et al. 2019b), allowing the extraction of oils and other high-value compounds such as polysaccharides, proteins, and pigments.

Despite the multiple benefits of these novel sources of high-value compounds, according to the European legislation, products that were not significantly consumed in the European Union (EU) before May 15, 1997 have to follow the novel foods Regulation (EC) No 258/97 (EC 1997) and will only be marketed in the EU following a safety assessment and approval. The Regulation (EU) 2015/2283 (EU 2015) also indicates that whole animals, and thus whole insects, are covered as "novel foods."

I. GENERAL FOOD LAW. Regulation EC 178/200

Art. 1. Food: "any substance or product, whether processed, partially processed or unprocessed, intended to be, or reasonably expected to be ingested by humans".
Art. 2. Feed: "any substance or product, including additives, whether processed, partially processed or unprocessed, intended to be used for oral feeding to animals".
Art. 17. Responsibilities of food/feed business operators (FBO)
Art. 14-15. Food/feed should not be placed in the market if they are believed unsafe.
Art. 16. Presentation of the product should not mislead consumers.
Art. 18. Food/feed traceable by FBO
Art. 19-20. Withdraw a product from the market and inform authorities.

II. FOOD/FEED HYGIENE

II.1. Food hygiene regulations
Regulation EC 853/2004. Hygiene rules for food of animal origin.
Regulation EC 2073/2005. Regulation EC 1441/2007. Microbiological criteria for foodstuffs.

II.2. Feed hygiene regulations
Regulation EC 183/2005. Requirements for feed hygiene.
Annex I. Obligations in primary production.
Annex II. Obligations on feed after primary production
- Monitor undesirable substance and contaminants.
- HACCP (hazard analysis and Critical control point).
- Guides of good practice (national/community).

Regulation EC 852/2004. Hygiene foodstuff.
Annex I. Hygiene and records in primary feed production.
- Measures to control contamination of the products.
- Keep records on nature and origin of feed fed to animals.
- Guidelines on good hygiene practices (national).

Annex II. Requirements for non-primary producers.

III. NOVEL FOOD LEGISLATION

Regulation EC 258/1997. The Novel Food Regulation.
Novel food/food ingredients.
Seaweed as Novel Food in art. 1(2) d.
Recommendation 97/618/EC. Data requirements applicants of novel food product.

IV. NOVEL FEED LEGISLATION

Regulation EC 767/2009. On the placing on the market and use of feed.
Definition. Notification to place a novel feed for the first time after 1 September 2010.
Annex III. List of materials prohibited for feed.
Recommendation 2011/25/EU. Definition on feed (materials, additives and others).

Regulation EU 68/2013. Feed Catalogue (marketing names).

V. LEGISLATION ON CHEMICAL FOOD/FEED SAFETY

Regulation EC 396/2005. Maximum residue levels (MRL) pesticides food/feed of plant/ animal origin.
Annex I replaced by **Regulation (EU) 212/2013.** MRL primary agricultural products (seaweed as food).

Regulation EC 1881/2006. Maximum levels for certain contaminants in foodstuffs.

Directive 2002/32/EC. Undesirable substances in animal feed.

Figure 3.5 Scheme from Garcia-Vaquero and Hayes (2016) summarizing the main regulations and criteria to market food and feed products and ingredients in the European Union. *Source:* Image reproduced with permission from Taylor & Francis.

Moreover, the imports of insects (live or dead), insect parts, and other insect-derived ingredients coming from third countries are subject to special import procedures, including veterinary checks and border inspections as described in detail in the Directive 97/78/EC (EC 1997), Directive 91/496/EC (EEC 1991) and Decision 2007/275/EC (European Commission 2007). A summary of the main legislation applied to food and feed products in the EU, focusing on algal products is shown in Figure 3.5.

Moreover, the complexity of the EU regulations concerning novel sources of proteins such as insects, algae, duckweed, and rapeseed was reviewed by Van der Spiegel et al. (2013), emphasizing the need to clarify the European legislation concerning the expected entry into the EU of novel food and feed replacers for animal-derived proteins.

Acknowledgements

The authors thank the EU for financial support through Interreg Atlantic Area Project EAPA_338/2016 EnhanceMicroAlgae (High added-value industrial opportunities for microalgae in the Atlantic Area).

References

Abdo, M.S., Ali, G.H., and El-Baz, F.K. (2015). Potential production of omega fatty acids from microalgae. *International Journal of Pharmaceutical Sciences Review and Resarch* 32 (2): 210–215.

Adarme-Vega, T.C., Thomas-Hall, S.R., and Schenk, P.M. (2014). Towards sustainable sources for omega-3 fatty acids production. *Current Opinion in Biotechnology* 26: 14–18.

Bialek, A., Bialek, M., Jelinska, M. et al. (2017). Fatty acid composition and oxidative characteristics of novel edible oils in Poland. *CyTA Journal of Food* 15 (1): 1–8.

Bophimai, P. and Siri, S. (2010). Fatty acid composition of some edible dung beetles in Thailand. *International Food Research Journal* 17: 1025–1030.

Chaudhary, A., Gustafson, D., and Mathys, A. (2018). Multi-indicator sustainability assessment of global food systems. *Nature Communications* 9 (1): 848.

Chen, Y.-C. (2012). The biomass and total lipid content and composition of twelve species of marine diatoms cultured under various environments. *Food Chemistry* 131 (1): 211–219.

Cheng, C.-H., Du, T.-B., Pi, H.-C. et al. (2011). Comparative study of lipid extraction from microalgae by organic solvent and supercritical CO_2. *Bioresource Technology* 102 (21): 10151–10153.

Chew, K.W., Yap, J.Y., Show, P.L. et al. (2017). Microalgae biorefinery: high value products perspectives. *Bioresource Technology* 229: 53–62.

Converti, A., Casazza, A.A., Ortiz, E.Y. et al. (2009). Effect of temperature and nitrogen concentration on the growth and lipid content of *Nannochloropsis oculata* and *Chlorella vulgaris* for biodiesel production. *Chemical Engineering and Processing: Process Intensification* 48 (6): 1146–1151.

da Costa, E., Domingues, P., Melo, T. et al. (2019). Lipidomic signatures reveal seasonal shifts on the relative abundance of high-valued lipids from the brown algae Fucus vesiculosus. *Marine Drugs* 17 (6): 335.

Dawczynski, C., Schubert, R., and Jahreis, G. (2007). Amino acids, fatty acids, and dietary fibre in edible seaweed products. *Food Chemistry* 103 (3): 891–899.

Dominguez, H. and Loret, E.P. (2019). Ulva lactuca, a source of troubles and potential riches. *Marine Drugs* 17 (6): 357.

Durst, P., Johnson, D., Leslie, R. et al. (2010). Forest insects as food: humans bite back. In: *Proceedings of a Workshop on Asia-Pacific Resources and Their Potential for*

Development. Chiang Mai, Thailand, 19–21 February, 2008. Rome: Food and Agriculture Organization of the United Nations (FAO).
EC (1997). Council Directive 97/78/EC of 18 December 1997 laying down the principles governing the organisation of veterinary checks on products entering the community from third countries. Brussels: EC.
EC (1997). Regulation (EC) no 258/97 of the European Parliament and of the Council of 27 January 1997 concerning novel foods and novel food ingredients. Brussels: EC.
EEC (1991). Council Directive 91/496/EEC of 15 July 1991 laying down the principles governing the organization of veterinary checks on animals entering the community from third countries and amending directives 89/662/EEC, 90/425/EEC and 90/675/EEC. Brussels: EEC.
Elhassan, M., Wendin, K., Olsson, V. et al. (2019). Quality aspects of insects as food – nutritional, sensory, and related concepts. *Food* 8 (3): 95.
Elorduy, J.R., Pino, J.M., and Correa, S.C. (1998). Insectos comestibles del estado de méxico y determinación de su valor nutritivo. *Anales del Instituto de Biología. Serie Zoología* 69 (1): 65–104.
EU (2015). Regulation (EU) 2015/2283 of the European Parliament and of the Council of 25 November 2015 on novel foods, amending regulation (EU) no 1169/2011 of the European Parliament and of the Council and repealing Regulation (EC) no 258/97 of the European Parliament and of the Council and Commission Regulation (EC) no 1852/2001. Brussels: EU.
European Commission 2007. Commission Decision 2007/275/EC of 17 April 2007 concerning lists of animals and products to be subject to controls at border inspection posts under council directives 91/496/EEC and 97/78/EC. Brussells: European Commission.
FAO (2014). *Building a Common Vision for Sustainable Food and Agriculture: Principles and Approaches*. Rome: FAO.
FAO (2019). Edible insects: a solution for food and food security. http://www.fao.org/edible-insects/en (accessed October 3, 2019).
Fradique, M., Batista, A.P., Nunes, M.C. et al. (2013). *Isochrysis galbana* and *Diacronema vlkianum* biomass incorporation in pasta products as PUFAs source. *LWT – Food Science and Technology* 50: 312–319.
García-Vaquero, M. (2019). Seaweed proteins and applications in animal feed. In: *Novel Proteins for Food, Pharmaceuticals and Agriculture: Sources, Applications and Advances* (ed. M. Hayes), 139–161. Hoboken, NJ: John Wiley & Sons.
Garcia-Vaquero, M. and Hayes, M. (2016). Red and green macroalgae for fish and animal feed and human functional food development. *Food Reviews International* 32 (1): 15–45.
Garcia-Vaquero, M., Lopez-Alonso, M., and Hayes, M. (2017). Assessment of the functional properties of protein extracted from the brown seaweed Himanthalia elongata (Linnaeus) sf gray. *Food Research International* 99: 971–978.
Garcia-Vaquero, M., Rajauria, G., O'Doherty, J. et al. (2017). Polysaccharides from macroalgae: recent advances, innovative technologies and challenges in extraction and purification. *Food Research International* 99: 1011–1020.
Garcia-Vaquero, M., Rajauria, G., Tiwari, B. et al. (2018). Extraction and yield optimisation of fucose, glucans and associated antioxidant activities from Laminaria digitata by applying response surface methodology to high intensity ultrasound-assisted extraction. *Marine Drugs* 16 (8): 257.
Garcia-Vaquero, M., Mora, L., and Hayes, M. (2019a). in vitro; and in silico approaches to generating and identifying angiotensin-converting enzyme i inhibitory peptides from green macroalga *Ulva lactuca*. *Marine Drugs* 17 (4): 204.
Garcia-Vaquero, M., O'Doherty, J.V., Tiwari, B.K. et al. (2019b). Enhancing the extraction of polysaccharides and antioxidants from macroalgae using sequential hydrothermal-assisted extraction followed by ultrasound and thermal technologies. *Marine Drugs* 17 (8): 457.

Garcia-Vaquero, M., Ummat, V., Tiwari, B. et al. (2020). Exploring ultrasound, microwave and ultrasound–microwave assisted extraction technologies to increase the extraction of bioactive compounds and antioxidants from brown macroalgae. *Marine Drugs* 18 (3): 172.

Gong, M. and Bassi, A. (2016). Carotenoids from microalgae: a review of recent developments. *Biotechnology Advances* 34 (8): 1396–1412.

Gosch, B.J., Magnusson, M., Paul, N.A. et al. (2012). Total lipid and fatty acid composition of seaweeds for the selection of species for oil-based biofuel and bioproducts. *GCB Bioenergy* 4 (6): 919–930.

Guil-Guerrero, J.L., Ramos-Bueno, R.P., González-Fernández, M.J. et al. (2018). Insects as food: fatty acid profiles, lipid classes, and sn-2 fatty acid distribution of lepidoptera larvae. *European Journal of Lipid Science and Technology* 120 (6): 1700391.

Hayes, M., Bastiaens, L., Gouveia, L. et al. (2019). Microalgal bioactive compounds including protein, peptides, and pigments: applications, opportunities, and challenges during biorefinery processes. In: *Novel Proteins for Food, Pharmaceuticals and Agriculture: Sources, Applications and Advances* (ed. M. Hayes), 239–255. Hoboken, NJ: John Wiley & Sons.

Hernández, D., Solana, M., Riaño, B. et al. (2014). Biofuels from microalgae: lipid extraction and methane production from the residual biomass in a biorefinery approach. *Bioresource Technology* 170: 370–378.

Hildebrand, G., Poojary, M.M., O'Donnell, C. et al. (2020). Ultrasound-assisted processing of Chlorella vulgaris for enhanced protein extraction. *Journal of Applied Phycology* 32 (3): 1709–1718.

Holdt, S. and Kraan, S. (2011). Bioactive compounds in seaweed: functional food applications and legislation. *Journal of Applied Phycology* 23 (3): 543–597.

Imbs, T., Shevchenko, N., Sukhoverkhov, S. et al. (2010). Seasonal variations of the composition and structural characteristics of polysaccharides from the brown alga Costaria costata. *Chemistry of Natural Compounds* 45 (6): 786–791.

Kendel, M., Wielgosz-Collin, G., Bertrand, S. et al. (2015). Lipid composition, fatty acids and sterols in the seaweeds Ulva armoricana, and Solieria chordalis from Brittany (France): an analysis from nutritional, chemotaxonomic, and antiproliferative activity perspectives. *Marine Drugs* 13 (9): 5606–5628.

Kim, D.-Y., Vijayan, D., Praveenkumar, R. et al. (2016). Cell-wall disruption and lipid/astaxanthin extraction from microalgae: *Chlorella* and *Haematococcus*. *Bioresource Technology* 199: 300–310.

Kraan, S. (2013). Pigments and minor compounds in algae. In: *Functional Ingredients from Algae for Foods and Nutraceuticals* (ed. H. Dominguez), 205–251. Philadelphia, PA: Woodhead.

Kwak, H.S., Kim, J.Y.H., Woo, H.M. et al. (2016). Synergistic effect of multiple stress conditions for improving microalgal lipid production. *Algal Research* 19: 215–224.

Lafarga, T. (2019). Effect of microalgal biomass incorporation into foods: nutritional and sensorial attributed of the end products. *Algal Research* 41: 101566.

Lafarga, T. (2020). Cultured microalgae and compounds derived thereof for food applications: strain selection and cultivation, drying, and processing strategies. *Food Reviews International* 36 (6): 559–583.

Lafarga, T., Clemente, I., and Garcia-Vaquero, M. (2019a). Carotenoids from microalgae. In: *Carotenoids: Properties, Processing and Applications* (ed. C.M. Galanakis), 149. London: Academic Press.

Lafarga, T., Mayre, E., Echeverria, G. et al. (2019b). Potential of the microalgae *Nannochloropsos* and *Tetraselmis* for being used as innovative ingredients in baked goods. *LWT - Food Science and Technology* 115: 108439.

Lafarga, T., Acién-Fernández, F.G., Castellari, M. et al. (2019c). Effect of microalgae incorporation on the physicochemical, nutritional, and sensorial properties of an innovative broccoli soup. *LWT - Food Science and Technology* 111: 167–174.

Lafarga, T., Acién-Fernández, F.G., and García-Vaquero, M. (2020). Bioactive peptides and carbohydrates from seaweed for food applications: natural occurrence, isolation, purification, and identification. *Algal Research* 48: 101909.

Liau, B.-C., Shen, C.-T., Liang, F.-P. et al. (2010). Supercritical fluids extraction and anti-solvent purification of carotenoids from microalgae and associated bioactivity. *The Journal of Supercritical Fluids* 55 (1): 169–175.

Mariod, A.A., Abdel-Wahab, S.I., and Ain, N.M. (2011). Proximate amino acid, fatty acid and mineral composition of two Sudanese edible pentatomid insects. *International Journal of Tropical Insect Science* 31 (3): 145–153.

Mariod, A.A., Saeed Mirghani, M.E., and Hussein, I. (2017). Principles of oil extraction, processing, and oil composition. In: *Unconventional Oilseeds and Oil Sources* (eds. A.A. Mariod, M.E. Saeed Mirghani and I. Hussein), 347–350. London: Academic Press.

Mariod, A.A., Mirghani, M.E.S., and Hussein, I.H. (2017). *Unconventional Oilseeds and Oil Sources*. London: Academic Press.

Mathimani, T., Uma, L., and Prabaharan, D. (2018). Formulation of low-cost seawater medium for high cell density and high lipid content of *Chlorella vulgaris* BDUG 91771 using central composite design in biodiesel perspective. *Journal of Cleaner Production* 198: 575–586.

Matos, Â.P., Feller, R., Moecke, E.H.S. et al. (2016). Chemical characterization of six microalgae with potential utility for food application. *Journal of the American Oil Chemists' Society* 93 (7): 963–972.

Maurer, N.E., Hatta-Sakoda, B., Pascual-Chagman, G. et al. (2012). Characterization and authentication of a novel vegetable source of omega-3 fatty acids, sacha inchi (*Plukenetia volubilis* L.) oil. *Food Chemistry* 134 (2): 1173–1180.

Mba, A.R.F., Kansci, G., Viau, M. et al. (2019). Edible caterpillars of *Imbrasia truncata* and *Imbrasia epimethea* contain lipids and proteins of high potential for nutrition. *Journal of Food Composition and Analysis* 79: 70–79.

Megido, R.C., Gierts, C., Blecker, C. et al. (2016). Consumer acceptance of insect-based alternative meat products in western countries. *Food Quality and Preference* 52: 237–243.

Mišurcová, L., Ambrožová, J., and Samek, D. (2011). Seaweed lipids as nutraceuticals. *Advances in Food and Nutrition Research* 64: 339–355.

Mubarak, M., Shaija, A., and Suchithra, T. (2015). A review on the extraction of lipid from microalgae for biodiesel production. *Algal Research* 7: 117–123.

Musundire, R., Osuga, I.M., Cheseto, X. et al. (2016). Aflatoxin contamination detected in nutrient and anti-oxidant rich edible stink bug stored in recycled grain containers. *PLoS One* 11 (1): e0145914.

Nowak, V., Persijn, D., Rittenschober, D. et al. (2016). Review of food composition data for edible insects. *Food Chemistry* 193: 39–46.

Omotoso, O. and Adedire, C. (2007). Nutrient composition, mineral content and the solubility of the proteins of palm weevil, *Rhynchophorus phoenicis* f. (coleoptera: Curculionidae). *Journal of Zhejiang University. Science. B* 8 (5): 318–322.

Oranut, S., Subhachai, B., Shen, L.-R. et al. (2010). Lipids and fatty acid composition of dried edible red and black ants. *Agricultural Sciences in China* 9 (7): 1072–1077.

Patil, V., Källqvist, T., Olsen, E. et al. (2007). Fatty acid composition of 12 microalgae for possible use in aquaculture feed. *Aquaculture International* 15 (1): 1–9.

Patra, J., Das, G., and Baek, K.-H. (2015). Chemical composition and antioxidant and antibacterial activities of an essential oil extracted from an edible seaweed, *Laminaria japonica* L. *Molecules* 20 (7): 12093–12113.

Patra, J.K., Lee, S.-W., Park, J.G. et al. (2017). Antioxidant and antibacterial properties of essential oil extracted from an edible seaweed *Undaria pinnatifida*. *Journal of Food Biochemistry* 41 (1): e12278.

Pieber, S., Schober, S., and Mittelbach, M. (2012). Pressurized fluid extraction of polyunsaturated fatty acids from the microalga *Nannochloropsis oculata*. *Biomass and Bioenergy* 47: 474–482.

Ramadan, M.F., Asker, M.M.S., and Ibrahim, Z.K. (2008). Functional bioactive compounds and biological activities. *Czech Journal of Food Sciences* 26 (3): 211–222.

Ramesh Kumar, B., Deviram, G., Mathimani, T. et al. (2019). Microalgae as rich source of polyunsaturated fatty acids. *Biocatalysis and Agricultural Biotechnology* 17: 583–588.

Ramos-Elorduy, J., Moreno, J.M.P., Prado, E.E. et al. (1997). Nutritional value of edible insects from the state of Oaxaca, Mexico. *Journal of Food Composition and Analysis* 10 (2): 142–157.

Rubio-Rodríguez, N., Beltrán, S., Jaime, I. et al. (2010). Production of omega-3 polyunsaturated fatty acid concentrates: a review. *Innovative Food Science & Emerging Technologies* 11 (1): 1–12.

Ruiz del Castillo, M.L., Dobson, G., Brennan, R. et al. (2002). Genotypic variation in fatty acid content of blackcurrant seeds. *Journal of Agricultural and Food Chemistry* 50 (2): 332–335.

Rumpold, B.A. and Schlüter, O.K. (2013). Nutritional composition and safety aspects of edible insects. *Molecular Nutrition & Food Research* 57 (5): 802–823.

Saini, R.K. and Keum, Y.-S. (2018). Omega-3 and omega-6 polyunsaturated fatty acids: dietary sources, metabolism, and significance — a review. *Life Sciences* 203: 255–267.

Sánchez-Machado, D., López-Cervantes, J., Lopez-Hernandez, J. et al. (2004). Fatty acids, total lipid, protein and ash contents of processed edible seaweeds. *Food Chemistry* 85 (3): 439–444.

Sheng, J., Vannela, R., and Rittmann, B.E. (2011). Evaluation of methods to extract and quantify lipids from *Synechocystis* PCC 6803. *Bioresource Technology* 102 (2): 1697–1703.

Siegrist, M. (2008). Factors influencing public acceptance of innovative food technologies and products. *Trends in Food Science & Technology* 19 (11): 603–608.

da Silva Vaz, B., Moreira, J.B., de Morais, M.G. et al. (2016). Microalgae as a new source of bioactive compounds in food supplements. *Current Opinion in Food Science* 7: 73–77.

Skriptsova, A., Shevchenko, N., Tarbeeva, D. et al. (2011). Comparative study of polysaccharides from reproductive and sterile tissues of five brown seaweeds. *Marine Biotechnology* 14 (3): 304–311.

Suh, S.-S., Kim, S.J., Hwang, J. et al. (2015). Fatty acid methyl ester profiles and nutritive values of 20 marine microalgae in Korea. *Asian Pacific Journal of Tropical Medicine* 8 (3): 191–196.

Tang, S., Qin, C., Wang, H. et al. (2011). Study on supercritical extraction of lipids and enrichment of DHA from oil-rich microalgae. *The Journal of Supercritical Fluids* 57 (1): 44–49.

Taufeeque, M., Malik, A., and Sherwani, M. (2015). Screening of seed oils from four species of genus Ipomoea. *International Journal of Current Research and Review* 7 (21): 25.

Tzompa-Sosa, D.A., Yi, L., van Valenberg, H.J. et al. (2014). Insect lipid profile: aqueous versus organic solvent-based extraction methods. *Food Research International* 62: 1087–1094.

Van der Spiegel, M., Noordam, M., and Van der Fels-Klerx, H. (2013). Safety of novel protein sources (insects, microalgae, seaweed, duckweed, and rapeseed) and legislative aspects for their application in food and feed production. *Comprehensive Reviews in Food Science and Food Safety* 12 (6): 662–678.

Van Huis, A. (2013). Potential of insects as food and feed in assuring food security. *Annual Review of Entomology* 58: 563–583.

Vannuccini, S., Kavallari, A., Bellù, L.G. et al. (2019). Understanding the impacts of climate change for fisheries and aquaculture: global and regional supply and demand trends and prospects. In: *Impacts of Climate Change on Fisheries and Aquaculture* (eds. M. Barange, T. Bahri, M.C.M. Beveridge, et al.), 41–62. Rome: FAO.

Ventura, S., Nobre, B., Ertekin, F. et al. (2018). Extraction of added-value compounds from microalgae. In: *Microalgae-Based Biofuels and Bioproducts* (eds. R. Muñoz and C. Gonzalez-Fernandez), 461–484. Duxford, UK: Elsevier.
Ward, O.P. and Singh, A. (2005). Omega-3/6 fatty acids: alternative sources of production. *Process Biochemistry* 40 (12): 3627–3652.
Wei, L. and Huang, X. (2017). Long-duration effect of multi-factor stresses on the cellular biochemistry, oil-yielding performance and morphology of *Nannochloropsis oculata*. *PLoS One* 12 (3): e0174646.
Wilkinson, K., Muhlhausler, B., Motley, C. et al. (2018). Australian consumers' awareness and acceptance of insects as food. *Insects* 9 (2): 44.
Xue, Z., Wan, F., Yu, W. et al. (2018). Edible oil production from microalgae: a review. *European Journal of Lipid Science and Technology* 120 (6): 1700428.

4 Oils Extracted from Nuts and Grains

Nirupama Gangopadhyay
Kusuma School of Biological Sciences, Indian Institute of Technology Delhi, New Delhi, India

4.1 Introduction

Recent interest in specialty oils, i.e. oils from nuts, grains, and oilseeds has gained momentum owing to health-promoting components present in these oils. The global specialty oils market is expected to grow at a compound annual growth rate (CAGR) of 6.5% in the forecast period of 2020–2025, and high-income countries are projected to be at forefront of this growth. These specialty oils differ from regular commodity oils in terms of their health and nutritional quotient, which in turn drives their market price. The biggest difference lies in the presence of natural antioxidants like tocols and phenolic compounds in specialty oils which are absent in mainstream oils. These specialty oils are demanded as health supplements and food ingredients by consumers, who are also willing to pay a higher price for them. There is also an increased awareness with regards to "processing" of these oils. Oils that are minimally processed in presence of less or no chemicals are preferred due to maximum retention of the bioactive components present in them.

This chapter discusses the composition, distribution, processing, and possible applications of specialty oils from nuts and grains.

4.2 Oils

Oils, synonymous with lipids or fat, constitute a small portion in grains while they are major components in nuts and seeds. Not only do lipids determine the energy content of nuts and grains, they are also associated with a number of health-promoting bioactive components termed "lipophilic bioactives." Both nuts and grains show presence of two main lipid classes – neutral/non-polar lipids and phospholipids. The third major class of lipids present in grains, but absent in nuts, are glycolipids. The polar class of

Oil and Oilseed Processing: Opportunities and Challenges, First Edition.
Edited by Tomás Lafarga, Gloria Bobo, and Ingrid Aguiló-Aguayo.

lipids, also known as phospholipids, and glycolipids are minor lipid constituents that fulfill functional and physiological roles in the cell (e.g. defining properties of cell membrane). Neutral or non-polar lipids form the bulk in nuts and grains consisting of free fatty acids and triacylglycerols which are saponifiable, i.e. can be hydrolyzed under basic conditions due to the presence of ester group. Non polar lipids also constitute free sterols, steryl esters, and tocols which are mostly unsaponifiable. Triacylglycerols, which function as storage lipids and energy sources, exist as the most prominent class of lipids in plants. However, the health-promoting properties of nuts and grains are majorly linked to the presence of unsaturated fatty acids, sterols, and tocols.

The saponifiable lipid fraction of nuts and grains is a good source of essential fatty acids (Figure 4.1). Oils from nuts and grains are considered desirable because of much higher unsaturated fatty acids as compared to saturated fatty acids (Ryan et al. 2007; Venkatachalam and Sathe 2006). Monounsaturated fatty acids (MUFAs) and polyunsaturated fatty acids (PUFAs) have been shown to reduce plasma cholesterol levels in hypercholesterolemic subjects (Griel et al. 2008; Lupton et al. 1994). Whereas a part of the cholesterol-lowering property of oils from nuts and grains arises from PUFAs, minor non-fatty acid components, i.e. unsaponifiable fatty acids in oils also play a considerable role toward their hypolipidemic effect (Wilson et al. 2000).

Two components that constitute the bulk of the unsaponifiable lipid fraction of nuts and grains are phytosterols and tocols. Phytosterols is a collective term used for both plant "sterols" and "stanols," which are the saturated form of sterols. They are related in structure to cholesterol (Figure 4.2), differing mainly in the side chain group, and are produced as a secondary metabolite in nuts, oilseeds, and cereals. They are located in the cell membrane of plants either existing as free sterols, or in conjugated form as steryl esters or steryl glucosides. The health claim that sterols and stanols in foods help maintain normal blood low-density lipoprotein (LDL)-cholesterol levels has even been passed and approved by European Food Safety Authority (EFSA Journal n.d.). The role of phytosterols in cancer prevention (Ramprasath and Awad 2015), as well as in reducing absorption of LDL cholesterol and prevention of coronary heart diseases (Jones et al. 1997) has been well documented.

Other unsaponifiable components of oils are tocols/tocochromanols, which include four tocopherol (T) and four tocotrienol (T3) homologs, all categorized under the term Vitamin E. Tocopherols have a saturated side chain in their structure, while tocotrienols possess a tri-unsaturated side chain (Figure 4.3). Tocols are referred to as fat-soluble antioxidants that are synthesized by plants for quenching free radicals and for protection against oxidative spoilage of oils and fats through stages of germination and plant growth (Dörmann 2007). The antioxidant capacities of Ts and T3s has been extensively reported in *in vitro*, liposomal, and biological models (Yoshida et al. 2003).

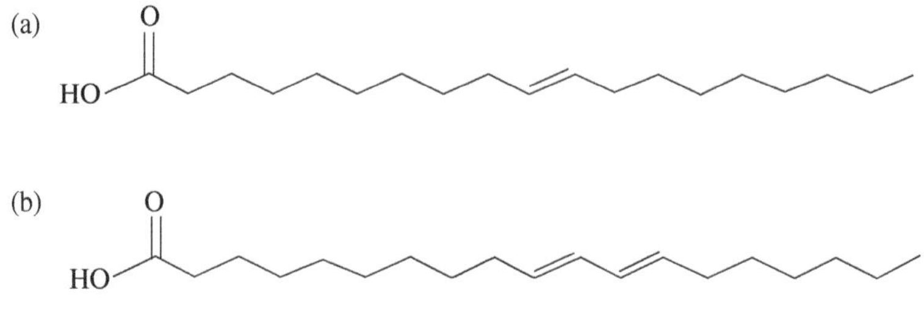

Figure 4.1 Structures of (a) oleic acid and (b) linoleic acid, the major monounsaturated fatty acid (MUFA) and polyunsaturated fatty acid (PUFA) in nuts and grains respectively.

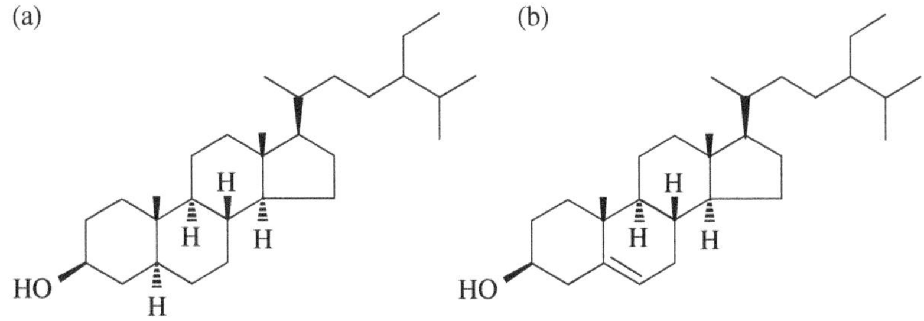

Figure 4.2 Structures of (a) β-sitosterol – one of the most prominent sterols in oils from nuts and grains, and (b) β-sitostanol.

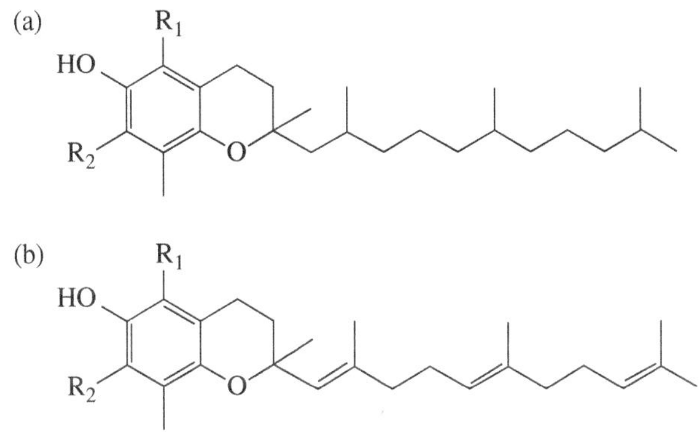

Type of tocol	R1	R2
α	CH3	CH3
β	CH3	H
γ	H	CH3
δ	H	H

Figure 4.3 Structure of tocols: (a) Tocopherols; (b) Tocotrienols.

Possible cancer-preventive effects of tocopherols and tocotrienols in animal and human models have also been explored (Ju et al. 2010). The tocopherol homolog, γ-tocopherol, has been proposed to be a very promising cancer preventive agent.

4.3 Nut Lipids

Edible nuts are commercially important as high-energy food items that are globally marketed as popular snacks as well as a part of various food products like nut butters and trail mixes. Edible nuts are considered valuable oil crops due to their sensorial, nutritional, and health attributes. Not only are they rich sources of lipids and proteins, edible nuts are also a source of vitamins, minerals, and fiber. A few globally popular and edible nuts include peanuts (*Arachis hypogaea*), several tree nuts namely cashew (*Anacardium occidentale*), almond (*Prunus dulcis*), Brazil nut (*Bertholetia excelssa*), hazelnut (*Corylus avellana*), pecan (*Carya illinoinensis*), pistachio (*Pistachia vera*), pine nut (*Pinus pinea*), and walnut (*Juglans regia*).

Edible nuts and their oils are known to contain bioactive and health-promoting components and have long been considered an important component of the Mediterranean diet (Hu and Stampfer 1999). In 2003, the United States Food and Drug Administration (FDA) approved a qualified health claim stating that the consumption of 1.5 oz/day of most nuts may reduce the risk of heart diseases. Epidemiological evidence indicates that several cardio-protective benefits are associated with consumption of edible nuts, which are speculated to arise from their lipid components (Hu and Stampfer 1999). In fact, recent investigations have also exerted higher benefits of consumption of nut oils as compared to whole nuts, possibly due to replacement of dietary carbohydrates with other components in oil extracts (Hu and Stampfer 1999).

4.3.1 Composition of Nut Lipids

Edible nuts provide a rich source of food lipids (Table 4.1), comprising up to 75% of their weight on a wet weight basis in many cases (USDA 2005).

The exact percentage of lipids in various nuts differs in different reports, which can be explained by differences in employed solvent systems and the method of extraction and analysis used by these reports. For instance, comparison of lipid extraction using two different solvent systems – hexane and chloroform/methanol demonstrated that for all nuts studied, the latter solvent system provided a higher yield of oil, whereas the hexane solvent system provided a more refined oil with a higher clarity (Miraliakbari and Shahidi 2008a). Among the examined nuts, pine nuts yield highest amount of oil of 75.1% with chloroform/methanol and 73.9% with hexane, whereas almonds have lowest oil yield of 53.5% with chloroform/methanol and 51.2% with hexane Miraliakbari and Shahidi 2008a). In another study using petroleum ether as extraction solvent (Venkatachalam and Sathe 2006), lipid content of the analyzed samples ranged from 42.88% for peanuts to 66.71% for Brazil nuts; wherein Brazil nut, hazelnut, macadamia, pecan, pine nut, and walnut registered a higher lipid content of more than 60%, while almond, cashew nut, pistachio, and peanut registered lipid contents ranging between 43 and 45%. A wide range of lipid content between 50 and 74% in various edible nuts (almonds [four cultivars, 52.5–57%], pistachios [one cultivar, 56.1%], hazelnuts [one cultivar, 64.1%], walnuts [four cultivars, 61.3–73.8%], and pecans [one cultivar, 71.7%]) grown in Italy and one commercial sample of pine nuts (50.3%) was reported in a study (Ruggeri et al. 1998). In general, edible nuts can be divided into two broad categories of: (i) nuts with a higher lipid content such as

Table 4.1 Total lipid content (%) of nuts.
Source: Miraliakbari and Shahidi (2008a).

Nut	Total lipid content (%)
Almond	53
Brazil nut	69
Pecan nut	73
Hazelnut	62
Pine nut	75
Pistachio	54
Walnut	72

pecans nuts, Brazil nuts, walnuts, and hazelnuts and (ii) nuts with a lower lipid content such as almonds, cashews, and peanuts.

Lipid classes in examined nut oils include triacylglycerols, sterols, sterol esters, phospholipids (phosphatidylserine, phosphatidylcholine), and sphingolipids, among which triacylglycerols are highest, while sterol esters are the smallest detectable lipid class present (Table 4.2) Miraliakbari and Shahidi (2008a).

The highest triacylglycerol content is found in almond oil, while pistachio and pecan oils contained the lowest amount. The following section will discuss compositional details of three important components viz. fatty acids, sterols, and tocols (Vitamin E) that are most commonly associated with cardio-protective benefits of nut oils.

4.3.1.1 Fatty Acids

Most edible nuts are low in saturated fatty acids, with the exceptions of Brazil nut, cashew nut, and pine nut, which have relatively high saturated fats of more than 20% (Robbins et al. 2011). Almonds, pecans, and hazelnuts have the lowest saturated fats with less than 10% (Venkatachalam and Sathe 2006). Nuts are a good source of unsaturated fatty acids, including MUFAs and PUFAs, however percentages of MUFAs and PUFAs are highly variable depending on type of nut (Table 4.3). MUFAs outweigh PUFAs in most nuts like peanut and hazelnut which contain more than 80% MUFAs of total fatty acids (Venkatachalam and Sathe 2006). Exceptions to this are Brazil nut, pine nut, and walnut where MUFAs contribute less than 30% of total fatty acids. Fatty acid composition analysis indicates that regardless of the type of nut, oleic acid (C18:1) is the major constituent of MUFAs, while linoleic acid (C18:2) is the main constituent of PUFAs in nuts (Robbins et al. 2011). Walnuts show quite a distinctive fatty acid profile among the other nuts (Amaral et al. 2003; Maguire et al. 2004; Venkatachalam and Sathe 2006). Linoleic acid is the predominant fatty acid in walnuts with levels as high 60% of total fatty acids, as compared to other nuts like macadamia nut (2% linoleic acid) or hazelnut (7% linoleic acid) (Table 4.3).

Although an overall range of the percentages of individual fatty acids in various nuts can be established, a consensus on exact amounts of these fatty acids is far from possible due to differences in cultivar and techniques used in different studies.

4.3.1.2 Non-Saponifiables

Sterols, stanols, and Vitamin E (tocopherols and tocotrienols) are regarded as the major components of the unsaponifiable fraction of nut lipids. The highest sterol contents have been found in oils from walnuts (2.9 mg/g), almonds (2.7–2.8 mg/g) and

Table 4.2 Composition of major lipid classes – triacylglycerols, sterols, phosphatidylserine, and sphingolipids in nuts (presented as % total lipids) *Source:* Miraliakbari and Shahidi (2008a).

Nut	Triacylglycerols	Sterols	Phosphatidyl-serine	Sphingolipids
Almond	98	0.25	0.32	0.63
Brazil nut	97	0.19	0.32	0.91
Pecan nut	96	0.28	0.47	0.55
Hazelnut	98	0.22	0.36	0.32
Pine nut	97	0.16	0.33	0.57
Pistachio	96	0.21	0.59	0.82
Walnut	97	0.28	0.46	0.68

Table 4.3 Composition of four major fatty acids – palmitic (C16:0), oleic (C18:1), linoleic (C18:2), and linolenic (C18:3) acid in nuts (presented as % of total lipids). *Source:* Kamal-Eldin and Moreau (2009).

Nut	C16:0	C18:1	C18:2	C18:3
Almond	6.8	69	22	0.2
Brazil nut	14	29	43	0.2
Pecan nut	7.6	50	38	1.5
Hazelnut	5.8	79	10	0.5
Pine nut	5	29	44	0.1
Pistachio	7.4	58	30	0.4
Walnut	6.7	21	58	12

Table 4.4 Composition of two major sterols (β-sitosterol and campesterol) and two major tocols (α-T and γ-T) in nuts (presented as μg/g). *Source:* Miraliakbari and Shahidi (2008a).

Nut	β-sitosterol	Campesterol	α-T	γ-T
Almond	2300	90	257	18
Brazil nut	1120	150	14	168
Pecan nut	1750	240	18	473
Hazelnut	1100	170	372	136
Pine nut	1120	220	166	247
Pistachio	1190	210	327	48
Walnut	2250	190	38	376

pecans (2.6–2.7 mg/g), while the lowest amount of sterol of 1.2 mg/g has been found in pine nuts Miraliakbari and Shahidi (2008a). Among the examined sterols and stanols, β-sitosterol is the predominant sterol in all the examined nuts, while stigmasterol, campesterol, and Δ5-avenasterol have also been found in nuts (Table 4.4). Small amounts of stanols (≤0.1 mg/g oil) are present in Brazil nut, hazelnut, pine nut and pistachio oils.

While the presence of tocopherol homologs has been reported in edible nuts, no exclusive mention of tocotrienols in nut oils exists in literature (Maguire et al. 2004; Miraliakbari and Shahidi 2008a, b). The highest amount of tocopherol is found in hazelnut oil (462–508 mg/kg oil), followed by pecan oil (454–490 mg/kg) and pine nut oil (399–458 mg/kg), while Brazil nut oil has the lowest amount of tocopherols (169 mg/kg oil) (Table 4.4) (Miraliakbari and Shahidi 2008b). α-tocopherol and γ-tocopherol are found in most nuts; however, their levels vary. While α-tocopherol is predominant in almond and hazelnut oils, both contain smaller amounts of γ-tocopherol (Maguire et al. 2004). On the other hand, nuts like walnut and pecan nuts predominantly contain γ-tocopherol. Nut oils contain smaller amounts of other homologs, δ-tocopherol and β-tocopherol.

4.3.2 Processing of Nuts

The high lipid content of nuts, a favorable sensory profile, together with the associated health benefits of nut oils has led to a demand for specialty nut oils in the market place. This demand has warranted an investigation into suitable methodologies and processes to enhance production of nut oils. If extracted and processed efficiently, nut oils can retain much of their nutritional and flavor profiles. Similar to oilseeds, harvested edible nuts must be cleaned and sorted to remove extraneous material and debris, and shelled before oil expression (Gong and Pegg 2015). Nuts like cashews require an additional process for removal of cashew nut shell liquid (CNSL) present in the cashew kernel. A hot oil bathing method for extraction of CNSL is employed, following which in-shell nuts are discharged from the oil bath, shelled and further prepared for oil extraction (Gedam and Sampathkumaran 1986). There are two leading methods for commercial oil extraction, mechanical pressing and solvent extraction:

1. Mechanical pressing

   Mechanical pressing of oil-bearing materials for oil expression has been a common practice by specialty oil industries. This process avoids contact with hydrocarbon solutions; so, the resulting oil is free from further refining processes. Cold pressing is one of the oldest methods that has developed over time, and involves the use of only hydraulic pressure with processing temperature being monitored and controlled below 45–50 °C. Process of cold pressing should involve very little or no heat development. While the legal definition of "cold-pressed" is different or nonexistent in some countries, in the United Kingdom, the temperature of oil exiting the press should be less than 50 °C to be labeled "cold pressed" (Panfilis et al. 1998).

   Screw pressing, another form of mechanical pressing involves a metal barrel housing a line of various screw elements. Application of a high shear force in screw pressing makes it a heat generating process, with temperatures reaching around 60–100 °C (Gong and Pegg 2015). While the application of heat improves oil yield of screw pressed oils (around 75–90% oil from oilseeds), it might negatively affect the lipid constituents of the oil. Optimizing moisture content of raw materials at the time of screw pressing improves oil recovery significantly without the need to increase the extraction temperature too much (Martínez et al. 2013; Singh et al. 2002; Wiesenborn et al. 2001). the highest oil recovery of 89% was obtained from walnuts at moisture content of 7.5% and 50 °C (Martínez et al. 2008), while 79% oil recovery could be achieved from almonds at 8% seed moisture content and pressing temperature of 40 °C (Martínez et al. 2013).
2. Solvent Extraction

   Although conventional extraction of oils using solvents like hexane or chloroform/methanol offers almost 100% extraction yield, it is not a method of choice in the specialty oil industry; where mechanical pressing is more preferable. A combination approach in which edible nuts are first pressed to obtain superior quality oil, and then subjected to solvent extraction for recovery of residual oil from oilcake is used in many cases (Karabulut et al. 2005).

   Nut oil obtained using solvent extraction needs to undergo a process of refining before it is bottled and marketed. Refining is done to get rid of free fatty acids, phospholipids, color pigments from oil without damaging its bioactive components (fatty acids, sterols, Vitamin E). Chemical refining of oil in which it is first neutralized using alkali treatment, bleached, and then deodorized, is considered to be a harsh process which strips the oil of its beneficial components. Industrially refined hazelnut oil using a chemical process showed a marked decrease in its phytosterols (from 168 to 142 mg/100 g oil) and tocols (from 52 to 47 mg/100 g oil) post the process of alkali treatment.

   Physical refining, which involves reduced use of chemicals, has replaced the traditional process of chemical refining. In physical refining, the oil must be thoroughly degummed to remove phospholipids, and further treated with superheated steam at low pressure and high temperatures of more than 220 °C to remove free fatty acids and volatile compounds (Čmolík and Pokorný 2000). Partial loss of tocopherols is a disadvantage of physical refining (Čmolík and Pokorný 2000). Physically refined oils and chemically refined oils are very similar in terms of their quality, but loss of neutral oil and the level of environment pollution is lower with the physical process.

Supercritical fluid extraction (SFE) is a widely adopted extraction or separation technique that has gained attention over traditional techniques of solvent extraction and steam distillation. SFE is popularly used for recovery of edible oils in the food industry and essential oils in the cosmetic industry. SFE involves the use of fluids in their supercritical state, which is achieved when its temperature and pressure is raised over its critical value, which gives the substance characteristics of both gases and liquids. Carbon dioxide is the most popularly used supercritical fluid (Sahena et al. 2009). Oxygen-free and refining-free extraction process employed by SFE might be beneficial for oxidative stability of edible nut oils that are rich in oxygen labile components like MUFA, PUFA, and Vitamin E. Oil obtained by SFE are completely free of solvent residues. Moreover, the CO_2 can be removed from the resulting oil and recycled without leaving any residue in the oil. The process of SFE using CO_2 has been optimized for extracting oils from various nuts like walnut (Crowe and White 2003), macadamia nut (Silva et al. 2008), and hazelnut (Özkal et al. 2005).

Between pressed and solvent extracted oils, pressed oils are considered to be of higher nutritional quality because they retain a lot of bioactive lipid components as compared to solvent extracted oils, where much of these components are lost during the process of refining (Gunstone 2011). However, compared to refined oils, their pressed counterparts have a shorter shelf life, and need to be stored in the dark and at lower temperatures to avoid degradation of their lipid components leading to rancidity.

4.3.3 Application/Utilization of Nut Lipids

Compared to cereal and other vegetable oils, nut oils show better oxidative stability due to higher level of MUFAs as compared to PUFAs. This along with the unique flavor profile and health-promoting lipid composition has created a market space and subsequent consumer demand for nut oils. Use of nut kernel oils like almond or hazelnut oil in salads or in everyday cooking, to enhance flavor of foods is becoming common – especially in the western world. Other than food applications, nut oils are used as moisturizing agents in cosmetic products including skin lotions, shampoos, and hair conditioning products. Defatted nut meals that are by-products of the process of nut oil extraction, are good source of proteins, and several anti-oxidative compounds. Thus, they could serve as functional food ingredients. For instance, defatted cashew nut meal contains about 42% crude protein and a better profile of amino acids as compared to soybean meal, and thus can be used to enhance livestock health and growth (Piva et al. 1971).

4.4 Grain Lipids

Cereal grains contain relatively small quantities of lipids usually between 2 and 7%. Wholegrain wheat contains about 3% lipids and wholegrain oats 7%. Even then, cereals comprise an important source of beneficial lipid associated phytochemicals due to the amount of cereals consumed in the diet of people all over the world. For example, rice (*Oryza sativa*) is an important grain in Asian countries, wheat (*Triticum aestivum*) is popular in America, barley (*Hordeum vulgare*) is consumed in European countries and North Africa, maize (*Zea mays*) is consumed in South Africa and Latin America. Other than these major grains, oats (*Avena sativa*), rye (*Secale cereale*), sorghum (*Sorghum bicolor*), triticale (*Triticosecale*) are some of the minor grains consumed in different parts of the world. Other grains that are recently gaining popularity like buckwheat (*Fagopyrum esculentum*), wild rice (*Zizania*), and amaranth (*Amaranthus viridis*) are not botanically true grains but are associated with the grain family due to their close composition to true grains. Wide consumption of these cereal grains makes them a significant source of lipophilic bioactives like fatty acids, phytosterols, and tocols along with other macronutrients and dietary fiber.

Lipids in cereals can be classified based on solubility, chemically based on structural considerations, or according to their location in the grain. These classifications are not mutually exclusive. Classification of cereals based on location in the grain is of special interest from the point of view of cereal processing.

4.4.1 Composition of Cereal Grains

Cereals have not yet been used as a source of edible oil, as the amount of oil in commercial cereal cultivars is lower compared to nuts and oil seed crops. Nevertheless, identification of health promoting components in cereal lipids has triggered interest and investigation into the composition, distribution, and processing of cereal oils. Among the popular cereal grains, oats have much higher lipid content reaching up to 11% in some cultivars (Morrison 1978). The lipid content in other cereals like barley, rice, wheat, maize, rye, sorghum lie in the range of 2 to 5%.

Cereal lipids are complex and moreover differences in analytical techniques used for their characterization, in particular the extracting solvents and purification procedures used, make it difficult to compare the lipid content and compositional data reported by different researchers. Nevertheless, they can be grouped into neutral/non-polar lipids, glycolipids, and phospholipids (Table 4.5) (Morrison 1978).

Neutral lipids consist of free fatty acids and triglycerides which form the saponifiable fraction, along with free sterols, wax esters, steryl esters, tocopherols, and carotenoids constituting the unsaponifiable fraction (Figure 4.4).

The major glycolipids in cereals are digalactosyl diglyceride and monogalactosyl diglyceride, along with other minor glycolipids. Major phospholipids in cereals are phosphatidyl choline, N-acyl phosphatidyl ethanolamine, and N-acyl lysophosphatidyl

Table 4.5 Composition of the three main lipid classes – neutral lipids, phospholipids, and glycolipids in cereal grains (presented as % of total lipids).

Grain	Neutral Lipids	Phospholipids	Glycolipids
Wheat (Piva et al. 1971)	60–70	24–26	17–18
Barley (Knuckles and Chiu 1995)	65–80	9–18	7–26
Oat (Kamal-Eldin and Moreau 2009)	65–80	6–26	7–11
Rye (Knuckles and Chiu 1995; Piva et al. 1971)	60–70	18–25	10–12
Sorghum (Zhou et al. 1999)	77–87	11–13	2–3

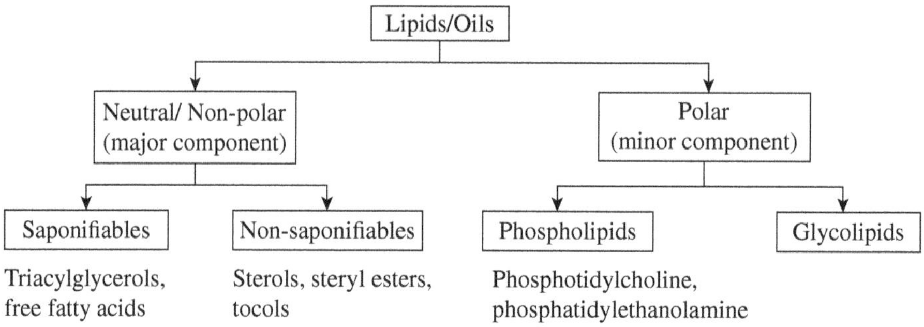

Figure 4.4 Classification of lipids in nuts and cereals.

ethanolamine, along with other minor phospholipids. All cereal grains are rich in neutral lipids than the other classes: 60–70% of total lipids in wheat, triticale, and rye are neutral lipids; 65–80% for barley and oats; 77–87% for rice and sorghum; 75–96% for corn and millet (Chung and Ohm 2002). Although, cereal lipids have not been explored as much as nut lipids, there are three lipid-associated components in cereals that have attracted attention of researchers and nutritionists – (i) Fatty acids, (ii) Phytosterols, and (iii) Tocols.

1. Fatty acids
   The saponifiable lipid fraction of cereals is a useful source of essential fatty acids (Table 4.6). Cereals contain both saturated and unsaturated fatty acids, however the level of unsaturated fatty acids is much higher, constituting approximately more than 70% of total fatty acids. Both saturated and unsaturated fatty acids in cereals are important from technological and nutritional standpoints. For instance, saturated fatty acids impart stability to the oil against oxidation while unsaturated fatty acids are important for mammalian nutrition.

   Principal fatty acids in cereals are the saturated palmitic acid (C16:0), and unsaturated fatty acids viz. oleic acid (C18:1), and linoleic acid (C18:2) (Table 4.6). Unlike nuts, PUFAs outweigh MUFAs in cereal grains. Oleic acid is the most abundant MUFA, while linoleic acid is the most abundant PUFA in cereals. Linoleic (C18:2) and linolenic (C18:3) acids are considered essential fatty acids as they are required for good health and proper functioning but cannot be synthesized by the body. These two fatty acids comprise a significant portion of the total fatty acids in cereals ranging from 39% in rice to 58% in sorghum (Adeyeye and Ajewole 1992). Thus, in spite of their low oil content, cereal lipids are good sources of essential fatty acids.
2. Non-saponifiables
   Phytosterols, tocopherols, and tocotrienols are the major health-promoting components in the non-saponifiable fraction of cereals (Table 4.7). Most important natural sources of plant sterols are oils and margarines. However, cereals have a higher contribution toward dietary phytosterols due to higher consumption of cereals in general. Among Finnish varieties of rye, wheat, barley, and oats, the highest plant sterol content was found in rye (95.5 mg/100 g), followed by barley (76.1 mg/100 g), wheat (69 mg/100 g), and the lowest amount was found in oats (44.7 mg/100 g) (Piironen et al. 2002).

Table 4.6 Composition of four major fatty acids – palmitic (C16:0), oleic (C18:1), linoleic (C18:2), and linolenic (C18:3) in cereal grains (presented as % of total lipids).

Grain	C16:0	C18:1	C18:2	C18:3
Wheat (Piva et al. 1971)	17–24	8–21	55–60	3–5
Barley (Osagie 1987)	21–27	4–9	56–65	4–6
Oat (Kamal-Eldin and Moreau 2009)	13–26	22–47	25–52	1–3
Rye (Piva et al. 1971)	12–19	20–27	46–57	7–10
Sorghum (Zhou et al. 1999)	17–30	17–29	42–56	1–5

Table 4.7 Composition of two major sterols (β-sitosterol and campesterol) and five major tocols (α-T, α-T3, β-T, β-T3, and γ-T) in cereal grains (presented as μg/g).

Grain	β-sitosterol	Campesterol	α-T	α-T3	β-T	β-T3	γ-T
Wheat (Adeyeye and Ajewole 1992; Gangopadhyay et al. 2017)	364	114	8	7	5	40	–
Barley (Adeyeye and Ajewole 1992; Gangopadhyay et al. 2017)	460	170	9	40	0.9	8.7	5.6
Oat (Adeyeye and Ajewole 1992; Gangopadhyay et al. 2017)	274	40	15	56	3	5	–
Rye (Price and Parsons 1974)	470	172	7	20	4	11	–
Sorghum (Panfili et al. 2003)	496	278	n.a.	n.a.	n.a.	n.a.	n.a.

n.a. – data not available.

In all cereals, β-sitosterol is the dominant sterol varying between 50 and 65%, followed by campesterol varying between 12 and 21%, and stigmasterol between 2 and 4% (Table 4.7) (Gangopadhyay et al. 2018; Normen et al. 2002; Piironen et al. 2002). Cereals are also a source of phytostanols like β-sitostanol and campestanol, but unlike sterols, the level of stanols varies considerably depending on cultivars and location. Some studies reported stanols in Finnish rye and wheat varieties, but could not detect them in oat and barley (Adeyeye and Ajewole 1992), while others detected significant amounts of β-sitostanol in Irish barley varieties (Gangopadhyay et al. 2017).

Highest levels of total tocols have been detected in soft wheat and barley varieties (Panfili et al. 2003). Various combinations of the eight tocol homologs are found in different cereals. Grains such as wheat, rye, and triticale contain significant proportion of β-tocols (β-T and β-T3) but γ-T and γ-T3 are not detectable in them (Table 4.7) (Barnes 1983). Conversely, in maize and rice, γ-tocols are detectable, but the corresponding β-tocols are not present. It could be possible because β-tocols and γ–tocols represent components of alternate biosynthetic routes to tocol production. Barley is a notable exception that shows presence of all eight tocol homologs, including all the four tocopherols and four tocotrienols (Morrison 1978). The homolog α-T3 is the most prominent tocol in barley grain followed by α-T and γ-T3 (Gangopadhyay et al. 2017). Only α- and β-tocols are present in wheat, rye, and oats, wherein β-T3 and α-T are major tocols in wheat, while α-T3 is the prominent tocol in rye and oats (Barnes 1983). In maize and rice, α-T and γ-T are the major components respectively.

4.4.2 Distribution of Lipids in Cereal Grains

All whole cereal grains have a bark-like hull on the outside, which forms a protective outer layer and constitutes about 10% of dry weight of the grain. Beneath the hull lie three major components: bran, endosperm, and germ. The bran fraction envelops the groat of the grain, and is composed of multiple layers of outermost pericarp, seed coat, aleurone, and sub-aleurone layers. The bran functions as a protection for the grain from its environment, including weather, insects, mold, and bacteria. Enclosed within the bran is the endosperm, the largest part of the grain. Endosperm is considered metabolically inactive, most of it consists of starch, along with a small proportion of storage proteins (Slavin 2004). The main function of the endosperm is to supply food to the growing seedling. Within the endosperm lies germ, the metabolically active fraction of a grain. Germ is a relatively minor contributor to the dry weight of the grain; however, it contains the embryo and plays a vital role during the process of germination by transporting nutrients from the endosperm to the embryonic axis. The germ in wheat and barley is small and contributes about 4–5% to the weight of grains, while the germ of maize is larger, and contributes a much higher proportion to the structure of the grain as compared to wheat or barley.

Nutrients and phytochemicals are not evenly distributed throughout the grain. Processing of cereal grains like milling provides fractions that are differentially rich in various phytochemicals. Relatively few nutrients, fiber, and phytochemicals are located in the endosperm fraction. A higher concentration of phytonutrients is located in the fiber-rich bran and germ of grains, although germ is a relatively small contributor to weight of the whole grain (Slavin et al. 1999). The outer bran of barley is a concentrated source of phenolic compounds and dietary fiber such as β-glucan (Gangopadhyay et al. 2018).

The distribution of lipids in various grains is varied, but bran and germ are two main sources of lipids/fats in grains. Composition and distribution of lipids in wheat, barley, rye, and triticale are generally similar, wherein lipids are mainly concentrated in outer bran, and germ of the grain, while a very small amount is localized in the endosperm (Morrison 1978). Triacylglycerols, the prominent class of lipids in these grains, are majorly located in the scutellum (outer layer of germ) and embryonic axis, while most glycolipids and phospholipids are in the endosperm. Oats are unique among cereals because a considerable amount of lipids are located in the endosperm. In oats, triacylglycerols are uniformly distributed between bran, endosperm, and germ fraction. Maize has a fairly large germ, and over 90% of the oil in maize is located in its germ.

Among lipophilic bioactives, the presence of essential fatty acids is majorly detected in the outer bran of cereals. In case of phytosterols, whole grains, bran, and germ fractions are richest sterol sources (Piironen et al. 2002). In cereals like rice and barley, sterols are concentrated in the bran, while in wheat the germ has been recognized as a rich source of sterols (Gangopadhyay et al. 2018; Jiang and Wang 2005). Like sterols, stanols are also concentrated in the outer fiber-rich fraction of cereals, while their presence in the endosperm or "flour" fraction of cereals is negligible (Normen et al. 2002). The process of "pearling" yields fractions from outer layers of grains called "pearling fines," which are an excellent source of phytosterols in barley and rye (Lampi et al. 2004). Among tocol homologs, tocotrienols are located in outer parts of the grain, while tocopherols are preferentially localized in the germ (Ko et al. 2003; Panfili et al. 2008). Germ fractions from wheat, barley, oat, triticale, and rye prepared by milling have been found to contain tocopherols but none of the four tocotrienols have been detected. The differential location of tocopherols and tocotrienols in grains is probably due to different functions that these components serve in the grain. Thus, increasing the use of whole meal flours, bran, and germ fractions in cereal products would strengthen the position of cereals as natural sources of lipophilic phytochemicals. The next section will discuss the various methods of processing employed to generate lipid-rich fractions of cereal grains with further applications.

4.4.3 Processing of Cereals

In light of the role played by cereal lipids and its components (fatty acids, sterols, and tocols) in human health and their uneven distribution in the grain, several processes have been employed to generate enriched fractions of grains that can be used as valuable ingredients or additives in food. Most processes in cereal grains start with "dehulling" the grain which separates the outer inedible hull from grain kernel without affecting the grain's process output or germination potential. In case of wheat, dehulling has helped lower the levels of mycotoxin in the grain. Dehulling of grains can be achieved mechanically by abrasion milling or by a less popular method of acid treatment. Post-dehulling processing of grains can be broadly divided into – (i) mechanical processing and (ii) solvent extraction.

1. Mechanical processing-
   The most commonly employed technique for grain processing is milling. Pearling is a traditional way of milling grains. It is an abrasive technique wherein the outermost layers of seed coat (testa and pericarp), aleurone and subaleurone layers, and germ are gradually removed as "pearling fines" to produce polished pearled grains (Lampi et al. 2004). "Pearled grains" thus obtained are high-starch products. Pearling is commonly used to fractionate grains like barley. Pearled barley grains are either used as a thickener or for texture in soups and stews, or milled to produce milled pearled barley. Pearling fines obtained after separation of high starch pearled grains are treated as by-products of the process of milling. In recent reports, these pearling fines have been explored as enriched sources of phytosterols in barley and rye (Lampi et al. 2004). A "barley pearling fine oil" contains 3% phytosterols which is several fold higher than phytosterols in commercial vegetable oils (Lampi et al. 2004). Pearling by-products of barley also show a five to sevenfold increase in levels of tocopherols and tocotrienols (Panfili et al. 2008). In the case of wheat, pearled by-products with high antioxidant activity have been reported (Liyana-Pathirana et al. 2006). Commercially milled fractions of wheat, rice, and oats have shown rice bran and wheat germ to have the highest oil and phytosterols content (Jiang and Wang 2005).

   Scarification of grains is a technique of milling that uses a scarifier, to abrade surface of grain kernels, instead of a pearler. Scarified fine fractions are further passed through a sieve to achieve a particle-based separation. Highest level of tocopherols has been reported in the scarification derived largest particle size fraction, mostly consisting of germ fraction, while the highest level of tocotrienols have been reported in smallest particle size fraction consisting of the outer layers (Moreau et al. 2007). Therefore, it might be possible to control the amount of tocopherols and tocotrienols in barley oil by controlling particle size of fractions used to extract oil.

Another process that is used for enrichment of cereal fractions is air classification which employs air to separate milled heterogeneous particulates of cereals into fairly uniform sized fractions based on particle density and mass. Air classification followed by sieving generates flours of varying sizes. Three varieties of barley were milled and fractionated by air classifiers to produce an enriched product with enhanced levels of dietary fiber, β-glucan, and tocols, which potentially lower cholesterol and glycaemic response in body and add functional properties to food systems (Knuckles and Chiu 1995).

Roller milling is another size-based fractionation technique that is commonly used for processing wheat and oats. This process generates three main fractions – bran, middlings, and flour (endosperm). The flour fraction of roller milling is commonly used for preparing baked goods and extruded products, while bran and middlings are used as animal fodder. Bran and middlings of roller milled barley, that are usually treated as processing by-products, contain several times higher levels of the phytochemicals β-glucan, polyphenols, and tocols, than the commonly used flour fraction (Gangopadhyay et al. 2018). It can thus be suggested that milling fragments isolated from grains hold potential to be used as functional food ingredients or extracted to yield oils enriched in health-promoting unsaturated fatty acids, phytosterols, tocopherols, and tocotrienols.

2. Extraction

Whole grains contain small amount of lipids, owing to which grain oils are not yet sold commercially like nut oils. However, oil enriched fractions of grains can be extracted to yield health-enhancing oils. Solvent extraction is a method of choice for extraction of analytes from grains. The conventional and officially approved method for extraction of non-starch free lipids has been Soxhlet extraction using non-polar solvents like diethyl ether or hexane. Soxhlet extraction recovers small amounts of phospholipids, while binary solvent mixtures like chloroform-methanol, dichloromethane-methanol, and hexane-isopropanol are used for total lipid extraction. All three classes of lipids namely neutral lipids, glycolipids, and phospholipids have been extracted using chloroform : methanol : water (1 : 1 : 0.9) from hulless barley and oat varieties (Price and Parsons 1979). The laborious and time-consuming setup of the Soxhlet apparatus along with its associated risks has made it imperative to search for alternative methods. Also, these widely used solvent mixtures suffer the disadvantage of having varying extraction capacities under different conditions, and require a fair amount of optimization to maximize the extraction potential.

An alternative method of solvent extraction called accelerated solvent extraction (ASE) is a process of extracting chemicals from a matrix using high temperature and high pressure in lesser time, requiring less solvent, along with giving better analyte recovery than traditional methods. The elevated temperature enhances the extraction efficiency of the analyte of interest, while elevated pressure maintains extraction solvent mixtures in a liquid state as temperature is increased above boiling point of the solvent. An ASE system using isopropanol-hexane extracted highest amount of fatty acids from cereals as compared to conventional extraction using chloroform-methane (Schäfer 1998).

Environmental concerns arising from disposal of hazardous solvent waste, exposure to toxic vapors, and the time-consuming nature of these extractions have led to the development of other "less toxic" or "green" technological processes like SFE using CO_2 (Colombo et al. 1998; Panfili et al. 2003). Comparison of free lipids extracted from wheat using SFE-CO_2 and Soxhlet extraction shows a quantitative disadvantage of using SFE, which extracts only 0.61–0.71% free lipids from wheat flour as compared to 0.98–1.04% extracted using Soxhlet extraction (Hubbard et al. 2004). This is probably because supercritical CO_2 is not polar enough to extract polar lipids from a matrix. This issue can be tackled by adding an organic solvent such as methanol or ethanol as a co-solvent in a precise amount to increase polarity of supercritical CO_2. Lipids extracted from wheat using supercritical CO_2 containing 12.0% ethanol modifier was similar to that extracted by Soxhlet method, thus averting the problem of lower yield using supercritical CO_2 alone (Hubbard et al. 2004). The cost of the SFE method per sample is less than the Soxhlet method. In terms of solvent consumption, SFE with a modifier required only 7.4 ml of ethanol per extraction, compared with 150 ml of petroleum ether for a Soxhlet, bringing down the solvent usage by almost 20 times (Hubbard et al. 2004). In addition to reducing costs, ethanol is much less of a fire hazard and is less toxic. Thus, SFE method could be used as an alternative method to save time, cost, and solvent usage, and is more environmentally friendly. Also, the addition of a polar modifier to supercritical CO_2 facilitates the extraction of polar lipids as efficiently as the traditional Soxhlet extraction.

Supercritical CO_2 has also been used to extract tocopherols and tocotrienols, which are thermolabile lipid components from barley (Colombo et al. 1998). Higher yields, clearer and more purified extracts of Vitamin E components were obtained by SFE; compared with n-hexane extract. Use of CO_2 extraction for lipids from brewer's barley is being carried out on commercial scale to yield cosmetic ingredients.

4.4.4 Application/Utilization of Cereal Lipids

Unlike nut oils, cereal oils have not found many applications in the market mainly because of their physical properties especially low melting point, and stability issues. Refining the process of extraction and purification of these oils is necessary to fulfill specifications for use in the food industry. Wheat germ oil is a popular cereal oil in the market, which is marketed either as a health supplement or as cosmetic product for skin applications in lotions, or shampoos (Niu et al. 2013). Popular marketing of wheat germ oil is linked with its high nutritional value and mass production of wheat germ during process of milling. While a high degree of unsaturated fatty acids and Vitamin E in wheat germ oil add to its nutritive value, they also impart oxidative instability to the oil. Refining of wheat germ oil to match the quality of mainstream edible oils has been attempted, however the process strips wheat germ oil of its functional components (Kahlon 1989).

The experimental use of cereal oils for baking or extending shelf life of food products has been explored. Oat oil has been tested as a dough improver in bread baking. Addition of 3% crude oat oil or 0.7% oat oil polar fraction was found to significantly delay bread firming and starch retrogradation, processes often associated with staling of bread (Erazo-Castrejón et al. 2001). In another experiment, the effect of a commercial mixture of natural tocopherols on shelf life of extruded breakfast cereals was evaluated (Paradiso et al. 2008). The inhibiting action of tocopherols on lipid oxidation in extruded corn flakes, and its effectiveness in preventing subsequent development of off-flavors during storage was confirmed. New approaches of production and refinement of cereal oils that target the issue with their stability, without compromising their nutritional functions will help open newer opportunities for commercialization of these oils.

4.5 Conclusions

Oils from nuts and grains fall in the category of "specialty oils" and they differ from regular commodity oils in terms of their health and nutritional properties. Compositional analyses of these specialty oils have revealed the presence of three major lipid associated components namely fatty acids, phytosterols, and tocols. The nutritional, especially cardio-protective, benefits associated with these components have led to in-depth investigation into possible techniques of processing that yield superior quality and higher quantities of oils. With further increase in demand for these oils along with improvement of techniques employed for extracting them, we are sure to see an increase in production and sale of these specialty oils in the market in the near future.

References

Adeyeye, A. and Ajewole, K. (1992). Chemical composition and fatty acid profiles of cereals in Nigeria. *Food Chemistry* 44 (1): 41–44.

Amaral, J.S., Casal, S., Pereira, J.A. et al. (2003). Determination of sterol and fatty acid compositions, oxidative stability, and nutritional value of six walnut (*Juglans regia* L.) cultivars grown in Portugal. *Journal of Agricultural and Food Chemistry* 51 (26): 7698–7702.

Barnes, P.J. (ed.) (1983). *Lipids in Cereal Technology*, 33–55. New York: Academic Press.

Chung, O.K. and Ohm, J.B. (2002). Cereal lipids. In: *Food Science and Technology* (eds. K. Kulp and J. Ponte), 417–478. New York;: Marcel Dekker.

Čmolík, J. and Pokorný, J. (2000). Physical refining of edible oils. *European Journal of Lipid Science and Technology* 102 (7): 472–486.

Colombo, M.L., Corsini, A., Mossa, A. et al. (1998). Supercritical carbon dioxide extraction, fluorimetric and electrochemical high performance liquid chromatrographic detection of vitamin E from *Hordeum vulgare* L. *Phytochemical Analysis: An International Journal of Plant Chemical and Biochemical Techniques* 9 (4): 192–195.

Crowe, T.D. and White, P.J. (2003). Oxidative stability of walnut oils extracted with supercritical carbon dioxide. *Journal of the American Oil Chemists' Society* 80 (6): 575–578.

Dörmann, P. (2007). Functional diversity of tocochromanols in plants. *Planta* 225 (2): 269–276.

EFSA Panel on Dietetic Products, Nutrition and Allergies (NDA) (2010). Scientific opinion on the substantiation of health claims related to plant sterols and plant stanols and maintenance of normal blood cholesterol concentrations (ID 549, 550, 567, 713, 1234, 1235, 1466, 1634, 1984, 2909, 3140), and maintenance of normal prostate size and normal urination (ID 714, 1467, 1635) pursuant to Article 13(1) of Regulation (EC) No 1924/20061, EFSA Panel on Dietetic Products, Nutrition and Allergies (NDA). *EFSA Journal* 8 (10): 1813–1825.

Erazo-Castrejón, S.V., Doehlert, D.C., and D'Appolonia, B.L. (2001). Application of oat oil in breadbaking. *Cereal Chemistry* 78 (3): 243–248.

Gangopadhyay, N., Rai, D.K., Brunton, N.P. et al. (2017). Fatty acids, sterols and tocols in Irish barley varieties: profiling and correlation analysis. *European Journal of Lipid Science and Technology* 119 (4): 1600213.

Gangopadhyay, N., Harrison, S.M., Brunton, N.P. et al. (2018). Brans of the roller-milled barley fractions rich in polyphenols and health-promoting lipophilic molecules. *Journal of Cereal Science* 83: 213–221.

Gedam, P.H. and Sampathkumaran, P.S. (1986). Cashew nut shell liquid: extraction, chemistry and applications. *Progress in Organic Coatings* 14 (2): 115–157.

Gong, Y. and Pegg, R.B. (2015). Tree nut oils: properties and processing for use in food. In: *Specialty Oils and Fats in Food and Nutrition* (ed. G. Talbot), 65–86. Cambridge: Woodhead Publishing.

Griel, A.E., Cao, Y., Bagshaw, D.D. et al. (2008). A macadamia nut-rich diet reduces total and LDL-cholesterol in mildly hypercholesterolemic men and women. *The Journal of Nutrition* 138 (4): 761–767.

Gunstone, F. (ed.) (2011). *Vegetable Oils in Food Technology: Composition, Properties and Uses*. Chichester: Wiley-Blackwell.

Hu, F.B. and Stampfer, M.J. (1999). Nut consumption and risk of coronary heart disease: a review of epidemiologic evidence. *Current Atherosclerosis Reports* 3: 204–209.

Hubbard, J.D., Downing, J.M., Ram, M.S. et al. (2004). Lipid extraction from wheat flour using supercritical fluid extraction. *Cereal Chemistry* 81 (6): 693–698.

Jiang, Y. and Wang, T. (2005). Phytosterols in cereal by-products. *Journal of the American Oil Chemists' Society* 82 (6): 439–444.

Jones, P.J., MacDougall, D.E., Ntanios, F. et al. (1997). Dietary phytosterols as cholesterol-lowering agents in humans. *Canadian Journal of Physiology and Pharmacology* 75 (3): 217–227.

Ju, J., Picinich, S.C., Yang, Z. et al. (2010). Cancer-preventive activities of tocopherols and tocotrienols. *Carcinogenesis* 31 (4): 533–542.

Kahlon, T.S. (1989). Nutritional implications and uses of wheat and oat kernel oil. *Cereal Foods World (USA)* 34 (10): 872–875.

Kamal-Eldin, A. and Moreau, R.A. (2009). Tree nut oils. In: *Gourmet and Health-Promoting Specialty Oils* (eds. R.A. Moreau and A. Kamal-Eldin), 127–149. Urbana, IL: AOCS Press.

Karabulut, I., Topcu, A., Yorulmaz, A. et al. (2005). Effects of the industrial refining process on some properties of hazelnut oil. *European Journal of Lipid Science and Technology* 107 (7–8): 476–480.

Knuckles, B.E. and Chiu, M.C.M. (1995). β-Glucan enrichment of barley fractions by air classification and sieving. *Journal of Food Science* 60 (5): 1070–1074.

Ko, S.N., Kim, C.J., Kim, H. et al. (2003). Tocol levels in milling fractions of some cereal grains and soybean. *Journal of the American Oil Chemists' Society* 80 (6): 585–589.

Lampi, A.M., Moreau, R.A., Piironen, V. et al. (2004). Pearling barley and rye to produce phytosterol rich fractions. *Lipids* 39 (8): 783–787.

Liyana-Pathirana, C., Dexter, J., and Shahidi, F. (2006). Antioxidant properties of wheat as affected by pearling. *Journal of Agricultural and Food Chemistry* 54 (17): 6177–6184.

Lupton, J.R., Robinson, M.C., and Morin, J.L. (1994). Cholesterol-lowering effect of barley bran flour and oil. *Journal of the American Dietetic Association* 94 (1): 65–70.

Maguire, L.S., O'sullivan, S.M., Galvin, K. et al. (2004). Fatty acid profile, tocopherol, squalene and phytosterol content of walnuts, almonds, peanuts, hazelnuts and the macadamia nut. *International Journal of Food Sciences and Nutrition* 55 (3): 171–178.

Martínez, M.L., Mattea, M.A., and Maestri, D.M. (2008). Pressing and supercritical carbon dioxide extraction of walnut oil. *Journal of Food Engineering* 88 (3): 399–404.

Martínez, M.L., Penci, M.C., Marin, M.A. et al. (2013). Screw press extraction of almond (*Prunus dulcis* [Miller] DA Webb): oil recovery and oxidative stability. *Journal of Food Engineering* 119 (1): 40–45.

Miraliakbari, H. and Shahidi, F. (2008a). Lipid class compositions, tocopherols and sterols of tree nut oils extracted with different solvents. *Journal of Food Lipids* 15 (1): 81–96.

Miraliakbari, H. and Shahidi, F. (2008b). Oxidative stability of tree nut oils. *Journal of Agricultural and Food Chemistry* 56 (12): 4751–4759.

Moreau, R.A., Wayns, K.E., Flores, R.A. et al. (2007). Tocopherols and tocotrienols in barley oil prepared from germ and other fractions from scarification and sieving of hulless barley. *Cereal Chemistry* 84 (6): 587–592.

Morrison, W.R. (1978). Cereal lipids. *Advances in Cereal Science and Technology* 2: 221–348.

Niu, L.Y., Jiang, S.T., Pan, L.J. et al. (2013). Characterization of wheat germ oil in terms of volatile compounds, lipid composition, thermal behavior, and structure. *International Journal of Food Properties* 16 (8): 1740–1749.

Normen, L., Bryngelsson, S., Johnsson, M. et al. (2002). The phytosterol content of some cereal foods commonly consumed in Sweden and in the Netherlands. *Journal of Food Composition and Analysis* 15 (6): 693–704.

Osagie, A.U. (1987). Total lipids of sorghum grain. *Journal of Agricultural and Food Chemistry* 35 (4): 601–604.

Özkal, S.G., Salgın, U., and Yener, M.E. (2005). Supercritical carbon dioxide extraction of hazelnut oil. *Journal of Food Engineering* 69 (2): 217–223.

Panfili, G., Fratianni, A., and Irano, M. (2003). Normal phase high-performance liquid chromatography method for the determination of tocopherols and tocotrienols in cereals. *Journal of Agricultural and Food Chemistry* 51 (14): 3940–3944.

Panfili, G., Cinquanta, L., Fratianni, A. et al. (2003). Extraction of wheat germ oil by supercritical CO_2: oil and defatted cake characterization. *Journal of the American Oil Chemists' Society* 80 (2): 157–161.

Panfili, G., Fratianni, A., Di Criscio, T. et al. (2008). Tocol and β-glucan levels in barley varieties and in pearling by-products. *Food Chemistry* 107 (1): 84–91.

Panfilis, F.D., Lercker, G., and Toschi, T.G. (1998). Quality control for cold-pressed oils. *Inform* 9: 212–221.

Paradiso, V.M., Summo, C., Trani, A. et al. (2008). An effort to improve the shelf life of breakfast cereals using natural mixed tocopherols. *Journal of Cereal Science* 47 (2): 322–330.

Piironen, V., Toivo, J., and Lampi, A.M. (2002). Plant sterols in cereals and cereal products. *Cereal Chemistry* 79 (1): 148–154.

Piva, G., Santi, E., and Ekpenyong, T. (1971). Nutritive value of cashewnut extraction meal. *Journal of the Science of Food and Agriculture* 22 (1): 22–23.

Price, P.B. and Parsons, J.G. (1974). Lipids of six cultivated barley (*Hordeum vulgare* L.) varieties. *Lipids* 9 (8): 560.

Price, P.B. and Parsons, J. (1979). Distribution of lipids in embryonic axis, bran-endosperm, and hull fractions of hulless barley and hulless oat grain. *Journal of Agricultural and Food Chemistry* 27 (4): 813–815.

Ramprasath, V.R. and Awad, A.B. (2015). Role of phytosterols in cancer prevention and treatment. *Journal of AOAC International* 98 (3): 735–738.

Robbins, K.S., Shin, E.C., Shewfelt, R.L. et al. (2011). Update on the healthful lipid constituents of commercially important tree nuts. *Journal of Agricultural and Food Chemistry* 59 (22): 12083–12092.

Ruggeri, S., Cappelloni, M., Gambelli, L. et al. (1998). Chemical composition and nutritive value of nuts grown in Italy. *Italian Journal of Food Science* 10: 243–252.

Ryan, E., Galvin, K., O'Connor, T.P. et al. (2007). Phytosterol, squalene, tocopherol content and fatty acid profile of selected seeds, grains, and legumes. *Plant Foods for Human Nutrition* 62 (3): 85–91.

Sahena, F., Zaidul, I.S.M., Jinap, S. et al. (2009). Application of supercritical CO_2 in lipid extraction – a review. *Journal of Food Engineering* 95 (2): 240–253.

Schäfer, K. (1998). Accelerated solvent extraction of lipids for determining the fatty acid composition of biological material. *Analytica Chimica Acta* 358 (1): 69–77.

Silva, C.F., Mendes, M.F., Pessoa, F.L.P. et al. (2008). Supercritical carbon dioxide extraction of macadamia (*Macadamia integrifolia*) nut oil: experiments and modeling. *Brazilian Journal of Chemical Engineering* 25 (1): 175–181.

Singh, K.K., Wiesenborn, D.P., Tostenson, K. et al. (2002). Influence of moisture content and cooking on screw pressing of crambe seed. *Journal of the American Oil Chemists' Society* 79 (2): 165–170.

Slavin, J. (2004). Whole grains and human health. *Nutrition Research Reviews* 17 (1): 99–110.

Slavin, J.L., Martini, M.C., Jacobs, D.R. Jr. et al. (1999). Plausible mechanisms for the protectiveness of whole grains. *The American Journal of Clinical Nutrition* 70 (3): 459s–463s.

United States Department of Agriculture (USDA) (2005). USDA Nutrient Database Version 17. https://fdc.nal.usda.gov/download-datasets.html (accessed September 26, 2020).

Venkatachalam, M. and Sathe, S.K. (2006). Chemical composition of selected edible nut seeds. *Journal of Agricultural and Food Chemistry* 54 (13): 4705–4714.

Wiesenborn, D., Doddapaneni, R., Tostenson, K. et al. (2001). Cooking indices to predict screw-press performance for crambe seed. *Journal of the American Oil Chemists' Society* 78 (5): 467–471.

Wilson, T.A., Ausman, L.M., Lawton, C.W. et al. (2000). Comparative cholesterol lowering properties of vegetable oils: beyond fatty acids. *Journal of the American College of Nutrition* 19 (5): 601–607.

Yoshida, Y., Niki, E., and Noguchi, N. (2003). Comparative study on the action of tocopherols and tocotrienols as antioxidant: chemical and physical effects. *Chemistry and Physics of Lipids* 123 (1): 63–75.

Zhou, M., Robards, K., Glennie-Holmes, M. et al. (1999). Oat lipids. *Journal of the American Oil Chemists' Society* 76 (2): 159–169.

5
New Approaches to Detect Compositional Shifts in Fish Oils

Editha Giese[1,2] and Jan Fritsche[1]

[1] *Institute of Safety and Quality of Milk and Fish Products, Max Rubner-Institut, Federal Research Institute of Nutrition and Food, Kiel, Germany*

[2] *Department of Life Sciences, Hamburg University of Applied Sciences (HAW Hamburg), Hamburg, Germany*

5.1 Introduction

The global fish oil market has been growing rapidly in recent years. In 2018, it was worth US$ 3.28 billion (Reports And Data 2019). By 2026, it is expected to reach US$ 5.42 billion, which is mainly due to an increase in aquaculture (Reports And Data 2019). Most of the fish oil (75% in 2016) is used as aquaculture feed (Bachis 2017; Market Research Future 2019). Other applications include livestock and pet feed, pharmaceuticals, dietary supplements, and functional food (Reports And Data 2019). The 2012 National Health Interview Survey found that 7.8% of US adults had taken fish oil or omega-3 fatty acid supplements during the preceding 30 days (Clarke et al. 2015). The use of fish oil for direct human consumption, which accounted for over 9% of the global revenue in 2014, is expected to further increase in the coming years (Grand View Research 2016). Moreover, fish oil is becoming increasingly popular as an ingredient of cosmetic and skin care products (Grand View Research 2016). It is also used in paint production and leather making (Rizliya and Mendis 2014).

Geographically, the fish oil market is segmented across North America, Europe, Asia-Pacific, Latin America, and Africa (Prasad 2015). Global fish oil production stood at 991 600 Mt in 2017 (Auchterlonie 2018). Peru, Chile, and Scandinavia (Denmark, Norway, Iceland) are leaders in the production of fish oil, followed by the USA and China (Auchterlonie 2018). Norway is the biggest fish oil consumer (FAO 2018).

Fish oil is produced from whole fish or shellfish for which there is little or no demand for human consumption, from the liver of lean fish, or by-products of the fish filleting or fish meal industry (CODEX STAN 329-2017 2017; FAO 1986). Species mainly

Oil and Oilseed Processing: Opportunities and Challenges, First Edition.
Edited by Tomás Lafarga, Gloria Bobo, and Ingrid Aguiló-Aguayo.

include anchovy, salmon, herring, tuna, mackerel, sardine, and menhaden (Prasad 2015; Reports And Data 2019). Shark and cod are the most important species whose livers are extracted for oil (Rizliya and Mendis 2014). The fat content of fish varies from 2 to 30% (Kim and Mendis 2006).

This chapter will provide an introduction to fish oils with regard to their production and processing, their nutritional benefits, their oxidative stability, and analytical methods for fish oil analysis. The focus will be on new approaches to determine fish oil quality using rapid spectroscopic techniques in combination with machine learning.

5.2 Production and Processing

The production of fish oil generally employs the same techniques as the production of other edible oils. Methods used to extract fish oil from raw materials include high-speed centrifugation, low-temperature solvent extraction, supercritical fluid extraction, as well as wet and steam rendering methods (Kim and Mendis 2006). Typically, raw materials are minced and steam-heated to 95–100 °C within 15–20 minutes in order to liberate the oil from the fat depots of the fish and to condition the material for subsequent treatment. After a pressing step, the press cake is processed into fishmeal, while the liquid phase enters several (horizontal and vertical) centrifuges, where the oil is separated from water and suspended solids based on the different specific gravities. The resulting crude fish oil is stored in large tanks. If the fish oil is to be used for food applications, a refining process usually follows (FAO 1986).

The aim of refining is to remove components from the oil that are undesirable for human consumption or further processing: e.g. pigments, free fatty acids, phospholipids, lipid oxidation products, and contaminants such as polychlorinated dibenzodioxins, dibenzofurans, and biphenyls as well as heavy metals (Rubio-Rodríguez et al. 2010). Conventional oil refining in industry is carried out by chemical means and comprises steps such as degumming, de-acidification, bleaching, deodorization, winterization, and addition of additives (e.g. antioxidants, sequestrants, antifoaming agents, and emulsifiers) (CODEX STAN 329-2017 2017; Rizliya and Mendis 2014). In physical refining, de-acidification is achieved by steam distillation. More recent technologies for oil refining include supercritical fluid extraction as well as membrane and enzymatic processes (Rubio-Rodríguez et al. 2010).

Some fish oil producers do not submit the crude oil to the entire refining process but only apply a filtering step using activated charcoal. This saves the oil from high temperature exposure during deodorization (90–95 °C, often up to 180 °C), which may reduce levels of omega-3 fatty acids, vitamins, and antioxidants and can lead to the formation of *trans* fatty acids and 3-monochloropropanediol (BIOHAZ [EFSA Panel on Biological Hazards] 2010; John et al. 2013; van Duijn 2014).

Refined fish oil may be further processed into omega-3 concentrates by a physicochemical process comprising transesterification into ethyl or methyl esters, followed by concentration, deodorization/earth treatment, and antioxidant addition (BIOHAZ [EFSA Panel on Biological Hazards] 2010; Rubio-Rodríguez et al. 2010). Concentration is generally achieved by a combination of techniques such as molecular distillation (most commonly applied), urea complexation, low-temperature crystallization, adsorption chromatography, and supercritical fluid fractionation. Concentrates are sold as ethyl or methyl esters, triacylglycerols, or free fatty acids (BIOHAZ [EFSA

Panel on Biological Hazards] 2010; Rubio-Rodríguez et al. 2010). Re-esterification into triacylglycerols can be performed by chemical or enzymatic procedures (Cowan 2010).

5.3 Nutritional Benefits

Fish oil is a virtually unique natural source of the long-chain polyunsaturated omega-3 fatty acids eicosapentaenoic acid (EPA, 20:5), docosahexaenoic acid (DHA, 22:6), and docosapentaenoic acid (DPA, 22:5) (Pike and Jackson 2010). These fatty acids are produced by marine algae, which are directly or indirectly (via zooplankton) consumed by fish (Pike and Jackson 2010).

Long-chain omega-3 (LCn3) fatty acids are believed to provide a number of health benefits, in particular for the cardiovascular system. The European Commission has authorized nine health claims with regard to the consumption of DHA (and EPA) relating to the maintenance of normal brain function/development, vision/visual development, blood triacylglycerol levels, blood pressure, and function of the heart (European Commission 2018). A meta-analysis of randomized controlled trials published in 2020 concluded that increasing the intake of LCn3 fatty acids reduced serum triacylglycerol levels by approx. 15% in a dose-dependent manner (high-certainty evidence) (Abdelhamid et al. 2020). However, the authors found little or no effect on all-cause mortality and the risk of cardiovascular events as well as blood pressure and serum levels of total cholesterol, high-density lipoprotein, and low-density lipoprotein. Most of the included studies analyzed the effects of dietary supplements, some compared the intake of foods naturally rich in LCn3 fatty acids or enriched foodstuffs or dietary advice to placebo or the usual diet. Doses of LCn3 fatty acids ranged from 0.5 g/d to more than 5 g/d and the duration of the interventions varied between 12 and 88 months (Abdelhamid et al. 2020). The US Food and Drug Administration (FDA) considers that the consumption of EPA and DHA combined may reduce the risk of hypertension and coronary heart disease by lowering blood pressure (FDA 2019). However, qualified health claims must be accompanied by a disclaimer stating that the overall evidence is inconsistent and inconclusive (FDA 2019).

Because of the presumed health-promoting effects, the Food and Agriculture Organization of the United Nations (FAO) recommends a daily intake of 250 mg EPA plus DHA for men and nonpregnant/nonlactating women and a minimum daily intake of 300 mg EPA plus DHA for pregnant and lactating women (FAO 2008). Contents of EPA plus DHA in fish oils range from an average of 11% in herring oil to 26% in anchovy oil (Pike and Jackson 2010), while concentrated fish oil complying with the *Codex Standard for Fish Oils* contains between 35 and 50% (CODEX STAN 329-2017 2017).

Other compounds apart from lipids occurring in fish oils are the unsaponifiable matter, which includes hydrocarbons, sterols (cholesterol being the most abundant), glyceryl ethers, fatty alcohols (primarily derived from wax esters), and minor amounts of pigments (carotenoids), vitamins (A, D, and E), and oxidation products (Rizliya and Mendis 2014). Environmental contaminants (heavy metals and persistent organic pollutants), which may be present in crude fish oils and may pose a risk to human health, are virtually removed during refining. Maximum levels for fish oils intended for human consumption are laid down in the Council Regulation (EC) No. 2375/2001

(Council of the European Union 2001) and the *Global Organization for EPA and DHA Omega-3* (GOED) *Voluntary Monograph* (GOED 2012).

5.4 Oxidative Stability

A crucial aspect determining fish oil quality is oxidative stability. Due to the high number of double bonds in EPA and DHA, fish oils are highly prone to lipid oxidation, giving rise to the formation of secondary oxidation products, which are known to impart off-flavors including a fishy taste or odor (Serfert et al. 2010). In fact, volatile oxidation products resulting from the oxidation of LCn3 fatty acids have particularly low sensory threshold values in humans, which limit the consumption of these oils (Frankel 2012).

Crude fish oil contains a variety of compounds that may affect oxidative stability. The removal of prooxidant compounds (e.g. trace metals) during refining is accompanied by a loss of antioxidant compounds such as α-tocopherol, carotenoids (e.g. astaxanthin found in salmonids), retinol, and phospholipids (Oterhals and Vogt 2013). Measures used to limit lipid oxidation in fish oil include the use of fresh fish as raw material for the extraction of fish oil, minimizing exposure to ambient air during processing, refining oils under vacuum, flushing storage containers with nitrogen, and the addition of antioxidants (Ismail et al. 2016; Pike and Jackson 2010). Synthetic antioxidants such as butylated hydroxytoluene, tocopherols, ascorbic acid, and ascorbyl palmitate, and plant extracts like rosemary extract can reduce the rate of oxidation (Rubio-Rodríguez et al. 2010). Furthermore, encapsulation or microencapsulation with a coating material may retard lipid oxidation (Rizliya and Mendis 2014). Various physical and chemical processes are available for encapsulation. Spray-drying is the most common technique, alternatives include freeze-drying and ultrasonic atomization, which avoid the use of high temperatures during drying (Rubio-Rodríguez et al. 2010). Oil-in-water emulsions are increasingly used as delivery systems for fish oils in food applications, which, if properly designed, may reduce lipid oxidation in the fish oil before as well as after the addition to foods (Rizliya and Mendis 2014).

5.5 Methods for Quality Assessment

Lipid oxidation and hydrolysis provoke changes in color, odor, and taste, which is why sensory evaluation is considered one of the most important analyses to determine fish oil quality (Ritter and Budge 2012). Most studies apply descriptive sensory analyses where sensory attributes are assessed on intensity scales. Typical odor attributes include rancid, metallic, painty, pungent, buttery, and green. Descriptive sensory analyses require the product-specific training and maintenance of a sensory expert panel and are therefore cost-intensive and time-consuming (Ritter and Budge 2012; Serfert et al. 2010).

Apart from contents of omega-3 fatty acids, crude fish oils are typically tested for the following physicochemical quality parameters: moisture, unsaponifiable matter, appearance (color), free fatty acids/acid value, saponification value, iodine value, peroxide value, *p*-anisidine value, conjugated dienes and trienes (ultra violet extinction at 233 and 269 nm), heavy metals, sulfur, phosphorus, and persistent organic pollutants (FAO 1986; GOED 2012). The analysis of these parameters is part of a manufacturer's quality assurance but also serves to determine refining procedures (FAO 1986). The most common quality parameters currently used to measure lipid oxidation in fish oils for human consumption are peroxide and *p*-anisidine values (GOED and CRN [Council for Responsible Nutrition] 2015). Various national and international standards have set limits/guidelines for peroxide, *p*-anisidine, and acid values in refined fish oils. Most of them, including the *Codex Standard for Fish Oils* and the *GOED Voluntary Monograph*, define a peroxide value ≤5 meq O_2/kg, a *p*-anisidine value ≤20, and an acid value ≤3 mg KOH/g as acceptable (CODEX STAN 329-2017 2017; GOED and CRN 2015). However, the European Food Safety Authority (EFSA) pointed out that due to the lack of scientific data on potential negative health effects of oxidation products and free fatty acids, it is not possible to set maximum values for these parameters (BIOHAZ [EFSA Panel on Biological Hazards] 2010). In the following, an overview of conventional and new approaches to determine fish oil quality will be provided.

5.6 Conventional Methods

5.6.1 Wet-Chemical Methods

Moisture, free fatty acids, acid value, peroxide value, and *p*-anisidine value are typically determined by titration or spectrophotometric methods (AOCS [American Oil Chemists' Society] 2017). These methods have been criticized for their lack of specificity and sensitivity (BIOHAZ [EFSA Panel on Biological Hazards] 2010). Moreover, peroxide and *p*-anisidine values do not correlate well with the concentration of specific volatile oxidation products nor a sum of these nor the overall sensory quality (Ritter and Budge 2012; Sullivan and Budge 2012).

An aspect that is closely linked to fish oil composition and quality is authenticity. Due to its high market value, fish oil may become a target of fraudulent activities. In particular, it may be affected by adulteration or substitution with other marine oils or vegetable oils of lower value (Liverseege 1904). While molecular techniques are commonly used to determine the species and geographical origin of many fish products, these methods are not applicable to fish oils that have undergone high temperature treatment during refining, which causes the deoxyribonucleic acid (DNA) to degrade (Martinez et al. 2005). The authenticity of fish oils was historically determined by the amount of free fatty acids, the refractive index, or certain color reactions (Parry 1904). Even more recently, wet-chemical procedures are still used to authenticate fish oils in order to determine quality parameters such as iodine value, moisture, peroxide value, *p*-anisidine value, free fatty acids, and unsaponifiable matter (Tangendjaja 2015).

5.6.2 Instrumental Methods

Chromatographic methods have been used extensively in fish oil analysis in combination with traditional calibration approaches involving signal allocation by means of external standards or databases, determination of peak areas (or heights), and, if necessary, quantification using external calibration curves. Following this approach, Hartvigsen et al. 2000 determined the peroxide value in the oil phase of fish oil-enriched mayonnaises using high performance liquid chromatography (HPLC). Due to its high resolution, sensitivity, and reproducibility in quantitative analyses, gas chromatography (GC) coupled to a flame ionization detector (FID) is the standard method for determining fatty acid profiles of fish oils (CODEX STAN 329-2017 2017; GOED 2012).

Fatty acid profiles and the regiospecific distribution of fatty acids on the glycerol backbone of triacylglycerols provide information on fish oil authenticity. The *Codex Standard for Fish Oils* defines fish species-specific ranges for individual fatty acids determined by GC-FID (CODEX STAN 329-2017 2017), while the *European Pharmacopoeia* lays down ranges for the regiospecific distribution of DHA, EPA, and stearidonic acid in triacylglycerols for the authentication of salmon oil (Council of Europe 2005). The *European Pharmacopoeia* specifies ^{13}C nuclear magnetic resonance (NMR) spectroscopy as the standard method to determine triacylglycerol composition. The method is based on the integration of the respective carbonyl peaks in the ^{13}C NMR spectrum (Council of Europe 2005) and was found to be more reliable than traditional enzymatic hydrolysis methods (Tengku-Rozaina and Birch 2014). Fatty acid composition has also been determined by electrospray ionization-mass spectrometry (ESI-MS) to discriminate animal from vegetable fats and oils as well as marine oils from other animal fats and oils (Kurata et al. 2005).

GC-MS is used in the analysis of individual volatile oxidation products. While a large variety of volatiles can be detected and quantified, the interpretation of this complex data may be difficult as the relationship between the concentration of individual volatiles and their sensory impact is still poorly understood (BIOHAZ [EFSA Panel on Biological Hazards] 2010). The sensory properties of a sample are usually generated by a complex mixture of compounds (Benedetti et al. 2009; Serfert et al. 2010). Only very few studies combined GC with olfactometry, thereby identifying key aroma-active compounds in fish oil (Jónsdóttir et al. 2005; Sandgruber and Buettner 2012). Hexanal, 2,4-decadienals, 2,4,7-decatrienals, and 2,6-nonadienals have been described as quality indicators for monitoring fish oil oxidation, while *cis*-4-heptenal, *trans,trans*-2,4-heptadienal, and *trans,cis*-2,6-nonadienal have been associated with fishy flavor (Jónsdóttir et al. 2005; Karahadian and Lindsay 1989; Sandgruber and Buettner 2012; Sullivan and Budge 2012). There have also been attempts to replace the classical sensory analysis by electronic noses. Benedetti et al. (2009) studied lipid oxidation in LCn3-enriched microparticles by means of an electronic nose and solid-phase microextraction (SPME)-GC-MS. They found high correlations between certain sensor responses and several volatile compounds detected by SPME-GC-MS (propanal, hexanal, 2,6-nonadienal) and concluded that electronic noses could be a sensitive tool to evaluate lipid oxidation of LCn3-enriched microparticles (Benedetti et al. 2009).

Sullivan et al. (2009) developed an SPME-GC-MS method for the quantification of ethyl esters as contaminants/adulterants in fish oil. They calibrated peak area ratios of ethyl esters to an internal standard against ethyl ester concentrations determined by GC-FID. However, ethyl esters with low volatility and ethyl esters present in low concentrations could not be quantified by SPME-GC-MS (Sullivan et al. 2009).

While chromatographic methods may offer high sensitivity and specificity, they are generally laborious and time-consuming. Therefore, they are currently not implemented as routine methods for quality control in the fish oil industry (BIOHAZ [EFSA Panel on Biological Hazards] 2010). As an alternative, spectroscopic methods such as near-infrared (NIR), mid-infrared (MIR), Raman, and NMR spectroscopies have become increasingly popular. They are known to be nondestructive, rapid, and less laborious as they involve minimal or no sample preparation. Furthermore, they often require only very small amounts of sample and minimal or no chemicals. They can be automated and they provide comprehensive information in a single analysis (Wu et al. 2014).

In the field of fish oil quality assessment, vibrational and NMR spectroscopy are now most often used in combination with multivariate statistics (see Section 5.7). However, even without the application of statistical methods, they can provide a myriad of information. Fourier Transform (FT)-MIR spectroscopy has been used to monitor fish oil oxidation by the observation of changes in signal intensities and positions of bands/peaks (Belhaj et al. 2010; Daoud et al. 2019). Guillén et al. (2008) determined the molar percentage of polyunsaturated acyl groups in cod liver oils based on the frequency value of the band corresponding to =C–H stretching vibrations of *cis*-double bonds at around 3011 cm^{-1}. They also calculated the ratio of absorbance at 3011 cm^{-1} and 2854 (or 2925) cm^{-1}, which provides information on the total unsaturation degree (Guillén et al. 2008). Similarly, Wang et al. (2011) used FT-MIR spectroscopy to track changes in *trans* fatty acid and aldehyde contents during storage of fish oil at 30 °C by determining absorbance at wavenumbers that are characteristic of vibrations of carbonyl and *trans* double bonds.

Igarashi et al. (2002) validated a ^{1}H NMR method for determining absolute and molar concentrations of DHA and molar proportions of total omega-3 fatty acids in fish oils in an International Union of Pure and Applied Chemistry (IUPAC) interlaboratory study. The molar concentration was determined by comparing signal intensities at 2.36 ppm (C2 and C3 methylene protons of DHA) and 2.28 ppm (C2 methylene protons of other fatty acids) for DHA analysis and at 0.95 ppm (methyl protons of omega-3 fatty acids) and 0.85 ppm (methyl protons of other fatty acids) for omega-3 fatty acids analysis. Absolute quantification was performed using ethylene glycol dimethyl ether as an internal standard (Igarashi et al. 2002). Others have derived molar percentages of EPA, total omega-1, omega-6, saturated, and unsaturated fatty acids as well as *trans* fatty acids and phosphatidylcholine by integration of relevant signals in ^{1}H and ^{13}C NMR spectra (Burri et al. 2016; Dais et al. 2015; Vidal et al. 2015; Williamson and Hatzakis 2017). In addition, ^{13}C NMR spectroscopy can be used for the quantification of DPA, total omega-9 fatty acids as well as stearidonic, eicosatetraenoic, arachidonic, linoleic, and α-linolenic acids (Dais et al. 2015). Both ^{1}H and ^{13}C NMR spectroscopy can distinguish between triacylglycerols and ethyl esters (Siddiqui et al. 2003). Total sterols, 1,2-diacylglycerols, and 1,3-diacylglycerols can be quantified in ^{1}H NMR spectra, contents of 1,3-diacylglycerols and free fatty acids can be derived from ^{13}C NMR spectra (Dais et al. 2015; Williamson and Hatzakis 2017). ^{31}P NMR spectroscopy has been used to quantify mono- and diacylglycerols, vitamin E, and free fatty acids as well as classes of phospholipids occurring in krill oil (Burri et al. 2016; Dais et al. 2015; Monakhova et al. 2018).

Both ^{1}H and ^{13}C NMR spectroscopy have been applied to the study of fish oil oxidation. In the process of lipid oxidation, ^{1}H NMR spectra reveal changes in signal intensities of resonances related to omega-3 fatty acids, allylic, and olefinic protons, while new signals appear as a result of the formation of primary and secondary oxidation products (hydroperoxides, compounds containing conjugated double bonds, hydroxyl compounds, epoxides, aldehydes) (Siddiqui et al. 2003) (see Figure 5.1). These changes have been monitored by the analysis of signal intensity ratios, some of which correlated well with lipid oxidation parameters determined by traditional wet-chemical methods (Tyl et al. 2008). Vidal et al. (2015) integrated characteristic signals of alkanals, *trans*-2-alkenals, and *trans,trans*-2,4-alkadienals in the ^{1}H NMR spectrum of sea bass lipids in order to determine their molar concentrations. Other aldehydes like the genotoxic and cytotoxic 4-hydroxy-*trans*-2-alkenals could not be quantified due to the low intensity and resolution of their signals (Vidal et al. 2015).

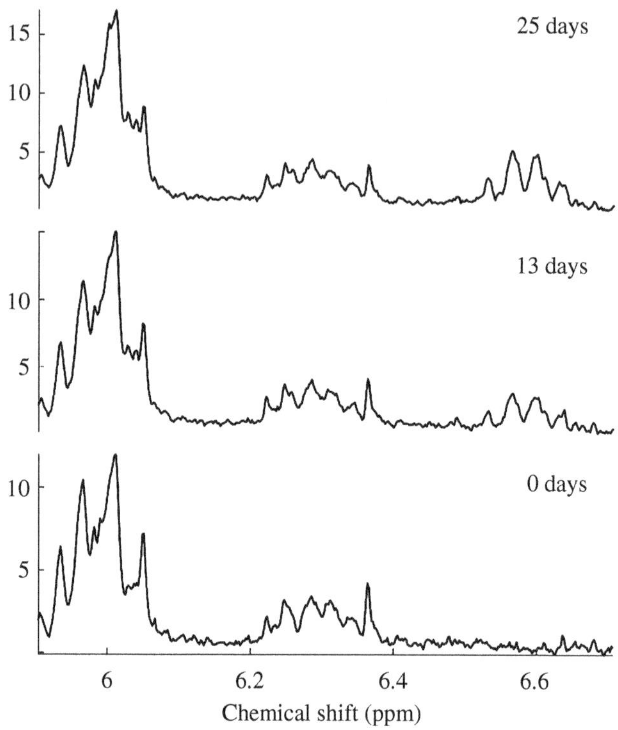

Figure 5.1 ^{1}H NMR spectra of a salmon oil stored at 40 °C (in closed 50 ml polypropylene centrifugation tubes positioned in a closed opaque plastic box inside a climate test chamber) for up to 25 days: resonances of *cis,trans*-conjugated dienes at around 6 ppm and 6.50–6.64 ppm (*Source:* adapted from Giese 2020).

Burri et al. (2016) and Monakhova et al. (2018) analyzed krill oil (and other fish oils) with regard to EPA and DHA contents and their positional distribution in phospholipids and triacylglycerols using ^{1}H and ^{13}C NMR spectroscopies. They found that in krill oil, EPA and DHA were mainly incorporated into phospholipids, whereas in other fish oils, they are concentrated in triacylglycerols. Adulterations of krill oil with other fish oil can be detected by an increased content of omega-3 fatty acids in the overall triacylglycerol fraction (^{1}H NMR) (Burri et al. 2016) and the *sn*-2 position of triacylglycerols (^{13}C NMR) (Monakhova et al. 2018). Monakhova et al. (2018) estimated that it was possible to detect fish oil in krill oil qualitatively at concentrations of as low as 10% based on 1 and 99% quantiles calculated for DHA/EPA resonances in the *sn*-2 region of triacylglycerols in 15 commercial krill oil supplements. Furthermore, Burri et al. (2016) identified phytyl wax esters in the ^{1}H NMR spectra, which are highly specific for krill oil and not found in other fish oils.

5.7 Machine Learning Approaches toward the Detection of Compositional Shifts

Spectroscopic and chromatographic instruments acquire a wealth of data that are complex and difficult to interpret, especially when multicomponent samples like fish oils are analyzed (Ropodi et al. 2016). Spectroscopic techniques are regarded as fingerprint methods, i.e. they provide holistic information on sample composition and can be used for nontargeted analyses. However, the acquired spectra may suffer from overlapping bands/peaks, which may make it impossible to integrate individual signals. Also, certain analytical problems cannot be solved by a univariate approach. This is where machine learning can be useful because it can help to extract the relevant information by statistical means. In general, machine learning does not require the integration of individual signals, but the acquired data is used directly (or after preprocessing/variable selection) as input for statistical models (Giese et al. 2019).

Machine learning is the scientific discipline of the development and application of algorithms designed to "learn" from training data (Ropodi et al. 2016). Unsupervised learning groups samples based on their similarities/differences, whereas supervised learning generates models that can be used to make predictions and/or decisions. The most common linear methods used in fish oil analysis are Principal Component Analysis (PCA; see Figure 5.2) for unsupervised learning and Partial Least Squares Regression (PLSR), Principal Component Regression (PCR), Multiple Linear Regression (MLR), and Linear Discriminant Analysis (DA) for supervised learning (see Section 5.7.1). Alternatively, more advanced techniques have increasingly attracted attention, including, e.g. Artificial Neural Networks (ANN) and Support Vector Machines (SVM) (see Section 5.7.2). Many multivariate statistical models do not only provide information on the identity of a sample of interest but also on the variables/features in the analytical data from which this information is derived. The typical process of developing a prediction model in supervised learning is depicted in Figure 5.3. A crucial step is the partitioning of the data into a calibration, a validation, and a test set. Some studies apply cross-validation and do not include an independent test set (Ropodi et al. 2016).

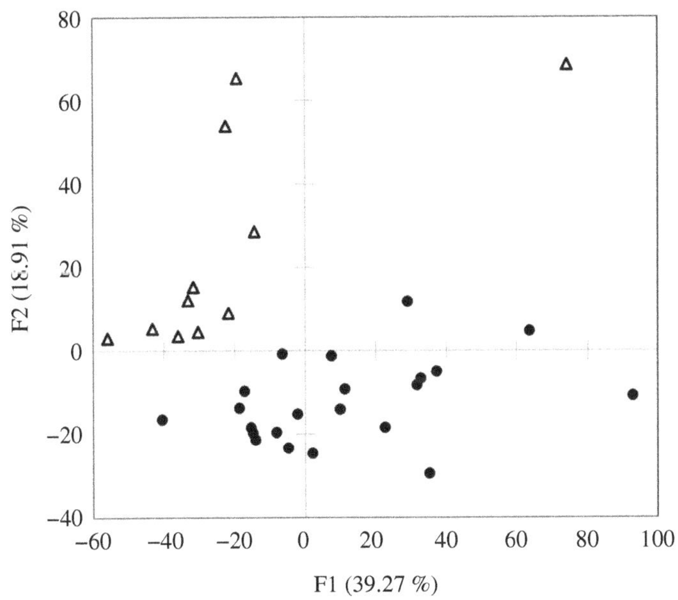

Figure 5.2 PCA score plot of salmon oil (circles) and other fish oil samples (triangles) analyzed by FT-MIR and FT-NIR spectroscopy (fused spectra were submitted to PCA) (*Source:* E. Giese, University of Hamburg, Hamburg, unpublished results).

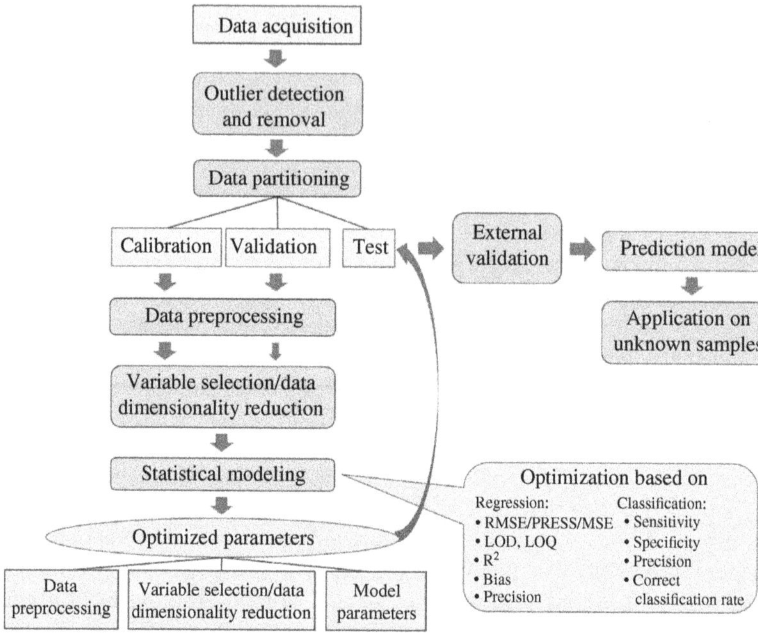

Figure 5.3 Process of how to develop a machine learning model for regression/classification.

5.7.1 Standard Methods

Two studies applied machine learning to SPME-GC-MS data (peak area percentages of volatiles) in order to predict the sensory quality of fish oil (Ritter and Budge 2012; Sullivan and Budge 2012). Ritter and Budge (2012) used Principal Components (PCs) as input variables in MLR to predict sensory scores, while Sullivan and Budge (2012) used Stepwise Discriminant Function Analysis to classify oils as acceptable or unacceptable based on sensory analysis. Key lipid oxidation products with a high impact on predictions were identified, the majority of which were aldehydes (e.g. *trans,cis*-2,4-octadienal, 3,6-nonadienal) and ketones (e.g. 3,5-octadien-2-one). While in the study of Sullivan and Budge (2012), 100% correct classification was achieved for cross-validation and the three test samples, Ritter and Budge (2012) found that it was more difficult to make quantitative predictions of the sensory scores. Both studies analyzed only a very small number of samples.

Various studies applied linear machine learning methods to determine fatty acid profiles. In most cases, models were developed by PLSR based on FT-MIR or NIR data. Karunathilaka et al. (2019) found that both FT-MIR and FT-NIR spectroscopy provided excellent or very good models for contents of EPA, DHA, and total omega-3 polyunsaturated fatty acids (PUFA), while FT-NIR spectroscopy was superior in predicting total monounsaturated fatty acids and omega-6 PUFA contents. Comparing FT-MIR, NIR, and Raman spectroscopies, Bekhit et al. (2014) developed the best model for EPA and total omega-3 fatty acids based on FT-MIR and NIR spectra, respectively, while the NIR model exhibited the best performance for DHA. In the study of Wu et al. (2014), FT-MIR spectroscopy provided the most reliable PLSR model for EPA, which was built on 551 spectral variables obtained after Uninformative Variable Elimination

(UVE) and exhibited good performance. ^{1}H NMR spectroscopy provided the most valid PLSR model for DHA, with very good performance. This model was built on 10 variables selected by Successive Projections Algorithm (SPA) after the application of UVE. Visible–short-wave NIR and long-wave NIR models were of moderate quality, except for the case of DPA modeling, where a very good model was generated by means of UVE-SPA-PLSR based on long-wave NIR spectra (Wu et al. 2014).

PLSR models of moderate quality were obtained for *trans* PUFA content based on FT-MIR spectra (Karunathilaka et al. 2017). In a preliminary study, PLSR predicted iodine and saponification values from FT-NIR spectra with high accuracy and even better reproducibility than the standard titration methods (Endo et al. 2005). Cozzolino et al. (2005) developed a PLSR model with very good performance to derive the content of free fatty acids from NIR spectra. They also modeled moisture as well as peroxide and *p*-anisidine values. However, these models exhibited poor accuracy (Cozzolino et al. 2005). In two preliminary studies, conjugated dienes in fish oil and fish oil emulsions were monitored by FT-Raman and FT-MIR spectroscopies applied in combination with PCA and PLSR, respectively (Daoud et al. 2019; Killeen et al. 2017).

Fatty acid and triacylglycerol profiles determined by GC-FID and LC-ESI-MS-MS, respectively, have been used in combination with PCA to distinguish between fish and marine mammal oils (GC-FID) (Araujo et al. 2010) as well as between pure cod liver or blue whiting oils and their mixtures with seal, salmon, or soybean oils at adulteration levels ranging from 25 up to 100% (LC-ESI-MS-MS) (Araujo et al. 2018; Zeng et al. 2010). Fish oils differ from marine mammal oils in that omega-3 fatty acids are mainly found in the *sn*-2 position of triacylglycerols, whereas in marine mammals, they are preferentially located in the *sn*-1/3 position (Araujo et al. 2018; Zeng et al. 2010). Rohman and Che Man (2011a) found that the contents of EPA, DHA, and gondoic acid separated beef fat, lard, mutton fat, and chicken fat from cod liver oil in the PCA score plot of fatty acid contents determined by GC-FID. The results of the aforementioned studies are regarded as preliminary because very small numbers of samples were analyzed.

A series of studies by the same research group determined cod liver oil adulteration based on FT-MIR spectra (Rohman and Che Man 2009, 2011a, b, c; Rohman et al. 2012, 2017). They used various animal fats and vegetable oils as adulterants and applied PLSR, PCR, and DA for modeling adulteration, which ranged from 1 to 50% in most cases. They identified the fingerprint region (1500–900 cm^{-1}) as well as the band at 3007 cm^{-1} (stretching vibrations of =C–H *cis*-olefinic groups) as particularly useful for modeling due to differences in peak absorbance and frequency values (Rohman and Che Man 2009, 2011a, b, c; Rohman et al. 2012, 2017) (see Figure 5.4). The Root Mean Square Error of Prediction (RMSEP) for the quantification of adulteration levels in the validation sets varied from 0.28% (corn oil) (Rohman et al. 2017) to 1.8% (canola oil) (Rohman and Che Man 2011b) and 100% correct classification of adulterated and genuine cod liver oils was obtained by DA (Rohman and Che Man 2009, 2011b, c; Rohman et al. 2012). Rohman and Che Man (2011a) to fatty acid profiles determined by GC-FID with regard to their suitability to determine cod liver oil adulteration with beef fat. They obtained a limit of detection (LOD) of 2% for the FT-MIR spectroscopy-based PLSR model compared to 5% for the GC-FID method. In the GC-FID method, the LOD was defined as the lowest concentration where a significant difference ($p<0.05$) was identified by one-way Analysis of Variance (Rohman and Che Man 2011a). Lingzhi et al. (2016) determined an LOD of 0.8, 0.6, 2, and 3% for the detection of lard, chicken oil, tallow, and suet, respectively, in fish oil by FT-MIR spectroscopy and PLS-DA. All these studies provided promising results

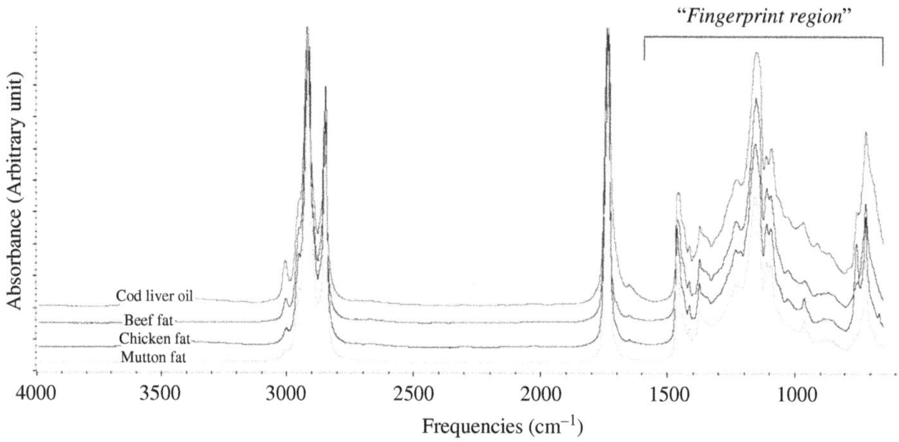

Figure 5.4 FT-MIR spectra of cod liver oil, beef fat, chicken fat, and mutton fat. *Source:* Rohman and Che Man 2011c, Taylor & Francis Ltd. (www.tandfonline.com).

regarding the detection of fish oil adulteration by FT-MIR spectroscopy. However, they exhibited certain limitations in that (i) their validation data sets consisted of blends containing the same original oils as the ones used in the calibration sets, (ii) the number of different pure oils was small, or (iii) the allocation of the different pure oils to calibration and validation sets was not reported.

Karunathilaka et al. (2019) classified marine oil supplements into triacylglycerols, ethyl esters, and mixtures thereof using PLS-DA based on FT-MIR and FT-NIR data. The sensitivity of the FT-MIR model was found to be higher compared to the FT-NIR model. The FT-MIR model successfully classified around 96% of the 95 products as belonging to the three classes and the LOD for the detection of ethyl esters in triacylglycerol oils was estimated at 11% (Karunathilaka et al. 2019).

5.7.2 Advanced Methods

To date, there is only a limited number of studies that applied advanced machine learning methods in fish oil analysis. Two studies predicted fish oil oxidation as measured by peroxide and *p*-anisidine values using ANN (Giese et al. 2018; Klaypradit et al. 2011). ANN are a set of mathematical methods mimicking the functioning of the human brain (Hastie et al. 2009). Figure 5.5 shows the structure of a typical multilayer feed-forward ANN. The major advantage of ANN is that they are capable of both linear and nonlinear modeling and can be flexibly adapted to a specific problem in terms of their architecture and learning procedure (Hastie et al. 2009).

In the study of Klaypradit et al. (2011), a menhaden oil was artificially oxidized and FT-MIR spectra were recorded at regular intervals. The FT-MIR data was used as input for ANN in order to model peroxide and *p*-anisidine values determined by wet-chemical methods. Hyperbolic tangent and linear activation functions were used for modeling (Klaypradit et al. 2011). Giese et al. (2018) compared ANN to PLSR with regard to their performance in modeling peroxide, *p*-anisidine, and acid values based on ^{1}H NMR, FT-MIR, and FT-NIR spectra. They found that in most cases, the accuracy of ANN models (using ramp and logistic activation functions) was superior to the

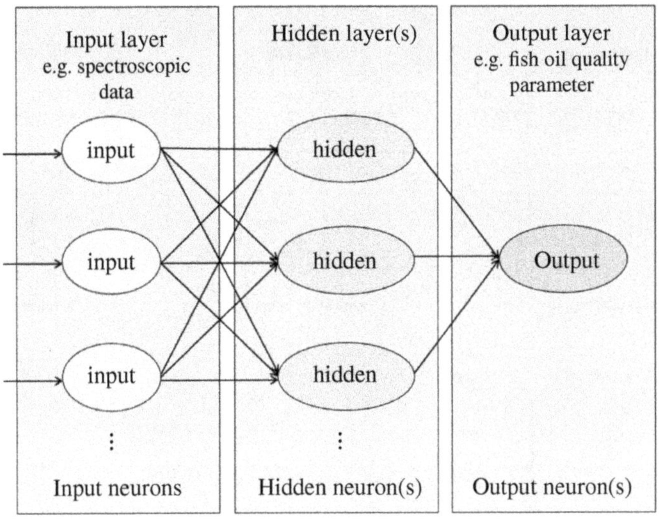

Figure 5.5 Example structure of a fully-connected multilayer feed-forward ANN.

respective PLSR models. ANN predicted peroxide, *p*-anisidine, and acid values with an RMSEP of 1.5 meq O_2 /kg (^{1}H NMR spectra), 0.74 (fused FT-MIR and FT-NIR spectra), and 0.43 mg NaOH/g (FT-MIR spectra), respectively, and with high precision. The ANN model for the peroxide value based on ^{1}H NMR data was regarded as suitable for quality screening, while the ANN model for the acid value based on FT-MIR data turned out to be adequate for quality control and the ANN model for the *p*-anisidine value based on the combined MIR/NIR data was even considered excellent for process control, development, and applied research (Giese et al. 2018).

As far as authentication is concerned, Aursand et al. (2007) used ^{13}C NMR spectroscopy in combination with Probabilistic Neural Networks to differentiate between fish oil samples of different species, geographical origins, and production methods (wild or farmed) and achieved >95% prediction accuracy. They also developed models for predicting the adulteration level of salmon oil with commercial South American fish oil (mainly sardine/anchovy oil) using ANN and PCR. Adulteration levels ranged from 5 to 40%. ANN quantified the degree of adulteration to within about 1%, whereas PCR yielded an RMSE of approx. 1.8% (Aursand et al. 2007).

Giese et al. (2019) investigated the adulteration of cod liver oil with canola and sunflower oils using ^{13}C NMR, ^{1}H NMR, and FT-MIR spectroscopies as well as fatty acid profiles determined by GC-FID in combination with various standard and more sophisticated machine learning methods such as MLR, ANN, SVM, Logistic Regression, k-Nearest Neighbors, and Multivariate Adaptive Regression Splines (MARS). Adulteration levels ranged from 1 to 50%. Exemplary spectra of pure cod liver and sunflower oils as well as a blend thereof (30% sunflower oil) are shown in Figure 5.6. The authors obtained the best regression models using ANN with sigmoid and linear activation functions. The most sensitive and accurate model was based on FT-MIR data with an LOD of 0.22% and an RMSEP of 0.86%. However, the ANN model using fatty acid profiles determined by the standard GC-FID method also yielded very promising results (LOD = 0.81%, RMSEP = 1.1%). The ANN models

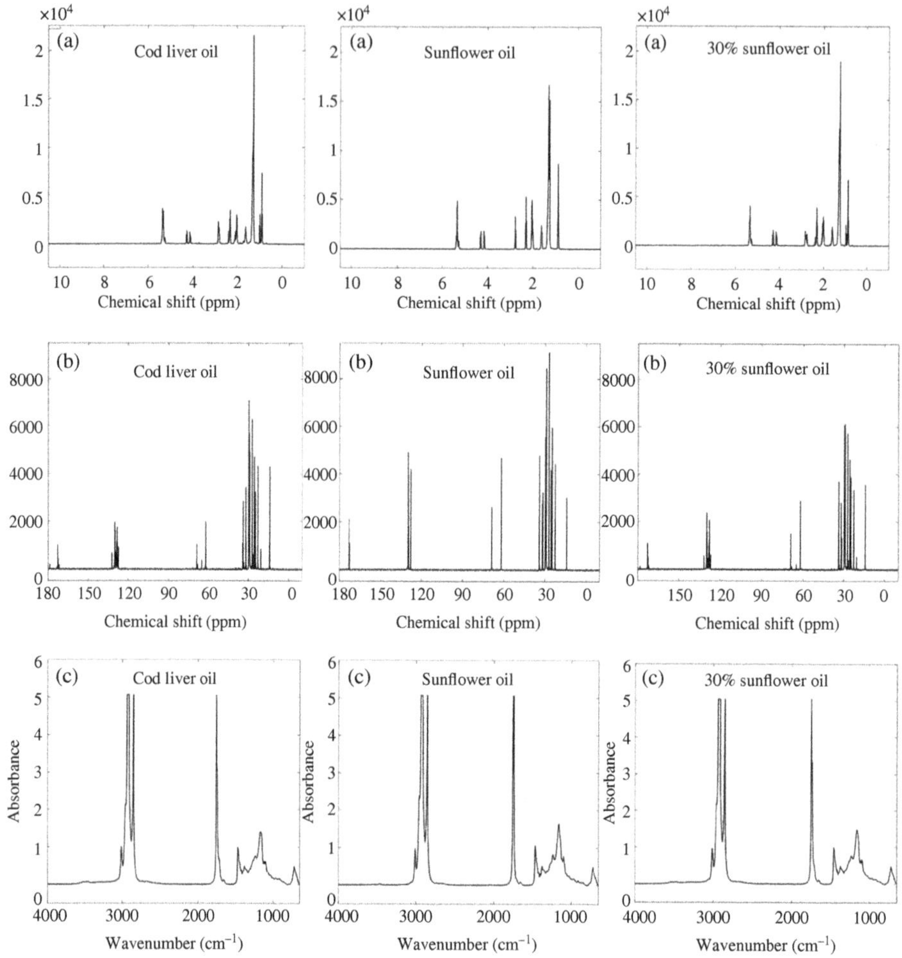

Figure 5.6 Exemplary spectra of a crude cod liver oil, a refined sunflower oil, and a blend thereof (crude cod liver oil containing 30% sunflower oil): (a) ^{1}H NMR; (b) ^{13}C NMR; (c) FT-MIR. *Source:* Giese et al. (2019), Springer Nature, distributed under the terms of the Creative Commons Attribution 4.0 International License (http://creativecommons.org/licenses/by/4.0).

based on ^{1}H NMR and ^{13}C NMR data exhibited an LOD of 3.0 and 1.8% and an RMSEP of 2.7 and 1.1%, respectively. In addition, the researchers generated classification models to distinguish between genuine and adulterated cod liver oils based on the same sample sets. Among the analytical and statistical techniques investigated, SVM (linear kernel) based on FT-MIR data and Flexible DA using MARS based on ^{13}C NMR data provided the most reliable models with 100% correct classification (Giese et al. 2019).

Nonlinear SVM using radial basis kernel functions were identified as being superior to PLSR in the quantification of DHA, EPA, and total omega-3 PUFA in marine oil dietary supplements based on FT-MIR or FT-NIR spectra, with all models exhibiting excellent performance (Karunathilaka et al. 2020).

5.7.3 Limitations

Although the approach of using instrumental, in particular spectroscopic techniques in combination with machine learning seems to be a powerful tool in fish oil analysis, its application in practice is still very limited. Apart from the fact that machine learning approaches are not included in any official control methods or guidance documents for fish oil analysis, a number of factors may be responsible for this phenomenon. The calibration of a prediction model requires the prior analysis by both the standard and the alternative method of a large number of fish oil samples (Ropodi et al. 2016; Xu et al. 2015). As the alternative methods are calibrated against a standard method, the resulting prediction models can only be as valid as the standard method. Furthermore, as the basis of a reliable and robust model, the calibration samples have to be authentic and representative of the samples of interest (capturing the fish oil variability caused by factors such as species, geographical origin, season, processing, and quality). It is vital that the calibration database is updated regularly (Ropodi et al. 2016; Xu et al. 2015). Another important hurdle for the application of machine learning approaches is the fact that expertise in statistics is required to develop and update prediction models (Ropodi et al. 2016). However, the routine application of the models does not necessitate any statistical knowledge.

ANN offer several advantages over standard machine learning methods, but they are also affected by limitations: If a large number of variables are used as inputs (e.g. all the data points of a spectrum), network training requires huge computational capacity and may take a long time (Hastie et al. 2009). Therefore, the application of variable selection or data dimensionality reduction techniques may be necessary prior to modeling. The downside of the high versatility of ANN and the large number of parameters that are optimized during the training process is that large calibration datasets are needed in order to ensure that the ratio of the number of training samples to the number of adjustable parameters is adequately high so that the network is able to generalize. The susceptibility of ANN to overfitting is also the reason why validation of the prediction models using an independent test set is crucial, especially if the networks are complex. Another drawback that does not only apply to ANN but also to other machine learning techniques such as SVM and k-Nearest Neighbors is that the generated models may be regarded as a black box because they are difficult to interpret (Hastie et al. 2009).

Moreover, certain limitations inherent to the respective analytical technique (e.g. with respect to sensitivity, resolution, and cost) continue to exist when machine learning is applied to the data generated by these techniques.

5.8 Future Perspectives

The fish oil market is growing, which is the result of an increased demand for aquaculture feed but also of a growing popularity of dietary supplements and functional foods containing fish oil. Fish oil quality parameters such as peroxide and *p*-anisidine values are traditionally determined by laborious wet-chemical methods, while GC-FID is the official method for determining fatty acid profiles. Spectroscopic techniques such as NMR, FT-MIR, and NIR spectroscopies exhibit many advantages over these traditional methods because they offer a rapid, simultaneous, nondestructive, less laborious, and potentially automated analysis of numerous quality parameters of complex samples like dietary supplements. Although NMR spectrometers are currently not widely used in the analysis of food products because of their elevated cost and the high expertise required for data analysis, they are gradually becoming more common due to the aforementioned advantages as well as advances in instrumentation and software making NMR spectroscopy more suitable for routine analysis (Dais et al. 2015).

While the determination of triacylglycerol profiles by ^{13}C NMR spectroscopy has been included in the *European Pharmacopeia* for salmon oil and the determination of *trans* fatty acid contents by FT-MIR spectroscopy has been standardized as an official AOCS method (AOCS [American Oil Chemists' Society] 2017), no machine learning approach has found its way into official control methods for fish oil or guidelines for fish oil analysis. Machine learning can help to extract the relevant information from spectroscopic data, which is generally complex and contains overlapping signals. To date, the application of machine learning is very limited in the field of fish oil analysis compared to vegetable oils. It has mainly been used to determine lipid oxidation and hydrolysis parameters, fatty acid profiles, and fish oil authenticity. The potential of more sophisticated machine learning methods such as ANN has not been fully explored yet, but it can be expected that these methods offer additional advantages in cases where nonlinear structures are present in the data. Although the existing literature provides promising results for spectroscopy-based machine learning approaches as high-throughput screening or even quality control methods, many of the studies involved only small numbers of samples and exhibited limitations in their validation, which is why they have to be interpreted with caution. In order to become part of official control methods and be widely used by fish oil manufacturers, food control authorities, and private laboratories, the developed statistical models have to be fully validated in interlaboratory trials. As spectroscopic techniques are considered to be fingerprint methods, they have a particularly high potential to be used in the growing field of authentication and may pave the way for a holistic quality assessment of fish oils.

Machine learning approaches also exhibit a great potential for the development of models that are able to predict food fraud risks based on data from various sources and of various types such as expert knowledge, measurement data, and feedback experience (Fritsche 2018). In order to identify known and emerging food fraud risks, preferably in real time, prediction models have to link all drivers of food fraud and their dependencies (e.g. global economy and trade, global cooperation and standard setting, governance, demography and social cohesion, consumer attitudes and behavior, new food chain technologies, competition for key resources, climate change, emerging food chain risks and disasters, and new agri-food chain structures). Models for the prediction of food fraud risks applying machine learning could enable adequate and timely resource allocations within the food control community as well as the agri-food business in order to prevent food fraud. As a result, they could contribute to improve the confidence of consumers in food quality and to minimize economic losses (Fritsche 2018).

References

Abdelhamid, A.S., Brown, T.J., Brainard, J.S. et al. (2020). Omega-3 fatty acids for the primary and secondary prevention of cardiovascular disease. *Cochrane Database of Systematic Reviews* 3: CD003177. https://doi.org/10.1002/14651858.CD003177.pub5.

AOCS (American Oil Chemists' Society) (2017). *The Official Methods and Recommended Practices of the AOCS*. Urbana: AOCS.

Araujo, P., Zeng, Y., Du, Z.Y. et al. (2010). Discrimination of n-3 rich oils by gas chromatography. *Lipids* 45: 1147–1158.

Araujo, P., Tilahun, E., and Zeng, Y. (2018). A novel strategy for discriminating marine oils by using the positional distribution (*sn*-1, *sn*-2, *sn*-3) of omega-3 polyunsaturated fatty acids in triacylglycerols. *Talanta* 182: 32–37.

Auchterlonie, N. (2018). Global Fish Oil Supply: inputs, outputs and markets. Presented at the Omega-3 Summit in Brussels, October 29–30, 2018.

Aursand, M., Standal, I.B., and Axelson, D.E. (2007). High-resolution ^{13}C nuclear magnetic resonance spectroscopy pattern recognition of fish oil capsules. *Journal of Agricultural and Food Chemistry* 55: 38–47.

Bachis, E. (2017). Fishmeal and fish oil: a summary of global trends. Presented at the 57th IFFO Annual Conference in Washington, October 23–25, 2017.

Bekhit, M.Y., Grung, B., and Mjøs, S.A. (2014). Determination of omega-3 fatty acids in fish oil supplements using vibrational spectroscopy and chemometric methods. *Applied Spectroscopy* 68: 1190–1200.

Belhaj, N., Arab-Tehrany, E., and Linder, M. (2010). Oxidative kinetics of salmon oil in bulk and in nanoemulsion stabilized by marine lecithin. *Process Biochemistry* 45: 187–195.

Benedetti, S., Drusch, S., and Mannino, S. (2009). Monitoring of autoxidation in LCPUFA-enriched lipid microparticles by electronic nose and SPME-GCMS. *Talanta* 78: 1266–1271.

BIOHAZ (EFSA Panel on Biological Hazards) (2010). Scientific opinion on fish oil for human consumption. Food hygiene, including rancidity. *EFSA Journal* 8 (10): 1874.

Burri, L., Hoem, N., Monakhova, Y.B. et al. (2016). Fingerprinting krill oil by ^{31}P, ^{1}H and ^{13}C NMR spectroscopies. *Journal of the American Oil Chemists' Society* 93: 1037–1049.

Clarke, T.C., Black, L.I., Stussman, B.J. et al. (2015). Trends in the use of complementary health approaches among adults: United States, 2002–2012. *National Health Statistics Reports* 79: 1–16.

CODEX STAN 329-2017 (2017). *Standard for Fish Oils*. Rome: Codex Alimentarius Commission.

Council of Europe (2005). *European Pharmacopoeia*. Strasbourg: Council of Europe.

Council of the European Union (2001). Council regulation (EC) no 2375/2001 of 29 November 2001 amending Commission regulation (EC) no 466/2001 setting maximum levels for certain contaminants in foodstuffs (text with EEA relevance). *Official Journal of the European Communities, L* 321: 1–5.

Cowan, D. (2010). Concentration of omega-3 fatty acids using enzymes. *Agro FOOD Industry Hi Tech* 21: 30–33.

Cozzolino, D., Murray, I., Chree, A. et al. (2005). Multivariate determination of free fatty acids and moisture in fish oils by partial least-squares regression and near-infrared spectroscopy. *LWT – Food Science and Technology* 38: 821–828.

Dais, P., Misiak, M., and Hatzakis, E. (2015). Analysis of marine dietary supplements using NMR spectroscopy. *Analytical Methods* 7: 5226–5238.

Daoud, S., Bou-Maroun, E., Dujourdy, L. et al. (2019). Fast and direct analysis of oxidation levels of oil-in-water emulsions using ATR-FTIR. *Food Chemistry* 293: 307–314.

van Duijn, G. (2014). Oils and fats. In: *Food Safety Management: A Practical Guide for the Food Industry* (eds. Y. Motarjemi and H. Lelieveld), 325–345. London: Academic Press.

Endo, Y., Tagiri-Endo, M., and Kimura, K. (2005). Rapid determination of iodine value and saponification value of fish oils by near-infrared spectroscopy. *Journal of Food Science* 70: C127–C131.

European Commission (2018). EU Register on nutrition and health claims. http://ec.europa.eu/food/safety/labelling_nutrition/claims/register/public/?event=search (accessed 13 September 2019).

FAO (1986). The production of fish meal and oil. FAO Fisheries Department. FAO Fisheries Tech. Paper 142.

FAO (2008). Fats and fatty acids in human nutrition – report of an expert consultation. FAO Food and Nutrition Paper 91.

FAO (2018). *The State of World Fisheries and Aquaculture 2018 - Meeting the Sustainable Development Goals*. Rome: FAO.

FDA (2019). FDA announces new qualified health claims for EPA and DHA omega-3 consumption and the risk of hypertension and coronary heart disease. https://www.fda.gov/food/cfsan-constituent-updates/fda-announces-new-qualified-health-claims-epa-and-dha-omega-3-consumption-and-risk-hypertension-and (accessed October 1, 2019).

Frankel, E.N. (2012). *Lipid Oxidation*, 2e. Cambridge: Woodhead Publishing.

Fritsche, J. (2018). Recent developments and digital perspectives in food safety and authenticity. *Journal of Agricultural and Food Chemistry* 66: 7562–7567.

Giese, E. (2020). Bewertung der Qualität von Fischölen mittels analytischer Schnellmethoden (Kernspinresonanz- und Infrarotspektroskopie) und multivariater Statistik unter besonderer Berücksichtigung von *Künstlichen Neuronalen Netzen*. PhD thesis. University of Hamburg.

Giese, E., Winkelmann, O., Rohn, S. et al. (2018). Determining quality parameters of fish oils by means of ^{1}H nuclear magnetic resonance, mid-infrared, and near-infrared spectroscopy in combination with multivariate statistics. *Food Research International* 106: 116–128.

Giese, E., Rohn, S., and Fritsche, J. (2019). Chemometric tools for the authentication of cod liver oil based on nuclear magnetic resonance and infrared spectroscopy data. *Analytical and Bioanalytical Chemistry* 411: 6931–6942.

GOED (2012). *GOED Voluntary Monograph*, 4e. Salt Lake City: GOED.

GOED and CRN (Council for Responsible Nutrition) (2015). Oxidation in omega-3 oils: an overview. A white paper prepared by the Global Organization for EPA and DHA omega-3s and the Council for Responsible Nutrition.

Grand View Research (2016). Fish oil market size and forecast report, 2012–2022. Report ID: MN17617439, San Francisco: Grand View Research.

Guillén, M.D., Carton, I., Goicoechea, E. et al. (2008). Characterization of cod liver oil by spectroscopic techniques. New approaches for the determination of compositional parameters, acyl groups, and cholesterol from ^{1}H nuclear magnetic resonance and Fourier transform infrared spectral data. *Journal of Agricultural and Food Chemistry* 56: 9072–9079.

Hartvigsen, K., Hansen, L.F., Lund, P. et al. (2000). Determination of neutral lipid hydroperoxides by size exclusion HPLC with fluorometric detection. Application to fish oil enriched mayonnaises during storage. *Journal of Agricultural and Food Chemistry* 48: 5842–5849.

Hastie, T., Tibshirani, R., and Friedman, J. (2009). *The Elements of Statistical Learning: Data Mining, Inference, and Prediction*, 2e. New York: Springer.

Igarashi, T., Aursand, M., Sacchi, R. et al. (2002). Determination of docosahexaenoic acid and n-3 fatty acids in refined fish oils by ^{1}H-NMR spectroscopy: IUPAC interlaboratory study. *Journal of AOAC International* 85: 1341–1354.

Ismail, A., Bannenberg, G., Rice, H.B. et al. (2016). Oxidation in EPA- and DHA-rich oils: an overview. *Lipid Technology* 28: 55–59.

John, T.E., Rebry, R.M., Yan, K.L. et al. (2013). Measurement of environmental contaminants in a globally-representative sample of fish oil supplements. White paper Nutrasource Diagnostics Inc. and GOED, Guelph, Salt Lake City.

Jónsdóttir, R., Bragadóttir, M., and Arnarson, G.O. (2005). Oxidatively derived volatile compounds in microencapsulated fish oil monitored by solid-phase microextraction (SPME). *Journal of Food Science* 70: C433–C440.

Karahadian, C. and Lindsay, R.C. (1989). Evaluation of compounds contributing characterizing fishy flavors in fish oils. *Journal of the American Oil Chemists' Society* 66: 953–960.

Karunathilaka, S.R., Mossoba, M.M., Chung, J.K. et al. (2017). Rapid prediction of fatty acid content in marine oil omega-3 dietary supplements using a portable Fourier transform infrared (FTIR) device and partial least-squares regression (PLSR) analysis. *Journal of Agricultural and Food Chemistry* 65: 224–233.

Karunathilaka, S.R., Choi, S.H., Mossoba, M.M. et al. (2019). Rapid classification and quantification of marine oil omega-3 supplements using ATR-FTIR, FT-NIR and chemometrics. *Journal of Food Composition and Analysis* 77: 9–19.

Karunathilaka, S.R., Yakes, B.J., Choi, S.H. et al. (2020). Comparison of the performance of partial least squares and support vector regressions for predicting fatty acids and fatty acid classes in marine oil dietary supplements by using vibrational spectroscopic data. *Journal of Food Protection* 83 (5): 881–889.

Killeen, D.P., Marshall, S.N., Burgess, E.J. et al. (2017). Raman spectroscopy of fish oil capsules: polyunsaturated fatty acid quantitation plus detection of ethyl esters and oxidation. *Journal of Agricultural and Food Chemistry* 65: 3551–3558.

Kim, S.-K. and Mendis, E. (2006). Bioactive compounds from marine processing byproducts – a review. *Food Research International* 39: 383–393.

Klaypradit, W., Kerdpiboon, S., and Singh, R.K. (2011). Application of artificial neural networks to predict the oxidation of menhaden fish oil obtained from Fourier transform infrared spectroscopy method. *Food and Bioprocess Technology* 4: 475–480.

Kurata, S., Yamaguchi, K., and Nagai, M. (2005). Rapid discrimination of fatty acid composition in fats and oils by electrospray ionization mass spectrometry. *Analytical Sciences* 21: 1457–1465.

Lingzhi, X., Fei, G., Zengling, Y. et al. (2016). Discriminant analysis of terrestrial animal fat and oil adulteration in fish oil by infrared spectroscopy. *International Journal of Agricultural and Biological Engineering* 9: 179–185.

Liverseege, J.F. (1904). Cod-liver oil and other fish oils. *The Analyst* 29: 210–215.

Market Research Future (2019). Fish oil market research report – forecast till 2023. Report ID: MRFR/F-B & N/5122-HCRR, WantStats Research and Media Pvt. Ltd, Pune.

Martinez, I., James, D., and Loréal, H. (2005). Application of modern analytical techniques to ensure seafood safety and authenticity. FAO Fisheries Tech. Paper 455.

Monakhova, Y.B., Zailer, E., and Diehl, B.W.K. (2018). Quality control of krill oil by nuclear magnetic resonance (NMR) spectroscopy: composition and detection of foreign species. *Analytical Letters* 51: 2551–2562.

Oterhals, Å. and Vogt, G. (2013). Impact of extraction, refining and concentration stages on the stability of fish oil. In: *Food Enrichment with Omega-3 Fatty Acids* (eds. C. Jacobsen, N.S. Nielsen, A.F. Horn, et al.), 111–129. Cambridge: Woodhead Publishing.

Parry, E.J. (1904). The adulteration of cod-liver oil. *The Lancet* 163: 378.

Pike, I.H. and Jackson, A. (2010). Fish oil: production and use now and in the future. *Lipid Technology* 22 (3)): 59–61.

Prasad, A. (2015). *Global fish oil market size, share, analysis and trends*. London: Allied Market Research.

Reports And Data (2019). Fish oil market by source (salmon, cod, anchovy, tuna, mackerel, sardines, herring and others), by application, by distribution channel and segment. Forecast, 2017–2026. Report ID: RND_001256, Marketysers Global Consulting LLP, Pune.

Ritter, J.C.S. and Budge, S.M. (2012). Key lipid oxidation products can be used to predict sensory quality of fish oils with different levels of EPA and DHA. *Lipids* 47: 1169–1179.

Rizliya, V. and Mendis, E. (2014). Biological, physical, and chemical properties of fish oil and industrial applications. In: *Seafood Processing By-Products: Trends and Applications* (ed. S.-K. Kim), 285–313. New York: Springer-Verlag.

Rohman, A. and Che Man, Y.B. (2009). Analysis of cod-liver oil adulteration using Fourier transform infrared (FTIR) spectroscopy. *Journal of the American Oil Chemists' Society* 86: 1149–1153.

Rohman, A. and Che Man, Y.B. (2011a). Authentication analysis of cod liver oil from beef fat using fatty acid composition and FTIR spectra. *Food Additives & Contaminants: Part A Chemistry, Analysis, Control, Exposure & Risk Assessment* 28: 1469–1474.

Rohman, A. and Che Man, Y.B. (2011b). Application of Fourier transform infrared (FT-IR) spectroscopy combined with chemometrics for authentication of cod-liver oil. *Vibrational Spectroscopy* 55: 141–145.

Rohman, A. and Che Man, Y.B. (2011c). Analysis of chicken fat as adulterant in cod liver oil using Fourier transform infrared (FTIR) spectroscopy and chemometrics. *CyTA – Journal of Food* 9: 187–191.

Rohman, A., Sunarminingsih, R., and Che Man, Y.B. (2012). The employment of FTIR spectroscopy and chemometrics for classification and quantification of mutton fat in cod liver oil. *American Journal of Food Technology* 7: 151–159.

Rohman, A., Widyaningtyas, R., and Amalia, F. (2017). Authentication of cod liver oil from selected edible oils using FTIR spectrophotometry and chemometrics. *International Food Research Journal* 24: 1362–1367.

Ropodi, A.I., Panagou, E.Z., and Nychas, G.J.E. (2016). Data mining derived from food analyses using non-invasive/non-destructive analytical techniques; determination of food authenticity, quality & safety in tandem with computer science disciplines. *Trends in Food Science & Technology* 50: 11–25.

Rubio-Rodríguez, N., Beltrán, S., Jaime, I. et al. (2010). Production of omega-3 polyunsaturated fatty acid concentrates: a review. *Innovative Food Science and Emerging Technologies* 11: 1–12.

Sandgruber, S. and Buettner, A. (2012). Comparative human-sensory evaluation and quantitative comparison of odour-active oxidation markers of encapsulated fish oil products used for supplementation during pregnancy and the breastfeeding period. *Food Chemistry* 133: 458–466.

Serfert, Y., Drusch, S., and Schwarz, K. (2010). Sensory odour profiling and lipid oxidation status of fish oil and microencapsulated fish oil. *Food Chemistry* 123: 968–975.

Siddiqui, N., Sim, J., Silwood, C.J.L. et al. (2003). Multicomponent analysis of encapsulated marine oil supplements using high-resolution ^{1}H and ^{13}C NMR techniques. *Journal of Lipid Research* 44: 2406–2427.

Sullivan, J.C. and Budge, S.M. (2012). Fish oil sensory properties can be predicted using key oxidative volatiles. *European Journal of Lipid Science and Technology* 114: 496–503.

Sullivan, J.C., Budge, S.M., and St-Onge, M. (2009). Determining ethyl esters in fish oil with solid phase microextraction and GC–MS. *Journal of the American Oil Chemists' Society* 86: 743–748.

Tangendjaja, B. (2015). Quality control of feed ingredients for aquaculture. In: *Feed and Feeding Practices in Aquaculture* (ed. D.A. Davis), 141–169. Cambridge: Woodhead.

Tengku-Rozaina, T.M. and Birch, E.J. (2014). Positional distribution of fatty acids on hoki and tuna oil triglycerides by pancreatic lipase and ^{13}C NMR analysis. *European Journal of Lipid Science and Technology* 116: 272–281.

Tyl, C.E., Brecker, L., and Wagner, K.-H. (2008). ^{1}H NMR spectroscopy as tool to follow changes in the fatty acids of fish oils. *European Journal of Lipid Science and Technology* 110: 141–148.

Vidal, N.P., Goicoechea, E., Manzanos, M.J. et al. (2015). ^{1}H NMR study of the changes in brine- and dry-salted sea bass lipids under thermo-oxidative conditions: both salting methods reduce oxidative stability. *European Journal of Lipid Science and Technology* 117: 440–449.

Wang, H., Liu, F., Yang, L. et al. (2011). Oxidative stability of fish oil supplemented with carnosic acid compared with synthetic antioxidants during long-term storage. *Food Chemistry* 128: 93–99.

Williamson, K. and Hatzakis, E. (2017). NMR spectroscopy as a robust tool for the rapid evaluation of the lipid profile of fish oil supplements. *Journal of Visualized Experiments* 123: e55547.

Wu, D., Chen, X., Cao, F. et al. (2014). Comparison of infrared spectroscopy and nuclear magnetic resonance techniques in tandem with multivariable selection for rapid determination of ω-3 polyunsaturated fatty acids in fish oil. *Food and Bioprocess Technology* 7: 1555–1569.

Xu, J.-L., Riccioli, C., and Sun, D.-W. (2015). An overview on nondestructive spectroscopic techniques for lipid and lipid oxidation analysis in fish and fish products. *Comprehensive Reviews in Food Science and Food Safety* 14 (4): 466–477.

Zeng, Y.X., Araujo, P., Du, Z.Y. et al. (2010). Elucidation of triacylglycerols in cod liver oil by liquid chromatography electrospray tandem ion-trap mass spectrometry. *Talanta* 82: 1261–1270.

6
Milk Fats

Rogelio Sánchez-Vega[1], América Chávez-Martínez[1], Juan Manuel Tirado-Gallegos[1], Néstor Gutiérrez-Méndez[2] and María Janeth Rodríguez-Roque[3]
[1] *Faculty of Zootechnics and Ecology, Autonomous University of Chihuahua, Chihuahua, México*
[2] *Faculty of Chemical Sciences, Autonomous University of Chihuahua, Chihuahua, México*
[3] *Faculty of Agrotechnological Sciences, Autonomous University of Chihuahua, Chihuahua, México*

6.1 Introduction

Milk is a very complex biological fluid where fat, proteins, minerals, and water form an oil-in-water emulsion. Factors such as stage of lactation, animal feeding, genetic variations, seasonal variations, age, and health affect the chemical composition of milk fat (Walstra et al. 2006).

The concentration of lipids in milk for domestic species, range between 19 to 183 g/l (Table 6.1). In the case of bovine milk, the most consumed and common source of milk, the content of fat is around 3.5%. As is shown in Table 6.2, milk fats encompass triglycerides, diglycerides, monoglycerides, individual fatty acids, cholesterol, milk fat globule membrane (MFGM), and phospholipids (Audic et al. 2003).

Milk fats are present as globules emulsified in the aqueous phase of milk with a range of diameter between 1 and 10 μm (Walstra et al. 2006; Jensen 2002). Milk fat globules are structurally complex constituted by a core of triglycerides, which are bounded by a monolayer of polar lipids and surrounded by a dense proteinaceous coat. The entire fat globule is coated by a bilayer plasma membrane. The MFGM is composed of the lipids and proteins of the epithelial cell plasma membrane, including significant quantities of cholesterol, phosphatidylcholine, and sphingomyelin. But it also contains glycolipids, gangliosides, membrane glycoproteins, and mucins. An essential role of membrane is to stabilize the fat globules as an emulsion in the milk (Ward et al. 2006).

Almost all the fat in milk is in fat globules and the triglycerides make up over 98% of the milk lipids. Therefore, triglycerides largely determine the properties of milk fat (Walstra et al. 2006). In addition, the water affinity of triglycerides (polar lipids) is in function of size chain: long chain triglycerides are insoluble in water; while short chain triglycerides are partially water soluble (Callegarin et al. 1997). Diglycerides and monoglycerides are also present in fresh milk fat; although their content increases as a consequence of

Oil and Oilseed Processing: Opportunities and Challenges, First Edition.
Edited by Tomás Lafarga, Gloria Bobo, and Ingrid Aguiló-Aguayo.

Table 6.1 Fat content of milks from domestic species.

Species	Concentration (g/l)
Cow	33–47
Sheep	40–99
Goat	41–45
Horse	19
Rabbit	183

Source: Fox et al. (2015).

Table 6.2 Lipids present in milk.

Lipid class	Amount (%, w/w)	Examples	
Triglycerides	98.3		
Diglycerides	0.3		
Monoglycerides	0.03		
Free fatty acids	0.1	4:0, Butyric	16:1, Palmitoleic
		6:0, Caproic	18:1 *cis*, Oleic
		8:0, Caprylic	18:2, Linoleic
		10:0, Capric	18:2 conj, Conjugated linoleic acid
		12:0, Lauric	
		14:0, Myristic	
		16:0, Palmitic	
		18:00 Stearic	
Phospholipids	0.8	Phosphatidylcholine	
		Phosphatidyl ethanolamine	
		Phosphatidylserine	
		Phosphatidylinositol	
		Sphingomyelin	
		Lysophospholipids	
Sterols	0.3	Cholesterol	
Carotenoids	Trace	β-carotene	
Fat-soluble vitamins	Trace	Timanins A, D, E and K	
Flavor compounds	Trace		

lipolysis or could represent incomplete synthesized lipids (Walstra et al. 2006). More than 400 fatty acids are present in milk fat. The fatty acid pattern is an important factor in determining lipid properties, such as melting range, chemical reactivity, and nutritional value. Other lipids present in milk are phospholipids (phosphatidylcholine, phosphatidylethanolamine and sphingomyelin) which represent <1% of total lipids (Park 2009).

The functions of milk fats are to serve as a source of energy for newborns, essential fatty acids (linoleic acid), and fat-soluble vitamins (A, D, E, K). Also, fatty acids are precursors of flavor compounds (methyl ketones and lactones) in some dairy processes (Fox et al. 2015).

From an economic point of view, the value of milk is based on fat content when it is used mainly or solely for butter production. Thus, there has been commercial pressure to increase the yield of milk fat content by nutrition or genetic means (Fox et al. 2015). Milk fat can be modified to expand its applications; however, so far, very few non-food uses of milk fat have been developed.

6.2 Health Effects of Milk Fats

Many foods contain components that have beneficial effects beyond those associated with their traditional nutrient content, and these are often referred to as "functional food" (Santini et al. 2018). The functional components or ingredients are non-conventional biomolecules that occur in food that possess the capacity to modulate one or more metabolic pathways to the benefit of the human organism (Abuajah et al. 2015).

Bovine milk and dairy products have been part of the human diet, even after weaning. Milk fat provides energy, fat soluble vitamins (A, D, E, and K), some of the essential fatty acids properties (Ba et al. 2005).

In this sense, dietary guidelines around the world have recommended daily consumption of milk and dairy products for the overall health of the population. *Dietary Guidelines for Americans* has highlighted that milk and dairy products provide important amounts of three of the five essential nutrients for children and four of the seven nutrients of concern for adults (USDA/USDHHS 2010).

6.2.1 Milk Fat Globule Membrane (MFGM)

MFGM is important from the physicochemical point of view, since it stabilizes the milk fat globule against coalescence and protects the core fat against chemical oxidation and the lipase attack (Le et al. 2015). Also, it has been proposed that MFGM promotes the growth of beneficial intestinal bacteria (Kvistgaard et al. 2014). Diverse studies have mentioned that the daily consumption of milk reduces serum cholesterol, and these effects were associated with MFMG (Ward et al. 2006).

6.2.2 Fatty Acids

So far, more than 400 fatty acids present in bovine milk have been identified, which occur as fatty acids esterified to the glycerol molecule. Saturated fatty acids comprise 60–70% of milk fatty acids (Jensen 2002; Lock et al. 2014).

For many years a low consumption of saturated fatty acids has been massively promoted. Recent estimates indicate that approximately 30% of the US dietary intake of saturated fat comes from dairy products (cheese, as major source) (Lock et al. 2014). However, that does not mean that milk or dairy products are unhealthy foods. In this sense, Taubes (2001) concluded that there was no real evidence to demonstrate that a diet low in saturated fat prolongs life. In fact, some studies concluded that there was no significant evidence to associate dietary saturated fat with an increased risk of cardiovascular disease (Siri-Tarino et al. 2010).

6.2.3 Oleic Acid

Some of unsaturated fatty acids are recognized as functional ingredients, that is, they bring benefits for disease prevention and health maintenance. For milk fat these include oleic acid, very long-chain omega-3 fatty acids. Milk is a rich source of oleic acid, a *cis*-9 18:1 fatty acid, being approximately 25% of total milk fatty acids (Lock et al. 2014).

6.2.4 Conjugated Linoleic Acid (CLA, 18:2 Conjugated)

Conjugated linoleic acid (CLA) is a group of conjugated dienoic 18:2s with double bonds at 9 and 11 or 10 and 12. The *cis*-9 and *trans*-11 CLAs (rumenic acid) are the most abundant isomer (up to 90% of total), which is produced by endogenous synthesis from vaccenic acid (Fox and McSweeney 2006; Jensen 2002). More than 20 different isomers of CLA have been identified in milk fat. The *cis*-9 and *trans*-11 are mainly produced by anaerobic fermentation of bacteria in the rumen via the biohydrogenation of the polyunsaturated fatty acids (Park et al. 2000).

It has been proposed that consumption of CLA is associated with many health benefits. CLA anticarcinogen activity has been tested in biomedical models, such as human cancer cell lines, transplanted cell lines, and *in situ* organ site carcinogenesis models (Le et al. 2015; Parodi et al. 2006).

Dietary supplements of soybean, sunflower, or linseed oils have been added to increment the CLA content in milk (Dhiman et al. 2000; Chouinard et al. 2001). Dietary supplements of CLA have been shown to be effective in inhibiting chemically-induced skin papilloma, fore-stomach neoplasia, and preneoplastic lesions and tumors in the colon and mammary gland (Ba et al. 2005; Parodi et al. 2006).

6.2.5 Sphingomyelin

It has been observed that in rats, sphingomyelin contributes to myelination of the central nervous system (Oshida et al. 2003), inhibits colon carcinogenesis (Dillehay et al. 1993; Schmelz et al. 1996), has antibacterial activity to various food-borne pathogens, and can protect against food-borne gastroenteritis (Sprong et al. 2002). In this context, a mixture of sphingolipids and MFGM also reduced the colon carcinogenesis in rats (Snow et al. 2010). Sphingolipids reduce the plasma cholesterol and triglycerides (Watanabe et al. 2011). Milk is an important source of exogenous sphingomyelin in the human diet (Burling and Graverholt 2008).

6.2.6 Phosphatidylcholine

Phosphatidylcholine is necessary for supporting the recovery of the liver after toxic or chronic viral damage (Niederau et al. 1998). Phosphatidylcholine is an important source of choline, a nutrient related to brain development (Blusztajn 1998). Also, the intake of phosphatidylcholine protects the gastrointestinal mucosa against the damage caused by anti-inflammatory drugs and help to maintain the defensive hydrophobic barrier (Dial et al. 2008; Ehehalt et al. 2010). In fact, phosphatidylcholine showed benefits against pharmaceutical and death cap mushroom poisoning, alcoholic liver damage, and the hepatitis B virus (Kidd 2002).

6.2.7 Phosphatidylserine

Phosphatidylserine is highly concentrated in brain, lungs, heart, liver, and skeletal muscles, where its functions are the regulation of immune and anti-inflammatory responses (Gaitonde et al. 2011), modulation of the activity of receptors, ion channels, enzymes, and signaling molecules, and involvement in governing membrane fluidity (Pepeu et al. 1996; Starks et al. 2008). Clinical studies reported that consumption of phosphatidylserine may reduce the risk of dementia and cognitive dysfunction (of the Alzheimer's type) in the elderly (Cenacchi et al. 1993; Pepeu et al. 1996). The mixture phosphatidylserine and omega-3 has beneficial effects in the attention-deficit hyperactivity disorder (Veereman-Wauters et al. 2012).

6.3 Pre-Treatment and Processing Technologies

The changes in milk occur during and immediately after milking (Le et al. 2015). In addition, physical changes can take place due to the inclusion of air during milking and agitation. Air bubbles cause damage to fat globule membranes (van Boekel and Walstra 1981). Chemical changes can occur through oxidation caused by the presence of oxygen and light. Milk fat is also susceptible to enzyme activity, such as lipases, proteases, and phosphatases (Walstra et al. 2006).

In this section, three important processing treatments are described: cooling, heat treatment, and homogenization.

6.3.1 Cooling

Cooling, reducing the milk temperature below 4 °C, is used to avoid microbial and biological changes. The cooling process commonly involves pumping and agitation, which may cause changes in the composition and stability of MFGM. Thus, the fat globules surface is damaged, causing the fat globules to coalesce with each other and causing losses of MFGM material to the serum (Le et al. 2015).

6.3.2 Heat Treatment

Thermal treatments are applied with the aim of inactivating microorganisms and enzymes, in order to increase the stability of food products (Ortega-Rivas and Salmerón-Ochoa 2014). However, physicochemical changes take place as consequence of thermal milk processing. Milk heated at 60–65 °C causes the adsorption of β-lactoglobulin and a lesser extent of α-lactalbumin to the fat globule surface (Corredig and Dalgleish 1996). This adsorption could occur when the hydrophobic regions of the lipids are accessible to interact with β-lactoglobulin. At higher temperatures of pasteurization (80 °C for 20 minutes), a notable loss of triglycerides, but no significant changes in polar lipids has been detected (Houlihan et al. 1992). Similarly, a reduction of 25% of total lipid content from the MFGM was detected when milk was thermally treated at 80 °C for 18 minutes (Lee and Sherbon 2002).

6.3.3 Homogenization

Homogenization is applied to counteract creaming and to prevent partial coalescence of fat globules, and to render desirable rheological properties for processed products through disruption of fat globules into smaller ones. In consequence, the surface of the fat globule increases 5–10 times (Walstra et al. 2006). Keenan et al. (1983) reported that about 10% of the surface of fat globules in homogenized milk was covered by natural MFGM.

6.4 Techniques for Obtaining Functionality of Milk Fats

As is shown in Table 6.3, milk fat components such as triglycerides, diglycerides, and individual fatty acids can be used to provide functional properties for traditional and nontraditional applications. For example, production of high- and low-melting fat fractions for pastries and cold-spreadable butter, respectively. In the case of monoglycerides, they can be used as emulsifiers in baked goods. Nontraditional applications involve use in cosmetics (Kaylegian 1995).

Table 6.3 Functional properties of milk fats.

Lipid	Functional characteristics	General definitions	Examples
Triglycerides	Structure formation	Form the structure of a fat ingredient or finished product	Formation of fat-stabilized foams in cakes and whipping cream for adequate structure
	Viscosity	Rheological properties (stress, deformation and yield)	Softening of chocolate
	Solution	Capability of milk fat to dissolve substances	Solutions can be formed between milk and cocoa butter. Milk can dissolve fat-soluble flavors and vitamins
	Dispersion	Milk fat is able to disperse substances	Milk can disperse
	Spreadability	The way a fat deforms and flows under stress	Spreadability of butter at refrigeration temperature
	Plasticity	Ability of milk fat to be pliable at selected temperature range	Milk fats used in pastries must be plastic to form a thing film without breaking
	Aeration	Air incorporation into crystal network and form a stable foam	Indispensable for the proper structure formation of cakes and whipping cream
	Coating	Coat food particles	Necessary for good shortening and lubricity properties
	Lubricity	Formation of a layer that allows particle to slide past each other	Important in proper mixing of ingredients, and the sensory properties of foods
	Heat transfer	Ability to conduct heat	Heating medium for cooking
	Compatibility with other fats	Formation of compatible systems with other fats	Compatibility of milk fat with cocoa butter in chocolate

Monoglycerides and diglycerides	Emulsification	Surface agents to reduce the tension at the emulsion interface	Monoglycerides are the most used in the food industry. The chain length and degree of saturation of the fatty acid can influence their suitability
	Dough conditioning and crumb softening properties	Monoglycerides interact with amylose and amylopectin to form insoluble complexes	Softer crumb and delayed staling effects in baked goods
	Crystalization properties	Monoglycerides tend to form β crystal structures and diglycerides form β′ crystals	Diglycerides can be used to stabilize β′ structures and provide the characteristic texture of margarines and low fat spreads
Fatty acids	Flavor and flavor precursors	Milk fat flavor is derived from free flavor compounds and flavor precursors that are released with heat, water, acid, and lipase	Low molecular weight, branched-chain, hydroxy and β-keto fatty acids
	Metabolic properties	Fatty acids with less than 10 carbons are metabolized quickly in liver to be metabolized as carbohydrates	C8 and C10 fatty acids are effective high energy nutritional supplements
	Fatty acid reservoir	Act as reservoir in the manufacture of food and nonfood ingredients	Production of flavors and structured lipids
	Antimicrobial	Antimicrobial properties of lauric acid	Lauric acid is present in small quantities in milk fat
	Antioxidant and anticarcinigenic	Associated with conjugated linoleic acid	Milk fat is a rich source of conjugated linoleic acid

Source: Adapted from Kaylegian (1995).

The changes of milk fats to modify their functionality can be classified in enzymatic, chemical or physical interactions.

Enzymatic: The use of enzymes offers gentler processing conditions and preserves the delicate desirable butter flavor for which milk fat is valued. The enzyme used is lipase (glycerol ester hydrolase, EC 3.1.1.3) that hydrolyses triglycerides, diglycerides, and monoglycerides and under certain circumstances may also catalyze the re-incorporation of free fatty acids onto the glycerol backbone (Augustin and Versteeg 2006).

Lipase from *Aspergillus niger* is used to hydrolyze butyric acid in butter-in-oil emulsions to augment the production of free fatty acids responsible for dairy flavor (García et al. 1991). Specific products in milk fat hydrolysis could be generated through an appropriate choice of enzyme. Lipase from *Candida rugosa* had an increased activity toward butyric acid compared to lipase from *Pseudomanas fluorescens* (Marangoni 1994).

Alternatively, the enzyme may be used in an encapsulated form as was suggested by Chen and Chang (1993), who reported the release of short-chain fatty acids by using a lipase from *Candida cyclindracea* encapsulated in reverse micelles.

Other enzymes may be used to transform cholesterol into other products (Boudreau and Arul 1993). The enzymes studied were cholesterol reductase, that converts cholesterol to coprostanol, a product that is poorly adsorbed by the body, and cholesterol oxidase, which oxidizes cholesterol to non-steroid compounds.

Chemical: Chemical composition refers to the type and concentration of fatty acids, triglycerides, as well as the positional distribution of the fatty acids on the glycerol backbone. A chemical factor that affects the functionality of milk fat is the flavor, which is a function of free flavor compounds and flavor precursors bound to the triglyceride. The compounds that contribute to milk fat flavor are volatile low molecular weight and branched-chain fatty acids, lactones, methyl ketones, and aldehydes. Therefore, flavor compounds are released in presence of heat, water, acid, or lipase (Kaylegian 1995).

Physical: This type of functionality is consequence of complex interactions between the solid and liquid phases combined with the effects of temperature. Also, physical functionality can be further affected by processing.

6.4.1 Melting

Milk fat has a melting temperature between −40 and 40 °C and melting is influenced by the chemical composition of milk fat. The melting point of triglycerides depends on the fatty acid profile and the position of the fatty acids in the triglyceride. The melting point of fatty acids augments with increasing length of the acyl chain and the number, position, and isomeric form of double bonds. That is, the melting point diminishes as the number of double bonds in the molecule raises, and *cis* isomers have lower melting points than those *trans* isomers. Symmetrical triglycerides have higher melting points than an asymmetrical triglyceride containing the same fatty acids (Fox et al. 2015; Kaylegian 1995).

6.4.2 Fractionation

Milk fat contains an elevated number of fatty acids (>400) and a very heterogeneous triacylglycerol content. Milk fat fractionation is used to obtain groups of triglycerides with distinct chemical and physical properties and altered rheological characteristics. Fractionation techniques include dry fractionation, solvent fractionation, supercritical fluid extraction, and short-path distillation (Walstra et al. 2006).

Through fractional crystallization the melting and spreading characteristics of butter can be altered. The crystals formed may be removed by physical methods (centrifugation or filtration), which shifts the melting point temperature curve to lower temperatures without significantly changing its shape. The resulting butter has acceptable spreadability at low temperatures, but its "stand-up" properties are unsatisfactory, for example, it becomes totally liquid at too low a temperature (Fox et al. 2015).

6.4.3 Crystallization

The solid phase of milk fat is present in the form of crystals. The functionality of milk fat crystals depends on the speed of melting and the size and amount of the crystal. The size of crystals, as well as the presence of mixed crystals, liquid milk fat in the crystal matrix, and crystal agglomerates are related to the conditions employed during crystallization, processing, chemical composition, cooling, and agitation rate (Kaylegian 1995). The crystalline form, α, β, β^1 has an influence on the melting point of milk fat (Park 2009).

6.4.4 Blending

Blending is applied to increase the level of unsaturation in milk fat. The blends of vegetable oils and milk fat reduce the butter hardness by mixing an emulsion of the oil with dairy cream for the manufacture of butter or by blending the oil directly with butter, increasing the content of polyunsaturated fatty acids, which has a nutritional advantage. Oils rich in omega-3 fatty acids, considered to have desirable nutritional properties, may be included in the blend although these oils may be susceptible to oxidative rancidity (Wright and Marangoni 2006).

6.4.5 Softening or Hardening of Milk Fat

To obtain soft and spreadable butters, milk fat fractions are blended with milk fat in cream or other fractions (Deffense 1993) or with liquid vegetable oils (canola oil) (Rousseau et al. 1996). Milk fat may be made harder by blending with high melting triglycerides or fats (Fox and McSweeney 2006) or by the addition of 10–20% of tripalmitin to increase the mechanical properties of coatings (Fairley et al. 1994).

6.4.6 Interesterification

Interesterification consists in the exchange of fatty acids within and between triglycerides to produce fats and oils with desired functionalities and health benefits. This redistribution of the fatty acids results in modification of the physical properties and nutritional properties of the fat (Augustin and Versteeg 2006). For example, it may be useful in incorporating healthful fatty acids, such as CLA, into milk and other fats (Garcia et al. 1998). Interesterification has been produced by chemical (Rodrigues and Gioielli 2003) and enzymatic methods (Liew et al. 2001). However, when milk fat is interesterified, its butter flavor is reduced (Kleyn 1992).

6.4.7 Hydrolysis

Hydrolysis of triglycerides may be developed at high temperatures and pressure with steam. Fatty acids can be obtained by saponification or chemically catalyzed hydrolysis. Enzymatic hydrolysis is more appealing since the reaction proceeds under mild conditions of pressure and temperature with specificity and reduced waste. This technology is currently employed in the production of fatty acids, diglycerides, monoglycerides, flavoring agents for dairy products and detergents for laundry and household uses (Villeneuve et al. 2000).

6.4.8 Hydrogenation

Hydrogenation is based on the addition of hydrogen atoms to the double bonds in the fatty acid chain. This process requires high temperature and a catalyst. The aim of hydrogenation is reducing the degree of unsaturation of fat and increasing its hardness, principally applied in the vegetable oil industry. Also, fats with higher melting points are obtained from hydrogenation of milk fat (Fox and McSweeney 2006). Hydrogenation increases the level of trans fatty acids, which are not desirable from a nutritional viewpoint (Timms 2005).

6.4.9 Cholesterol Reduction

The content of cholesterol in milk is around 0.25–0.46%. The interest in the elimination of cholesterol is based on the fact that consumers are more conscious about the possible relation between cholesterol and heart disease. Although there is no real evidence about the relationship between dietary cholesterol and cardiovascular disease, a market has been created for low-cholesterol products (Schlimme 1990). Physical, chemical, and biological processes have been used to reduce the level of cholesterol in milk fat (Boudreau and Arul 1993).

6.5 Current and Potential Applications in the Food Industry and Other Areas

Applications or uses of milk fats include both intact milk fat and milk fat fractions.

MFGM materials and polar lipids have an elevate potential to produce emulsions, liposomes, and emulsion-based foods, such as reconstituted cream, mayonnaise, and salad sauces. Indeed, reconstituted cream showed better whipping properties when mixed with MFGM material obtained from industrial whey buttermilk (Phan et al. 2014).

Liposomes are prepared from polar lipids of soybean or eggs with the objective to carry and control the release of various bioactive agents, nutrient delivery in food and feed, delivery of drugs and nutraceuticals, encapsulation of flavor, aroma, and natural coloring compounds in foods among others (Henna Lu et al. 2011).

Liposomes get great transcendence due to the health-beneficial properties of MFGM and the commercial availability of milk polar lipids. Liposomes made from MFGM are more stable than those prepared from lecithin (Thompson and Singh 2006). Through the use of MFGM and polar lipids, new dairy products with different functional and nutritional properties can be produced. For example, Le et al. (2011) replaced part of milk solids with MFGM material isolated from buttermilk using microfiltration to fortify a dairy product with milk polar lipids. They reported that the yogurt enriched with polar lipids had higher water-holding capacity and adhesiveness, because polar lipids have a polar nature and, as a consequence, they have high water-holding capacity. Gülseren and Corredig (2013) prepared nanoliposomes with high-pressure homogenization from milk polar lipids to encapsulate tea polyphenolic compounds. The authors reported a higher encapsulation efficiency, liposomes with smaller size and lower extent of release of tea polyphenols during storage in comparison with soy liposomes.

On the other hand, phospholipids are added to impair the stability of milk foams, particularly at a temperature at which the milk fat is partially crystalline (20 °C). An example is the reconstitution of milk powders or infant formula (Huppertz 2010).

6.5.1 Milk Fats in Foods

The design of new fat-fortified foods or modified milk fats needs to gain consumer acceptance. Diverse studies have reported that there were no differences from unenriched dairy products and naturally enriched dairy products (Ramaswamy et al. 2001).

6.5.2 Structured Lipids

Novel applications of milk fat as a fatty acid reservoir have been investigated in structured lipids and infant formulas. The structured lipids are lipid food ingredients designed for people with special nutritional needs and as frying oils and confectionery and bakery fats (Kennedy 1991; Stein 1999). The physical or chemical properties of structured lipids are in function of their use. For example, structured lipids and medium-chain triglycerides are produced by random transesterification of a blend of fatty acid reservoirs. Elsewhere, structured lipids are used as reservoir of C8 and C10 fatty acids, which increase the potential value of milk fat (Kaylegian 1995).

6.5.3 Edible Films

Milk fat has been suggested for the manufacture of edible and packaging films and protective coatings, since lipids are effective barriers to the transport of moisture (Fairley et al. 1994). Edible films made from milk fat or milk fat fractions exhibit low gas and water vapor permeability and improved hydrophobic properties (Audic et al. 2003).

6.6 Non-food Uses of Milk Fats

In this section non-food applications of milk fats are described. Polar lipids from dairy sources have been used for pharmaceutics, skin care creams, shampoo, and cosmetic applications (Miura et al. 2006).

Another example is the conversion of milk fat into more profitable products, such as fatty acids, diglycerides, and monoglycerides obtained by lipase, as well as flavoring agents and detergents. Milk fats are also used in the manufacture of emulsifiers. In addition, the emulsifying properties of whey can be improved by the addition of monoglycerides and diglycerides derived from milk fat hydrolysis (Kaylegian 1995).

Other alternative is for the manufacture of soapstock, a byproduct of the refining of vegetable oils and another low value feedstock for biodiesel production. Residual butter and related fat, may be used for soapstock production (Pinzi and Dorado 2012).

Milk fat sucrose polyesters have been used as fat substitutes in cheese and other dairy products. Sucrose polyesters are produced by transesterification reactions between sucrose and fatty acid methyl esters and are used as fat substitutes that provide the flavor and physical functionalities of fat, but that are not absorbed by the body and provide no caloric value. The flavor and physical properties of sucrose polyesters are influenced by the fat blend used, as well as the fatty acid source (Drake et al. 1992).

6.7 Future Trends

The chemical and physical complexity of milk fat can be viewed as an opportunity. Milk fat can be manipulated to meet specific functional needs for food and nonfood applications. Nowadays, milk fats receive much attention due to their health promoting properties and advantageous technological functionalities. Technological functionalities of milk fats are associated with their emulsifying properties and water-binding capacity, making them potential ingredients for the development of emulsion-based foods or emulsions for protection of sensitive bioactive nutrients. Health promoting properties of milk fats include the contribution to development of the brain and intestinal immune systems of infants and lowering cholesterol absorption, among others. Thus, milk fats can be used for food fortification to create functional foods for infants, adults, athletes, and the elderly.

References

Abuajah, C.I., Ogbonna, A.C., and Osuji, C.M. (2015). Functional components and medicinal properties of food: a review. *Journal of Food Science and Technology* 52 (5): 2522–2529.

Audic, J.-L., Chaufer, B., and Daufin, G. (2003). Non-food applications of milk components and dairy co-products: a review. *Lait* 83: 417–438.

Augustin, M.A. and Versteeg, C. (2006). Milk fat: physical, chemical and enzymatic modification. In: *Advanced Dairy Chemistry Volume 2 Lipids*, 3e (eds. P.F. Fox and P.L.H. McSweeney), 293–332. New York: Springer.

Bauman, D.E., Lock, A.L., Corl, B.A. et al. (2005). Milk fatty acids and human health: potential role of conjugated linoleic acid and trans fatty acids. In: *Ruminant Physiology: Digestion, Metabolism and Impact of Nutrition on Gene Expression, Immunology and Stress* (eds. K. Sejrsen, T. Hvelplund and M.O. Nielsen), 523–555. Wageningen, Netherlands: Wageningen Academic Publishers.

Blusztajn, J.K. (1998). Choline, a vital amine. *Science* 281 (5378): 794–795.

van Boekel, M.A.J.S. and Walstra, P. (1981). Stability of oil-in-water emulsions with crystals in the disperse phase. *Colloids and Surfaces* 3 (2): 109–118.

Boudreau, A. and Arul, J. (1993). Cholesterol reduction and fat fractionation Technologies for Milk fat: an overview. *Journal of Dairy Science* 76 (6): 1772–1781.

Burling, H. and Graverholt, G. (2008). Milk – a new source for bioactive phospholipids for use in food formulations. *Lipid Technology* 20 (10): 229–231.

Callegarin, F., Gallo, J.A.Q., Debeaufort, F. et al. (1997). Lipids and biopackaging. *JAOCS, Journal of the American Oil Chemists' Society* 74 (10): 1183–1192.

Cenacchi, T., Bertoldin, T., Farina, C. et al. (1993). Cognitive decline in the elderly: a double-blind, placebo-controlled multicenter study on efficacy of phosphatidylserine administration. *Aging Clinical and Experimental Research* 5 (2): 123–133.

Chen, J.P. and Chang, K.C. (1993). Lipase-catalyzed hydrolysis of milk fat in lecithin reverse micelles. *Journal of Fermentation and Bioengineering* 76 (2): 98–104.

Chouinard, P.Y., Corneau, L., Butler, W.R. et al. (2001). Effect of dietary lipid source on conjugated linoleic acid concentrations in milk fat. *Journal of Dairy Science* 84 (3): 680–690.

Corredig, M. and Dalgleish, D.G. (1996). Effect of different heat treatments on the strong binding interactions between whey proteins and milk fat globules in whole milk. *Journal of Dairy Research* 63 (3): 441–449.

Deffense, E. (1993). Milk fat fractionation today: a review. *Journal of the American Oil Scientists' Society* 70 (12): 1193–1201.

Dhiman, T.R., Satter, L.D., Pariza, M.W. et al. (2000). Conjugated linoleic acid (CLA) content of milk from cows offered diets rich in linoleic and linolenic acid. *Journal of Dairy Scienece* 83 (5): 1016–1027.

Dial, E.J., Rooijakkers, S.H.M., Darling, R.L. et al. (2008). Role of phosphatidylcholine saturation in preventing bile salt toxicity to gastrointestinal epithelia and membranes. *Journal of Gastroenterology and Hepatology* 23 (3): 430–436.

Dillehay, D.L., Webb, S.K., Andalfred, E.S. et al. (1994). Biochemical and molecular roles of nutrients dietary Sphingomyelin inhibits colon cancer in CF1 mice. *Journal of Nutrition* 124: 615–620.

Drake, M.A., Younce, F., Cleary, D.O. et al. (1992). Melting onset and hardness of milk fat blend sucrose polyesters. In: *Institute of Food Technologists Annual Meeting* (ed. IFT), 179. Chicago, IL: IFT.

Ehehalt, R., Braun, A., Karner, M. et al. (2010). Phosphatidylcholine as a constituent in the colonic mucosal barrier-physiological and clinical relevance. *Molecular and Cell Biology of Lipids* 1801 (9): 983–993.

Fairley, P., German, J.B., and Krochta, J.M. (1994). Phase behavior and mechanical properties of tripalmitin/butterfat mixtures. *Journa; of Food Science* 59 (2): 321–325.

Fox, P. and McSweeney, P. (eds.) (2006). *Advanced Dairy Chemistry – 2: Lipids*, 3e. New York: Springer.

Fox, P.F., Uniacke-Lowe, T., McSweeney, P.L.H. et al. (eds.) (2015). *Dairy Chemistry and Biochemistry*. Cham, Switzerland: Springer.

Fox, P.F., Uniacke-Lowe, T., McSweeney, P.L.H. et al. (2015). Milk lipids. In: *Dairy Chemistry and Biochemistry*, 2e (eds. P.F. Fox, T. Uniacke-Lowe, P.L.H. McSweeney, et al.), 69–144. Cham, Switzerland: Springer International Publishing.

Gaitonde, P., Peng, A., Straubinger, R.M. et al. (2011). Phosphatidylserine reduces immune response against human recombinant factor VIII in hemophilia a mice by regulation of dendritic cell function. *Clinical Immunology* 138 (2): 135–145.

García, H.S., Amundson, C.H., and Hill, C.G.J. (1991). Partial characterization of the action of an a. niger lipase on butteroil emulsions. *Journal of Food Science* 56 (5): 1233–1237.

Garcia, H.S., Storkson, J.M., Pariza, M.W. et al. (1998). Enrichment of butteroil with conjugated linoleic acid via enzymatic interesterification (acidolysis) reactions. *Biotechnology Letters* 20 (4): 393–395.

Gülseren, I. and Corredig, M. (2013). Storage stability and physical characteristics of tea-polyphenol-bearing nanoliposomes prepared with milk fat globule membrane phospholipids. *Journal of Agriculture and Food Chemistry* 61 (13): 3242–3251.

Henna Lu, F.S., Nielsen, N.S., Timm-Heinrich, M. et al. (2011). Oxidative stability of marine phospholipids in the liposomal form and their applications. *Lipids* 46 (1): 3–23.

Houlihan, A.V., Goddard, P.A., Kitchen, B.J. et al. (1992). Changes in structure of the bovine milk fat globule membrane on heating whole milk. *Journal of Dairy Research* 59 (3): 321–329.

Huppertz, T. (2010). Foaming properties of milk: a review of the influence of composition and processing. *International Journal of Dairy Technology* 63 (4): 477–488.

Jensen, R.G. (2002). The composition of bovine milk lipids: January 1995 to December 2000. *Journal of Dairy Science* 85 (2): 295–350.

Kaylegian, K.E. (1995). Functional characteristics and nontraditional applications of milk lipid components in food and nonfood systems. *Journal of Dairy Science* 78 (11): 2524–2540.

Keenan, T.W., Dylewski, D.P., Woodford, T.A. et al. (1983). Origin of milk fat globules and the nature of the milk fat globule membrane. In: *Developments in Dairy Chemistry*, vol. 2 (ed. P.F. Fox), 83–118. New York: Elsevier.

Kennedy, J.P. (1991). Structured lipids: fats of the future. *Food Technology* 45 (11): 76–83.

Kidd, P.M. (2002). Phosphatidylcholine. *Alternative Medicine. Review Monographs* 7: 310–315.
Kleyn, D.K. (1992). Textural aspects of butter. *Food Technology* 46 (1): 118–121.
Kvistgaard, A.S., Schroder, J.B., Jensen, E. et al. (2014). Milk ingredients as functional foods. In: *Milk and Dairy Products as Functional Foods*, 1e (ed. A. Kanekanian), 198–236. Chichester: Wiley Blackwell.
Le, T.T., van Camp, J., Pascual, P.A.L. et al. (2011). Physical properties and microstructure of yoghurt enriched with milk fat globule membrane material. *International Dairy Journal* 21 (10): 798–805.
Le, T.T., Phan, T.T.Q., Camp, J. et al. (2015). Milk and dairy polar lipids: occurrence, purification, and nutritional and technological properties. In: *Polar Lipids: Biology, Chemistry, and Technology*, 1e (eds. M.U. Ahmad and X. Xu), 91–143. Urbana, IL: AOCS Press.
Lee, S.J. and Sherbon, J.W. (2002). Chemical changes in bovine milk fat globule membrane caused by heat treatment and homogenization of whole milk. *Journal of Dairy Research* 69 (4): 555–567.
Liew, M.Y.B., Ghazali, H.M., Long, K. et al. (2001). Physical properties of palm kernel olein-anhydrous milk fat mixtures transesterified using mycelium-bound lipase from Rhizomucor miehei. *Food Chemistry* 72 (4): 447–454.
Lock, A.L., Givens, D.I., and Bauman, D.E. (2014). Dairy fat: perceptions and realities. In: *Milk and Dairy Products as Functional Foods*, 1e (ed. A. Kanekanian), 174–197. Chichester: Wiley Blackwell.
Marangoni, A.G. (1994). Candida and Pseudomonas lipase-catalyzed hydrolysis of butter-oil in the absence of organic solvents. *Journal of Food Science* 59 (5): 1096–1099.
Miura, S., Mutoh, T., Shiinoki, Y. et al. (2006). Emulsifying properties of phospholipids in the reconstitution of cream using butter oil. *European Journal of Lipid Science and Technology* 108 (11): 898–903.
Niederau, C., Strohmeyer, G., and Heintges, T. (1998). Polyunsaturated phosphatidyl-choline and interferon alpha for treatment of chronic hepatitis B and C: a multi-center, randomized, double-blind, placebo-controlled trial. *Hepatogastroenterology* 45 (21): 797–804.
Ortega-Rivas, E. and Salmerón-Ochoa, I. (2014). Nonthermal food preservation alternatives and their effects on taste and flavor compounds of beverages. *Critical Reviews in Food Science and Nutrition* 54 (2): 190–207.
Oshida, K., Shimizu, T., Takase, M. et al. (2003). Effects of dietary sphingomyelin on central nervous system myelination in developing rats. *Pediatric Research* 53 (4): 589–593.
Park, Y.W. (ed.) (2009). *Bioactive Components in Milk and Dairy Products*, 1e, 439. Ames, IA: Wiley Blackwell.
Park, Y., Storkson, J.M., Ntambi, J.M. et al. (2000). Inhibition of hepatic stearoyl-CoA desaturase activity by trans −10, cis −12 conjugated linoleic acid and its derivatives. *Biochim Biophys Acta* 1486: 285–292.
Parodi, P.W. (2006). Nutritional significance of milk lipids. In: *Advanced Dairy Chemistry Volume 2 Lipids*, 3e (eds. P.F. Fox and P.L.H. McSweeney), 601–639. New York: Springer.
Pepeu, G., Pepeu, I.M., and Amaducci, L. (1996). A review of phosphatidylserine pharmacological and clinical effects. Is phosphatidylserine a drug for the ageing brain? *Pharmacological Research* 33 (2): 73–80.
Phan, T.T.Q., Moens, K., Le, T.T. et al. (2014). Potential of milk fat globule membrane enriched materials to improve the whipping properties of recombined cream. *International Dairy Journal* 39 (1): 16–23.
Pinzi, S. and Dorado, M.P. (2012). Feedstocks for advanced biodiesel production. In: *Advances in Biodiesel Production Processes and Technologies*, 1e (eds. R. Luque and J.A. Melero), 69–90. Cambridge, UK: Woodhead Publishing.
Ramaswamy, N., Baer, R.J., Schingoethe, D.J. et al. (2001). Composition and flavor of milk and butter from cows fed fish oil, extruded soybeans, or their combination. *Journal of Dairy Science* 84 (10): 2144–2151.

Rodrigues, J.N. and Gioielli, L.A. (2003). Chemical interesterification of milkfat and milk-fat-corn oil blends. *Food Research International* 36 (2): 149–159.

Rousseau, D., Forestiere, K., Hill, A.R. et al. (1996). Restructuring butterfat through blending and chemical interesterification. I. Melting behavior and triacylglycerol modifications. *Journal of the American Oil Chemists' Society* 73 (8): 963–972.

Santini, A., Cammarata, S.M., Capone, G. et al. (2018). Nutraceuticals: opening the debate for a regulatory framework. *British Journal of Clinical Pharmacology* 84 (4): 659–672.

Schlimme, E. (1990). Removal of cholesterol from milk fat. *European Dairy Magazine* 4: 12–21.

Schmelz, E.M., Dilehay, D.L., Webb, S.K. et al. (1996). Sphingomyelin consumption suppresses aberrant colonic crypt foci and increases the proportion of adenomas versus adenocarcinomas in CF1 mice treated with 1,2-dimethylhydrazine: implications for dietary sphingolipids and colon carcinogenesis. *Cancer Research* 56 (404): 4936–4941.

Siri-Tarino, P.W., Sun, Q., Hu, F.B. et al. (2010). Meta-analysis of prospective cohort studies evaluating the association of saturated fat with cardiovascular disease. *American Journal of Clinical Nutrition* 91 (3): 535–546.

Snow, D.R., Jimenez-Flores, R., Ward, R.E. et al. (2010). Dietary milk fat globule membrane reduces the incidence of aberrant crypt foci in Fischer-344 rats. *Journal of Agricultural and Food Chemistry* 58 (4): 2157–2163.

Sprong, R.C., Hulstein, M.F.E., and Van Der Meer, R. (2002). Bovine milk fat components inhibit food-borne pathogens. *International Dairy Journal* 12 (2–3): 209–215.

Starks, M.A., Starks, S.L., Kingsley, M. et al. (2008). The effects of phosphatidylserine on endocrine response to moderate intensity exercise. *Journal of the International Society of Sports Nutrition* 5: 1–6.

Stein, J. (1999). Chemically defined structured lipids: current status and future directions in gastrointestinal diseases. *International Journal of Colorectal Disease* 14 (2): 79–85.

Taubes, G. (2001). The soft science of dietary fat. *Science* 291 (5513): 2536–2545.

Thompson, A.K. and Singh, H. (2006). Preparation of liposomes from milk fat globule membrane phospholipids using a microfluidizer. *Journal of Dairy Science* 89 (2): 410–419.

Timms, R.E. (2005). Fractional crystallisation – the fat modification process for the 21st century. *European Journal of Lipid Science and Technology* 107 (1): 48–57.

USDA/USDHHS (2010). Dietary Guidelines for Americans: Report of the Dietary Guidelines Advisory Committee. https://health.gov/sites/default/files/2020-01/DietaryGuidelines2010.pdf (accessed January 18, 2020).

Veereman-Wauters, G., Staelens, S., Rombaut, R. et al. (2012). Milk fat globule membrane (INPULSE) enriched formula milk decreases febrile episodes and may improve behavioral regulation in young children. *Nutrition* 28 (7–8): 749–752.

Villeneuve, P., Muderhwa, J.M., Graille, J. et al. (2000). Customizing lipases for biocatalysis: a survey of chemical, physical and molecular biological approaches. *Journal of Molecular Catalysis B: Enzymatic* 9 (4–6): 113–148.

Walstra, P., Wouters, J.T.M., and Geurts, T.J. (eds.) (2006). *Dairy Science and Technology*, 2e. Boca Raton, FL: CRC Press.

Ward, R.E., German, J.B., and Corredig, M. (2006). Composition, applications, fractionation, technological and nutritional significance of milk fat globule membrane material. In: *Advanced Dairy Chemistry: Vol. 2 Lipids*, 3e (eds. P.F. Fox and P.L.H. McSweeney), 213–244. New York: Springer.

Watanabe, S., Takahashi, T., Tanaka, L. et al. (2011). The effect of milk polar lipids separated from butter serum on the lipid levels in the liver and the plasma of obese-model mouse (KK-Ay). *Jounal of Functiomal Foods* 3 (4): 313–320.

Wright, A.J. and Marangoni, A.G. (2006). Crystallization and rheological properties of milk fat. In: *Advanced Dairy Chemistry: Vol. 2 Lipids*, 3e (eds. P.F. Fox and P.L.H. McSweeney), 245–291. New York: Springer.

7
Oils and Their Use Beyond the Food Industry

Douglas G. Hayes
Department of Biosystems Engineering and Soil Science, University of Tennessee, Knoxville, TN, USA

7.1 Introduction

Oilseed-bearing crops are well-known sources of fats and oils for edible applications, as addressed by the majority of the chapters in this book. However, the same materials are also employed in several non-food applications. For instance, the use of seed oils for preparing soaps via alkali-catalyzed saponification has been practiced for millennia, going back to the Mesopotamian civilization (Spitz 2009). Up to the early 1900s, seed oils were widely employed in numerous non-food applications, such as lubricants, fuels, and coatings. However, fossil fuels, particularly petroleum, replaced seed oils as the principal source of such products approximately one century ago. In recent years, there has been increased interest in expanding the employment of seed oils for non-food applications due to the association of fossil fuel usage with increased levels of atmospheric CO_2 and other greenhouse gases, which in turn is associated with climate change. In addition, the long-term sustainability of fossil fuel supplies for future generations is uncertain (Hayes and Smith 2019).

This chapter will focus on non-food and industrial applications of seed oils. Initially, oilseed sources for industrial utilization will be described, including common sources of triacylglycerols (TAG) composed of C_{16}–C_{18} saturated and unsaturated fatty acyl groups such as the oils of low-erucic acid rapeseed (LEAR; i.e. canola), palm, maize, and soybean (SBO), as well as industrial crop oils such as those derived from castor, jatropha, and camelina. Subsequently, an overview of different application areas for oilseeds will be reviewed, including biopolymers (employed as materials and in coatings), biofuels (particularly biodiesel), surfactants, lubricants, and plasticizers. Because this chapter covers a broad topic, with each of the subtopics listed above suitable for individual chapters (and indeed several review papers and chapters on these subtopics has been published), an overview will be given for each, along with new developments and approaches. This chapter will *not* address

Oil and Oilseed Processing: Opportunities and Challenges, First Edition.
Edited by Tomás Lafarga, Gloria Bobo, and Ingrid Aguiló-Aguayo.

application for oil crop co-products such as phospholipids, proteins, polysaccharides, extractives, and other minor components nor for animal or microbial sources of oil.

7.2 Seed Oils for Non-food and Industrial Applications

7.2.1 Common Oil Crops

The majority of seed oil production is attributable to vegetable oils, employed for foods, such as SBO and the oils from palm, LEAR, sunflower, and others, with global production in 2019 anticipated to be 203 Mt (USDA 2019). The common vegetable oils are enriched in C_{16} and C_{18} fatty acids, particularly palmitic (16:0), palmitoleic (16:1-9c), stearic (18:0), oleic (18:1-9c), linoleic (18:2-9c,12c) and α-linolenic (18:3-9c,12c,15c) acids (Table 7.1). Generally, competition between food and non-food applications for the same feedstock is not desirable for long-term economic viability. However, currently the industrial use of oils is a much smaller than for foods; therefore, under most situations, the use of common vegetable oils for oleochemicals would minimally impact the availability of fats and oils for foods. Value-added feedstocks that can replace virgin vegetable oils can be employed, particularly in the production of transportation fuels, such as used cooking oils and greases collected from restaurants (Tsoutsos et al. 2016). In addition, fatty acids can also be derived from inexpensive co-product streams resulting from refining of vegetable oils: free fatty acid (FFA; such as palm fatty acid distillate, PFAD [Abdul Kapor et al. 2017; Top 2010] or soapstock (a co-product formed during degumming [Cruz et al. 2018]). Also, oils derived from crops developed specifically for non-food applications mimic common vegetable oils in composition, such as jatropha and soapnut (*Sapindus*), which are commonly utilized for biodiesel preparation (Barros et al. 2015; Chen et al. 2013). An emerging source of oils for non-food applications is algae, plant-like organisms that contain chlorophyll, which converts sunlight and CO_2 into carbohydrates and oil through photosynthesis. Algae are versatile sources of oils that can be controlled through recombinant DNA technology to produce oils of specific fatty acyl chain length and unsaturation (Hayes and Smith 2019).

Generally, oils are limited in their chemical functionality to the carboxylic acid groups of FFA and ester bonds of TAG. However, as will be discussed later, carbon–carbon double bonds can be utilized to increase functionality, e.g.,. to add epoxy or hydroxy groups (or shorten the acyl chain). In addition, partial glycerides (mono- and diacylglycerols, MAG and DAG, respectively), readily formed through hydrolysis or

Table 7.1 Fatty acid composition of selected feedstocks used to prepare biobased surfactants.

Fatty Acid	Palm Kernel	Coconut	Cuphea	Palm	Jatropha	Beef Tallow	Palm Stearin
10:0	3–7	4.4	77–84				
12:0	40–52	44.5	2–3				0.10
14:0	14–18	18.6	2–4	0.5–2		1.0–1.5	1.1
16:0	7–9	12.0		32–45	13.0–16.0	24.0–28.0	55.9
18:0	1–3	4.8		2–7	6.0–8.0	20.0–24.0	3.7
18:1	11–19	11.0		38–52	39.0–41.3	38.0–43.5	32.4
18:2	0.5–2	2.2		5–11	37.0–38.0	2.0–4.0	6.7

Source: Hayes and Smith (2019). Reproduced with permission from AOCS Press.

alcoholysis of TAG, contain -OH groups on the glycerol backbone that can serve as conjugation sites. FFA can be readily reduced to fatty alcohols or fatty amines, e.g. via ozonolysis (Biermann et al. 2011; Egan 1968; Giraldo et al. 2010; Narine et al. 2007), or converted into acid chlorides (Bauer 1946; Busch et al. 2004), providing additional functionality. If oxidative stability is an important criterion of the industrial application, the double bonds can be hydrogenated (e.g. using Ni-based heterogeneous catalysts) to improve their oxidative stability (Veldsink et al. 1997).

7.2.2 Industrial Oil Crops

Nature provides us with several oils that possess uniquely functional fatty acids that can be utilized for industrial non-food applications. One important category of such oils is those possessing oxygenated fatty acids. Hydroxy acid-rich oils, particularly those derived from castor, *Wrightia sp*., and physaria (the latter formerly known as lesquerella), have several specific applications as surfactants, cosmetics agents, and lubricants, and possess biological activity (Ahmad and Lie Ken Jie 2008; Chen 2016; McKeon 2016; Mutlu and Meier 2010; Nekhavhambe et al. 2019; Sylwia and Józef 2016). The three oils are enriched in ricinoleic (*R*-18:1-9c, -OH-12), isoricinoleic (*S*-18:1-12c, -OH-9), and lesquerolic (*R*-20:1-11c, -OH-14) acids, respectively (Figure 7.1). Ricinoleic acid and castor oil have numerous industrial applications, several of which will be discussed later (reviewed in Kunduru et al. 2015; Mubofu 2016; Nautiyal 2018). 12-Hydroxysteararic acid, derived from ricinoleic acid via hydration, is a well-known gelator (Fameau and Rogers 2020). Dimorphotheca oil is a source of a unique hydroxy acid, domorphecolic acid (18:2-10t,12t-OH-9), which possesses conjugation, making it a useful building block for paints and coatings (Figure 7.1) (Cuperus et al. 1996). Similarly, epoxy fatty acids may serve similar applications as hydroxy acids, and can occur naturally (e.g. vernolic acid, 18:1–9, –epoxy-12*S*,13*R*) from *Vernonia galamensis* and *Euphorbia lagascae* oil (Grinberg et al. 1994; McKeon 2016) (Figure 7.1) or via chemical epoxidation of double bonds found in common oils such as soybean (discussed in Section 7.3.1) (Tan and Chow 2010).

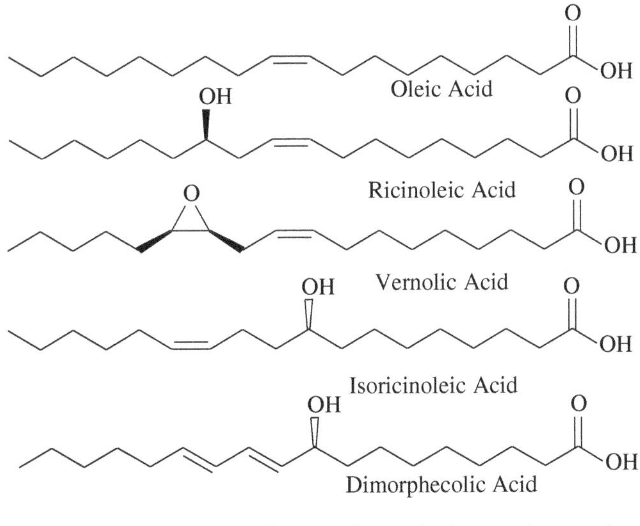

Figure 7.1 Oleic acid and naturally occurring C_{18} hydroxy and epoxy fatty acids.

A second category of industrial oils is highly polyunsaturated oils, which have been employed as "drying oils" for wood-based furniture, and linoleum floor covers for many years. Linseed (flaxseed) oil, enriched in α-linolenic acid (Figure 7.2), has fulfilled this purpose for centuries. In addition, camelina, chia, and perilla oils are emerging new crop oils that are also enriched in α-linolenic acid (Hayes and Hildebrand 2016; Murphy 2016; Serson et al. 2016). γ-Linolenic acid (18:3-6c, 9c, 12c) (Figure 7.2) is available through evening primrose and borage oils (Navarro-Herrera et al. 2018; Zhao et al. 2019). The performance of polyunsaturated oils as drying oils is enhanced when double bonds are conjugated. Nature provides a few conjugated oils: tung, pot marigold (*Calendia officinalis* L), and oiticica (*Licania rigida*) oils, which are enriched in α-eleostearic (18:3-9c, 11t, 13t), calendic (18:3-8t, 10t, 12c), and α-licanic (18:3-9c,11t,13t, 4-carboxy) acid, respectively (Figure 7.2) (Król and Paszko 2017; McKeon 2016; Melo et al. 2014; Shockey et al. 2016). Conjugated linoleic acid, a mixture of linoleic acid isomers (particularly, 18:2-9c,11t and 18:2-10t,12c) that is produced via alkaline isomerization of linoleic acid, serves as another source of conjugated fatty acids (Figure 7.2) (Moreno et al. 2012).

A third category of oils are those possessing long-chain fatty acids, which are useful in lubrication and cosmetics. Examples of long-chain fatty acids include erucic acid

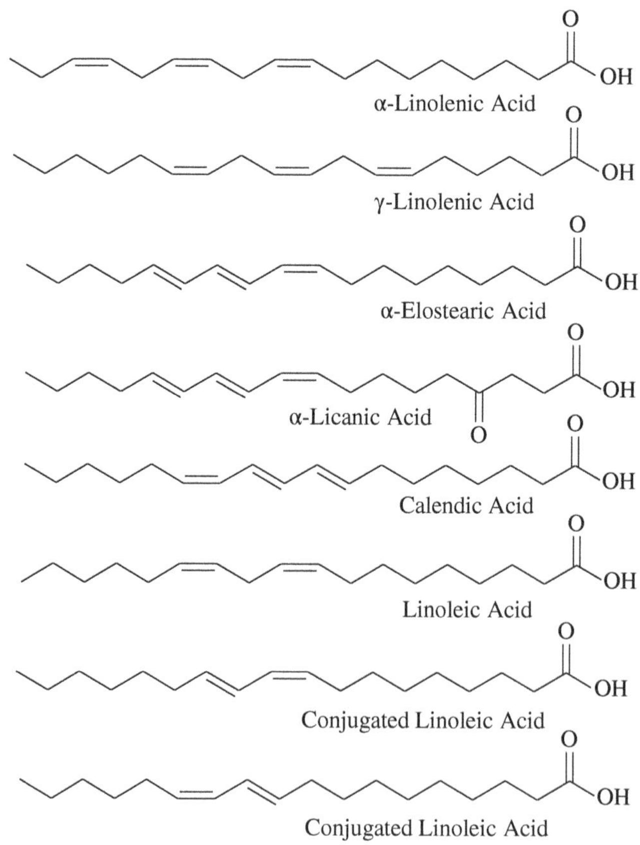

Figure 7.2 Naturally occurring C_{18} di- and tri-unsaturated fatty acids.

(C22:1-13c), found in high-erucic acid rapeseed, crambe, and pennycress oils (Chen 2016; McVetty et al. 2016; Zhu 2016), and 22:1-5c and 22:2-5c,13c found in meadowfoam oil (McKeon 2016). On the other end of the spectrum would be the fourth category, oils rich in medium-chain fatty acids, often referred to as "lauric oils." These oils are often used as feedstocks for industrial surfactants. Major sources include palm kernel and coconut oils, with cuphea oil serving as an emerging source (Hayes and Smith 2019; McKeon 2016). It should be also mentioned that common fatty acids can be modified to produce novel chemical building blocks, such as diacids (e.g. azelaic acid, $HOOC(CH_2)_7COOH$, produced via ozonolysis of oleic acid [Koppireddi et al. 2016]), medium-chain fatty acids with terminal unsaturation (e.g. 10-undecenoic acid produced via pyrolysis of ricinoleic acid [Van der Steen and Stevens Christian 2009] and methyl 9-dodecenate obtained from methyl oleate via cross-metathesis with ethylene [Rybak et al. 2008; Van der Steen and Stevens Christian 2009]).

7.3 Industrial Applications of Seed Oils

7.3.1 Biopolymers

Seed oils are attractive building blocks for polymer synthesis due to their possession of long alkyl chains, which can impart hydrophobicity, amphiphilicity, high oxidative stability, and other desirable properties to the polymers. Applications of oleochemical-based polymers are numerous, with many occurring in the paints, coatings and lubricant sectors. Several polymeric products derived from oilseeds are listed in Table 7.2. Several naturally occurring fatty acyl groups possess two or more functional groups, as needed for polymer synthesis, such as the hydroxy and epoxy fatty acyl groups depicted in Figure 7.1. Other fatty acyl groups possess multiple, and in some cases, conjugated double bonds, a desirable trait for free radical polymerization (Figure 7.2). The classical example of the latter is the use of linseed oil as a "drying oil" for wood varnish and floor finishing. However, most oils, particularly vegetable oils, do not possess sufficient functionality to serve as monomers, with only the carboxylic acid groups of fatty acids and the hydroxyls of MAG and DAG serving as reactive groups. To address the deficiency, C—C double bonds present in fatty acyl backbones have been utilized to introduce new functional groups. Examples are given in Figure 7.3. In addition, functional groups can be introduced via formation of diacids or terminal C=C bonds, as described earlier. In this section, major polymeric syntheses and products will be briefly reviewed. The reader can find more detailed information available through several excellent review articles (Gandini et al. 2016; Hayes and Dumont 2016; Llevot 2017; Zhang et al. 2017; Zia et al. 2016).

The most common approach to adding functional groups to double bonds is to introduce (secondary) hydroxy or epoxy groups via addition-type reactions (products **II, III,** and **XV** of Figure 7.3). Product **II**, epoxidized FFA or TAG (e.g. epoxidized soybean oil, or ESBO), is employed in lubrication, as a plasticizer, and in polyurethanes (thermosets) (Table 7.2). To incorporate primary hydroxyl groups, which are more reactive than secondary OH groups, hydroformylation (**I→VIII**) or addition (e.g. with 1,3-propanediol: **II→XV**, with $R_3 = CH_3(CH_2)_2OH$) can be employed (Figure 7.3). Ozonolysis of unsaturated TAG (e.g. triolein), via cleavage at C=C bonds, yields a triol with primary ω-OH medium-chain acyl groups attached to the glycerol backbone (1-nonanyl groups for triolein), useful reagents for polymerization and polyols for polyurethanes (**I → VII,** Figure 7.3)

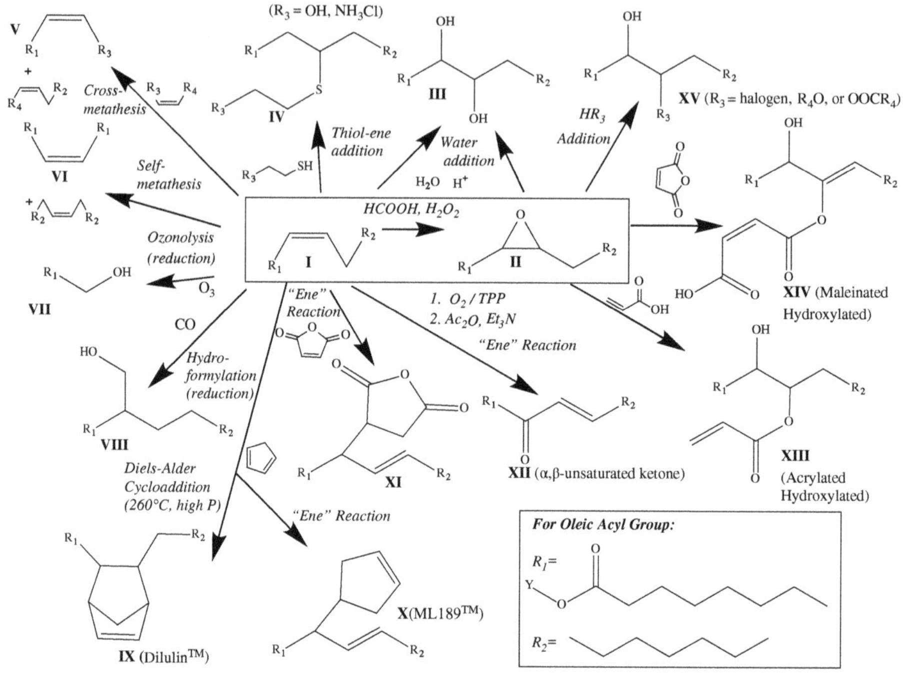

Figure 7.3 Modification of double bonds in fatty acyl groups to produce functional groups useful for polymerization. *Source:* Hayes and Dumont (2016). Reproduced with permission from AOCS Press.

Table 7.2 Commercially available non-food polymeric products from industrial oilseeds.

Product Name	Manufacturer	Application
Epoxidized Oils		
HiBond™ (Linseed)	Polar Industries (Fisher Branch, MB, Canada)	Paints, plasticizers, adhesives, coatings, polyols (polyurethanes), thermosets
Vikoflex® (Linseed, Soybean)	Arkema (King-of-Prussia, PA USA)	
PlastiSoy™ epoxidized soybean oil (ESO)	CHS (Inver Grove Heights, MN USA)	
ESO	Makwell (Maharashtra, India), The Chemical Company (Jamestown, RI USA), MultiPlus (Nonthaburi, Thailand), PolyMar Enterprises (Irvine, CA USA), FMC (Philadelphia, PA, USA)	
Ebecryl 860 Acrylated ESO	Advent International (Boston, MA, USA)	Surface coatings
Heloxy™ flexibilizers: castor oil and dimer acid glycidyl ether	Momentive Specialty Chemicals (Columbus, OH, USA)	Epoxy resins, adhesives

ERISYS™ GE-35 H castor oil and dimer acid glycidyl ethers	Emerald Performance Materials (Moorestown, NJ, USA)	Epoxy resins, adhesives
Biobased Polyurethanes		
Minwax® (Tung Polyurethanes)	Miniwax (Upper Saddle River, NJ, USA)	Wood finish
Polymerized Tung Oil	Sutherland Welles (North Hyde Park, VT, USA)	Coatings
BiOH® polyols	Cargill (Minneapolis, MN, USA)	
Biobased Polyurethanes	Synthetic Natural Polymers (Durham, NC, USA)	
Agrol® Biobased Polyols (Soy, castor)	BioBased Technologies (Springdale, AR, USA)	
Lupranol® Balance 50 (Ethoxylated Castor-based polyol)	BASF (Ludwigshafen, Germany)	
Renuva® Biobased Polyols	Dow Chemical (Midland, MI, USA)	
Linseed Oil – Cyclopentadiene Polymers		
Dilulin (via Diels-Alder)	Cargill (Minneapolis, MN, USA)	Drying oil
ML189 (via *ene* reaction)	Archer Daniels Midland (Decatur, IL, USA)	Varnish, enamel, aluminum paint, reinforced oil
Estolides		
LubriGreen® (Coco Estolide™)	Biosynthetic Technologies (Irvine, CA, USA)	Lubricants
Vorite® polymerized castor oil	Vertellus (Indianapolis, IN, USA)	Sealant, adhesive
Zenigloss® isostearic acid-grafted copolymer of castor oil and succinic acid	Zenitech (Toronto, ON, Canada)	Emollient, lip gloss agent, personal care applications
Polyamides		
Rilsan® Polyamide 11 (from 11-aminoundecanoic acid, derived from castor oil)	Arkema (King of Prussia, PA USA)	Electrical cables, automotive, and pneumatic and hydraulic hose

Source: Hayes and Dumont (2016). Reproduced with permission from AOCS Press.

(Narine et al. 2007). An additional approach to obtain primary OH groups is the thiol-*ene* reaction, where a thiolated alcohol such as mercaptoethanol adds across C=C, resulting in a thiol bond and a terminal primary OH group (**I → IV,** Figure 7.3) (reviewed in Resetco et al. [2017]). The thiol-*ene* reaction can also be employed to prepare a derivative with a primary amine (**IV,** Figure 7.3), as well as numerous other reactions, for applications in coatings (Resetco et al. 2017). A carboxylic acid group is introduced in the fatty acyl chain by ring-opening an epoxide group by maleic anhydride (**II→ XIV,** Figure 7.3). Malonylated epoxy FA (i.e. **XIV**) can serve as a cross-linking reagent for epoxy or hydroxy-TAG.

The occurrence of terminal C=C bonds increases the reactivity of seed oil-derived monomers toward free radical polymerization. Examples of ω-unsaturated FA include 10-undecenoic acid, which is derived from ricinoleic acid of castor oil (Van der Steen and Stevens Christian 2009) or those formed by ring-opening of epoxy FA by acrylic acid (**II→ XIII,** Figure 7.3). A frequently employed monomer along the lines of **XIII** is acrylated epoxidized soybean oil, used in surface coatings (e.g. Ebecryl 860; Table 7.2). The (Alder) *ene* reaction is another effective approach (reviewed in Lligadas et al. [2013]). A C=C bond undergoes reaction with an "enophile," i.e. a polyunsaturated molecule with conjugated double bonds and electron withdrawing groups that participate in the conjugation (e.g. C=O and C=N), resulting in formation of a new C–C (sigma) bond. During the reaction, the double bond of the alkene donor undergoes an allylic shift. Two examples of enophiles are shown in Figure 7.3, using cyclopentadiene and maleic anhydride as enophiles (**I → X** and **XI**, respectively). The former is employed in the commercial product ML 189™ as a co-monomer for varnishes and coatings (Table 7.2). The *ene* reaction can also be employed to introduce a ketone group in the fatty acyl chain (**I → XII,** Figure 7.3) or by oxyfunctionalization, and the carbonyl group is oxidized to form an amine group (Winkler and Meier 2014).

Diels-Alder reactions are "[4+2]" cycloaddition reactions that occur at elevated temperatures (>100 °C) between an alkene's double bond ("akeneophile") and a conjugated diene. The Diels-Alder reaction between calendic acid methyl ester (specifically, its *trans*-C_{10} and *cis*-C_{12} conjugated bonds) and acrylic acid is shown in Figure 7.4 (Huang et al. 2010). A six-member ring is formed consisting of C_{10}-C_{13} of methyl calendate and double bond of acrylic acid, and the carboxylic acid group of the latter is available for further conjugation or polymerization. A second example is shown in Figure 7.3, for the reaction between an alkene (**I**) and cyclopentadiene. The monomer is employed in the preparation of Dilulin™, used as a drying oil and resin (Table 7.2) and is used in metathesis and other polymeric reactions (**IX**). The Diels-Alder, thiol-*ene*, and *ene* reactions are all examples of "click chemistry," defined as quick and reliable reactions that involve the joining of smaller molecules to prepare a larger molecule.

Metathesis is a unique reaction that has expanded in use during the last two decades due to innovations with catalysts, specifically "second generation" Grubbs and Hoveyda-Grubbs (ruthenium-based) catalysts (reviewed in Dong et al. 2017; Montero de Espinosa and Meier 2012; Yelchuri et al. 2019). The reaction in its essence is the exchange of groups attached to C=C bonds between two molecules. Self- and cross-metathesis refer to metathesis between two identical and two different molecules, respectively (**I→ VI** and **I →V**, respectively, Figure 7.3). For example, self-metathesis of oleic (18:1-9c) acid yields 9-octadecedioic acid (a diacid that can be employed for nylon synthesis) and the alkene 9-octadecene. As shown for the cross-metathesis example in Figure 7.3 (**I→ VI**), *cis/trans* isomerism of the starting materials is not conserved. As mentioned above, cross-metathesis between a diacid and ethylene will produce a fatty acid with unsaturated in its terminal position; for example, oleic acid + ethylene → 9-decenoic acid (and 9-decene as co-product).

In addition to preparation of monomers, metathesis can be employed directly to oleochemical feedstocks to prepare polymers. Acyclic diene metathesis (ADMET) refers to a cascade of self-metathesis reactions (step-growth mechanism) involving a molecule with two double bonds (α,ω-dienes; e.g. 10-undecenoyl 10-undecenoate, which is produced via self-metathesis of 10-undecenoic acid) to produce a linear polymer (reviewed in Montero de Espinosa and Meier 2012). ADMET is terminated through addition of a chain-stopper, such as methyl acrylate. Acyclic triene metathesis

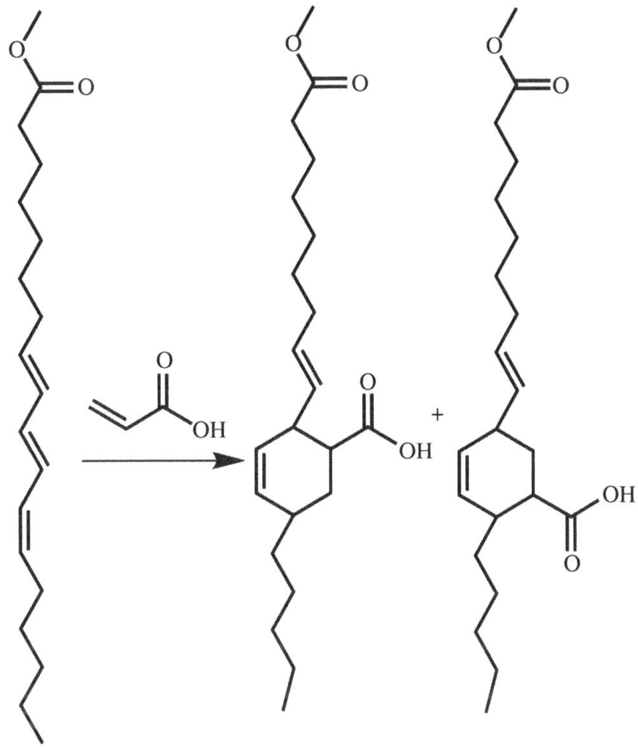

Figure 7.4 Diels-Alder reaction between calendic acid methyl ester (18:3-8t,10t,12c FAME) and acrylic acid.

(ATMET) is an analogous reaction, involving self-metathesis of a derivative possessing three double bonds (a "triene"). Figure 7.5 depicts the ATMET reaction of triolein, producing a hyperbranched polymer (Biermann et al. 2010). In this example, methyl acrylate is employed as the chain terminator, and the terminal carboxylate groups can serve as conjugation sites for further modification. A third example is ring opening metathesis polymerization (ROMP) (reviewed in Lu and Larock 2009; Montero de Espinosa and Meier 2011). The ROMP reaction applied to Dilulin is depicted in Figure 7.6, with dicyclopentadiene (a norborene) serving as initiator, acting via cross-metathesis and serving as a co-monomer. Dilulin undergoes self-metathesis, resulting in the cleavage of its cyclopentadiene ring, and producing a highly cross-linked polymer suitable for biobased composites with glass fibers (Henna et al. 2008) and ROMP of an analogous monomer to Dilulin has also been employed for the production of elasotomers (Wang et al. 2017).

Polyols derived from oleochemicals are valuable replacements for fossil fuel-derived polyols in the synthesis of polyurethanes (PUs), a large class of polymeric materials (typically thermosets; comprising ~5% of the global polymer market) that are employed for a wide variety of applications, such as adhesives, foam materials (e.g. automobile seats, couches and chairs, and mattresses), elastomers, and flooring (Alagi and Hong 2015; Hayes and Dumont 2016; Sawpan 2018; Zhang et al. 2017). The size of the global PU market was worth $ 95.13 billion in 2019 and is expected to grow at a rate

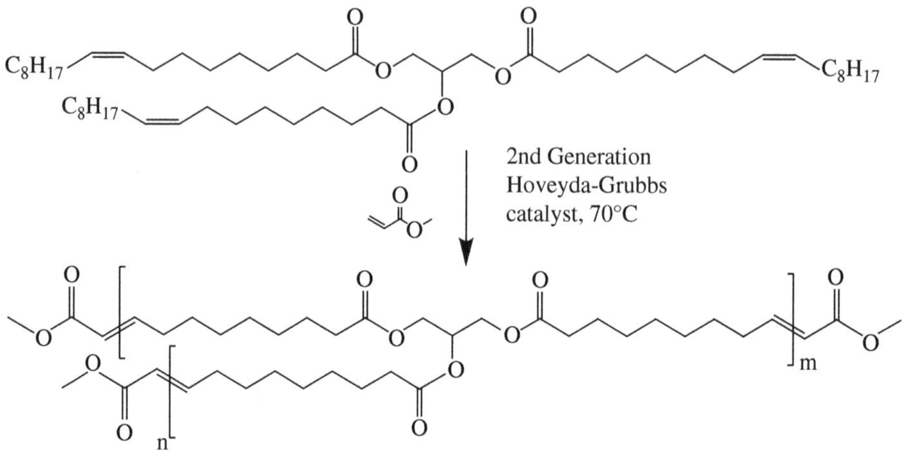

Figure 7.5 Acyclic triene metathesis (ATMET) of triolein, using methyl acrylate as chain stopper. *Source:* Adapted from Biermann et al. (2010). Redrawn in Hayes and Dumont (2016). Reproduced with permission from AOCS Press.

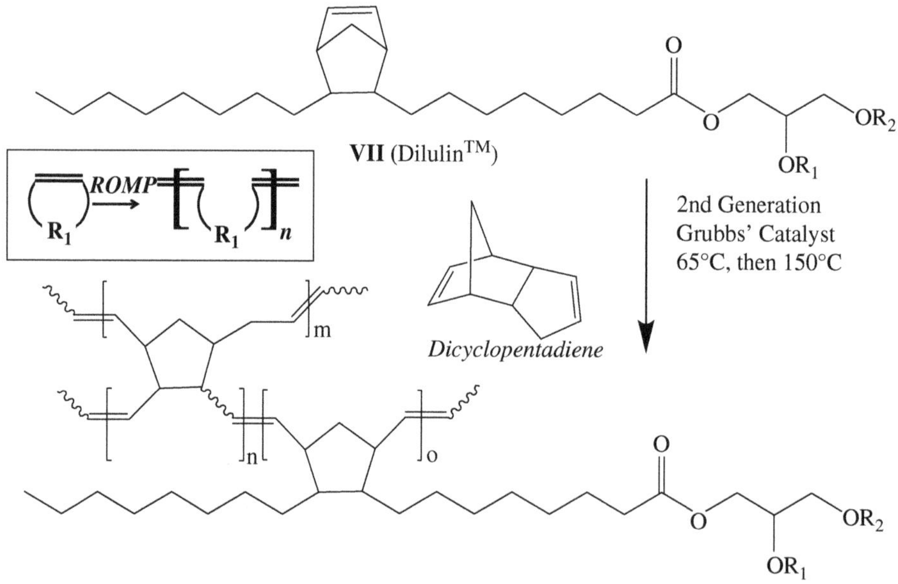

Figure 7.6 Ring-opening metathesis polymerization (ROMP) of Dilulin™ and discylopentadiene. Inset: mechanism of ROMP. *Source:* Adapted from Henna et al. (2008). Redrawn in Hayes and Dumont (2016). Reproduced with permission from AOCS Press.

of 12% per annum, reaching $149.91 billion by 2023 (Wood 2019). PUs are comprised of urethane (NH-COO-) linkages formed between molecules possessing two or more isocyanate (-N=C=O) groups and polyols in the presence of catalysts. The number of isocyanate building blocks commercially available for PU synthesis is much smaller than the number of polyol monomers. Nature provides several oleochemical (biobased) polyols, such ricinoleic acid in FFA or TAG form (the latter being castor oil) or derivatives of other hydroxy acids (Mutlu and Meier 2010; Nekhavhambe et al. 2019) (Figure 7.1). Epoxidized oils or their FFA or MAG can also serve as sources of polyols, either via vernonia oil (a natural source of vernolic acid; Figure 7.1) or epoxidation of unsaturated TAG (**I → II,** Figure 7.3). The ring-opening of the epoxy group produces OH groups (e.g. **XV, II** and **IV** of Figure 7.3). Another approach is to utilize synthetic polyols such as trimethylolpropane as acyl acceptors for hydroxy acids. Other methods of introducing OH groups into FA chains are described earlier in the chapter and reviewed elsewhere (Alagi and Hong 2015; Hayes and Dumont 2016; Sawpan 2018; Zhang et al. 2017). The preparation of biobased isocyanates and biobased PUs via other synthetic routes is reviewed elsewhere (Carre et al. 2019; Llevot 2017; Suryawanshi et al. 2019; Zhang et al. 2017).

As described earlier, oils enriched in polyunsaturated fatty acids, such as linseed and tung oils, are well known as "drying oils" that serve as varnishes for wood-based furniture and products, linoleum floor covers, printing inks, and alkyd resins employed in paints and coatings. The oils form a thin film coating on surfaces through autooxidation (via O_2), and then undergo free radical polymerization. Further detail on the underlying mechanism and operation of the reactions is described elsewhere (Llevot 2017; Xia et al. 2013). Cationic polymerization is an alternate approach to conduct free radical polymerization that often results in lower levels of artifacts in the coatings such as cracks and entrapped air bubbles (reviewed in Xia et al. 2013). The reaction is initiated by strong Brønsted-Lowry and Lewis acids, such as $AlCl_3$ and BF_3-$O(CH_3)_2$, and is conducted at temperatures >100 °C. The polymers that form are highly cross-linked and therefore possess high mechanical strength. Cationic polymerization has been employed to prepare thermosets via copolymerization of polyunsaturated oils with divinylbenzene, styrene, dicyclopentadiene, Dilulin, and ML189 (**X**, Figure 7.3).

An additional polymeric material prepared from seed oils (e.g. soybean, rapeseed, crambe, jojoba, and meadowfoam oils, or their mixtures) is factice, employed as processing aids and property modifiers in rubber (Erhan and Kleiman 1990a, b, 1993). In a reaction between oils with S_2Cl_2, a form that consists of cross-linked networks of TAG via sulfur atoms between C=C bonds arises. A final example, estolides, is described later (see Section 7.3.4).

7.3.2 Biofuels

Biofuels, of interest mainly for transportation and heating purposes, are receiving increased interest worldwide. Its market size, 168 billion USD in 2016, is projected to grow at 4.5% annually from 2017 to 2022, reaching 218.7 billion USD in 2022 (Zion Market Research 2018). The most widely produced biofuel from seed oils is biodiesel. With production at 36 billion L in 2017 and anticipated to reach 95 billion L by 2020, and its market size being 28.04 billion USD in 2016 (Grand View Research 2017; Rezania et al. 2019; World Bioenergy Association. Global Bioenergy Statistics 2019). Biodiesel is composed of fatty acid monoalkyl esters (typically methyl or ethyl esters, i.e. FAME and FAEE, respectively). Use of the term "biodiesel" infers that the fatty alkyl ester mixture meets standardized requirements designed to ensure good performance for compression-injection engines involving purity, viscosity, and temperature stability, such as ASTM D6751 (ASTM International 2020). (Seed oils cannot fulfill standardized requirements due to their high viscosities; therefore, modification of oils is required.) Biodiesel is often blended with petroleum-based diesel. Global interest in biodiesel as a transportation and heating fuel grew steadily in the first decade of the current century, as a result of rising petroleum prices. However, interest in North America waned during the previous decade due to the increased abundance and decreased cost of petroleum and natural gas. Biodiesel remains in use in places where government subsidies are in place. For example, Malaysia and Indonesia have mandated that the content of palm oil FAME in diesel blends be at least 20% (B20), to support domestic oil palm production. An additional motivating factor is sustainability. Biodiesel is reported to produce lower levels of pollutants in exhaust gases (e.g. carbon monoxide, particulate matter, and residual hydrocarbons) compared to petrodiesel due to its high cetane number (i.e. long acyl chain length), low sulfur content, and high flash point. The presence of O atoms in biodiesel produces more efficient combustion compared to petrodiesel. Unlike petroldiesel, biodiesel is biodegradable and possesses low ecotoxicity. Biodiesel is also attractive to many petroleum-poor countries to mitigate concerns for energy and fuel security. The main economic barrier to increased demand and usage for biodiesel remains the high feedstock cost compared to fossil fuels, with higher processing costs being an important secondary economic factor (Gebremariam and Marchetti 2018).

Biodiesel is most commonly produced via transesterification (alcoholysis) between TAG and acyl acceptor (e.g. methanol or ethanol) using homogeneous or heterogeneous catalysts, although the use of supercritical methanol negates the need for catalysts. Oleochemical feedstocks for "first generation" biodiesel consist of low-cost and abundant oils, e.g. from palm, LEAR, and soybean. However, a concern for such feedstocks is their prominence in foods; moreover, depletion from the food supply chain to support the transportation sector is undesirable. This concern has led to production of FAME from the oils of non-food resources such as cotton and tall oil (a co-product from wood pulp processing), and energy crops such as jatropha, which are non-edible, require minimal agricultural inputs and can grow on marginal land ("second generation biodiesel"). TAG feedstocks for "third-generation" biodiesel are less expensive and more economically feasible (e.g. use less agricultural inputs such as water and nutrients) than those used for previous biodiesel generations and include animal tallow, used cooking oils, and algal oils. Recently, "fourth generation" biodiesel has been developed, consisting of bioinspired fuel cells that efficiently harvest the energy of sunlight by applying scientific breakthroughs developed in synthetic biology. The different generations of biodiesel have been reviewed elsewhere (Rezania et al. 2019; Singh et al. 2019).

Selection of TAG feedstock is a strong determinant of the resultant properties and quality of the biodiesel produced (reviewed in Knothe et al. 2010; Singh et al. 2020). Oils rich in polyunsaturated fatty acids are infrequently employed due to the low thermostability of their FAME. A performance aspect directly associated with feedstock selection is fuel performance in cold temperatures. Under such conditions, biodiesel derived from oils enriched in long-chain and saturated FA (e.g. palm oil and tallow) perform poorly, resulting in poor fuel flow in engines and filter clogging. Oils rich in MAG and in steryl glucosides produce biodiesel that undergoes precipitation when stored in cold conditions.

The most common catalyst type is alkali, such as NaOH, KOH, or CaO, which is inexpensive and its employment is well-known. Alkali-catalyzed transesterification is usually conducted at ~60 °C under stirring with acyl acceptor (alcohol) present at ~six to ninefold excess for ~ one to four hours (Reviewed in Rezania et al. 2019; Knothe et al. 2010; Singh and Gaurav 2018). Microwave technology has recently made inroads into transesterification processing due to its enhancement of mixing and heating between phases (Singh and Gaurav 2018). The reaction medium is then allowed to settle (~eight hours), with a FAME-rich phase forming at the top and an alcoholic glycerol solution residing at the bottom. The top phase undergoes several purification steps, particularly neutralization (to remove any soaps) and distillation (to remove excess alcohol). Although alkali-catalyzed transesterification produces high yields (80–98%) and is simple and rapid, the approach is problematic for oils with high FFA content (e.g. rice bran oils and some used cooking oils), due to the formation of insoluble soaps. It is recommended that FFA must be present at ≤5%, but preferably at <0.5%. Often, an acid-catalyzed esterification step is employed prior to (or concurrently with) transesterification, to minimize soap formation. (Esterification has also been achieved using lipases or ion exchange resins [Singh and Gaurav 2018].) Another concern for alkali usage is the need for large amounts of water for removal of residual alkali and glycerol after transesterification is completed, which detracts from green manufacturing principles and increases operation costs. Moreover, the aqueous solution formed as a by-product of transesterification, enriched in alkali, excess alcohol reactant, and the reaction co-product glycerol, is expensive to purify, recycle, or upcycle.

Acidic catalysts have also been employed for transesterification, such as sulfuric acid. Their employment is usually limited to oils with high FFA content (due to their ability to simultaneously catalyze both esterification and transesterification) as a result of their lower catalytic efficiency (Singh and Gaurav 2018).

Several different heterogeneous catalysts, particularly nanocatalysts, have been developed in recent years, to improve the productivity and yield of alcoholysis, enable the utilization of low-cost high-FFA oils, and improve the environmental sustainability of production. The majority of the nanocatalyst technology development has focused upon metal oxides, zeolites, heteropolyacids, and activated carbon. Nanocatalysts provide rapid catalytic turnover (one to three hours) due to their high surface area and high yields (>95%). This topic is reviewed elsewhere (Singh and Gaurav 2018). A recent emphasis has been on developing low-cost heterogeneous catalysts from renewable resources or industrial by-products such as egg shells, biochar, and ash produced by lignocellulosic biorefineries (reviewed in Shan et al. 2018).

Enzymes, particularly lipases (acylglycerol hydrolases, EC 3.1.1.3), have been employed as biocatalysts for green synthesis of biodiesel (reviewed in Christopher et al. 2014). Lipases have several inherent advantages. They can catalyze transesterification and esterification of residual FFA concurrently, operate under mild reaction conditions (e.g. no extremes in pH, mild temperatures: 20–80 °C, and the absence of solvents) and residual water, and can utilize stoichiometric proportions of fatty acyl donor and acceptor substrates. Initial reports of biodiesel synthesis using immobilized lipases were reported in the 1990s, with biodiesel benchmarks for conversion and purity (e.g. per ASTM D6751) being met, in some cases without extensive purification required. In addition, the glycerol-rich co-product is of higher quality than produced using conventional alkali-catalyzed processing, with its isolation and purification being simplified and less expensive due to the reduced use of water. However, processing costs remained impractically high due to the high costs of the immobilized lipases, even when recycling or continuous processing through fixed beds was employed. Within the last 10–15 years, so-called "liquid lipases" have been employed as a less-expensive alternative to immobilized lipases (e.g. Callera™ Trans L and Eversa products, aqueous solutions composed of genetically modified *Thermomyces lanuginosa* lipase, manufactured by Novozymes, Inc., Franklinton, NC, USA) (Nielsen et al. 2016). Liquid lipases are stable in the presence of methanol and ethanol, unlike many immobilized lipases. The resultant reaction medium consists of three layers: FAMEs, a middle phase enriched in lipases (that is readily recycled), and a bottom phase enriched in glycerol.

Glycerol is widely recognized as an underutilized co-product from the preparation of biodiesel. Crude glycerol production is projected to reach 7.66 million tons per annum (Anuar and Abdullah 2016). However, recent research and development has led to novel applications for glycerol, such as the production of monomeric feedstocks for polymers (e.g. 1,3-propanediol and glycerol carbonate). This topic has been reviewed elsewhere (Gandini et al. 2016; Monteiro et al. 2018). But, the isolation of glycerol in a sufficiently pure form as needed for the applications is expensive, requiring neutralization and removal of alcohol (e.g. by distillation) and water.

An alternate approach is to rectify the underutilization of glycerol is to modify the reaction pathway to prevent the formation of free glycerol (reviewed in Estevez et al. 2019). Two approaches are outlined in Figure 7.7. The first involves using methyl acetate rather than methanol as acyl acceptor. The reaction (transesterification), in addition to producing 3 moles of FAME per mole of TAG, yields 1 mole of triacetin rather than free glycerol. Triacetate can undergo combustion in a diesel engine and also serves as an anti-knock agent and improves stability under cold conditions. The resultant homogeneous mixture of FAME and triacetin is referred to as "gliperol®." The second approach employs dimethyl carbonate rather than methanol as reactant, producing 2 (rather than 3) moles of FAME and 1 mole of fatty acid glycerol carbonate (Figure 7.7). The homogeneous mixture, referred to as "DMC-bioD," also contains small amounts of glycerol carbonate and glycerol decarbonate via side-reactions as well as unreacted dimethyl carbonate. The latter reactant is less toxic than methanol

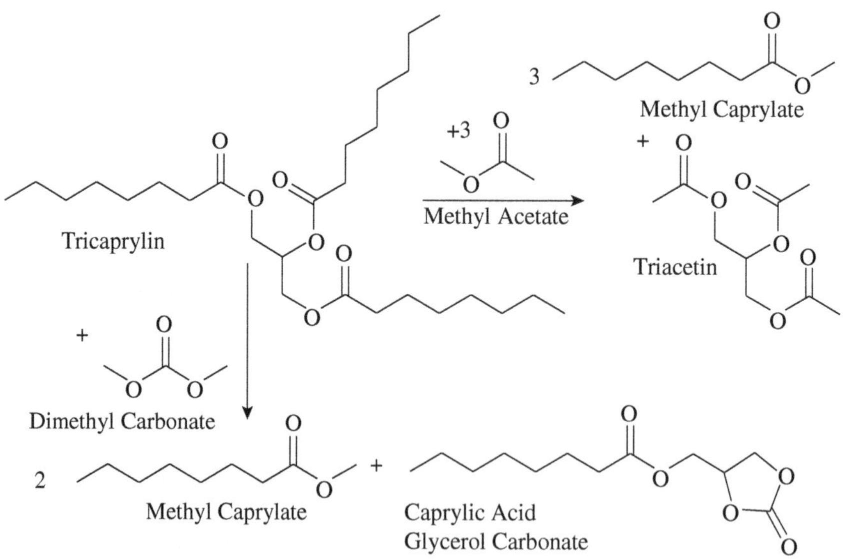

Figure 7.7 Formation of fatty acid methyl esters (FAME; i.e. biodiesel) in the absence of glycerol formation from tricaprylin via reactions with methyl acetate and dimethyl carbonate.

and is biobased, being derived from methanol and carbon monoxide. The presence of dimethyl carbonate produces less CO, particulate matter, and other undesirable compounds in engine exhaust compared to that generated by combustion of neat FAME. Another approach is to conduct transesterification using a stoichiometric deficiency of acyl acceptor (e.g. 2 moles of methanol per mole of TAG), yielding monoacylglycerol (MAG) and FAME at 1 mole and 2 moles per mole of TAG, respectively. Alternatively, 1,3-selective lipases can be employed as (bio)catalyst to achieve the same product distribution. The homogeneous mixture, referred to as "Ecodiesel," is equally as effective as FAME in terms of energy production. As an amendment to this approach, 1-(3-) MAG and 2-MAG co-product can be reacted with dimethoxymethane to produce acylated 1,3-dioxolane and 1,3-dioxane derivatives, respectively, that can undergo methanolysis to produce FAME and 1,3-dioxolane and 1,3-dioxane-based alcohols, respectively (Figure 7.8).

An additional biofuel product receiving great attention in recent years is the production of drop-in fuels, particularly jet fuel for commercial and military uses, via catalytic deoxygenation of vegetable oils using heterogeneous catalysis in the presence of H_2 (reviewed in Diaz-Perez and Serrano-Ruiz 2020; Khan et al. 2019; Wei et al. 2019). Jet fuel consumes approximately two billion barrels of crude petroleum per year and therefore contributes significantly to the production of atmospheric CO_2. Prior to the COVID-19 outbreak, air travel was on the rise, a trend that will likely continue in the years to come, accounting for 12% of transportation energy consumption. Therefore, the ability to partially replace petroleum consumption with biobased feedstocks for jet fuel is an attractive proposition. Several commercial and military trials of biobased jet fuel have been conducted. Approximately 80% of jet fuels are linear, branched (particularly iso-), and cyclic alkanes. Biobased jet fuel from vegetable

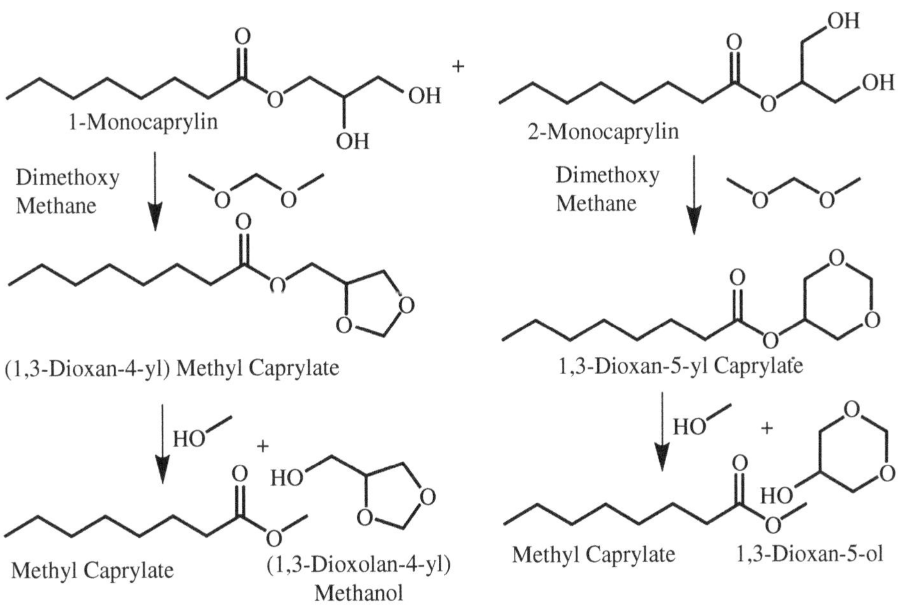

Figure 7.8 Conversion of monocaprylate into methyl caprylate and 1,3-dioxolane or 1,3-dioxane derivative via reaction with dimethoxymethane.

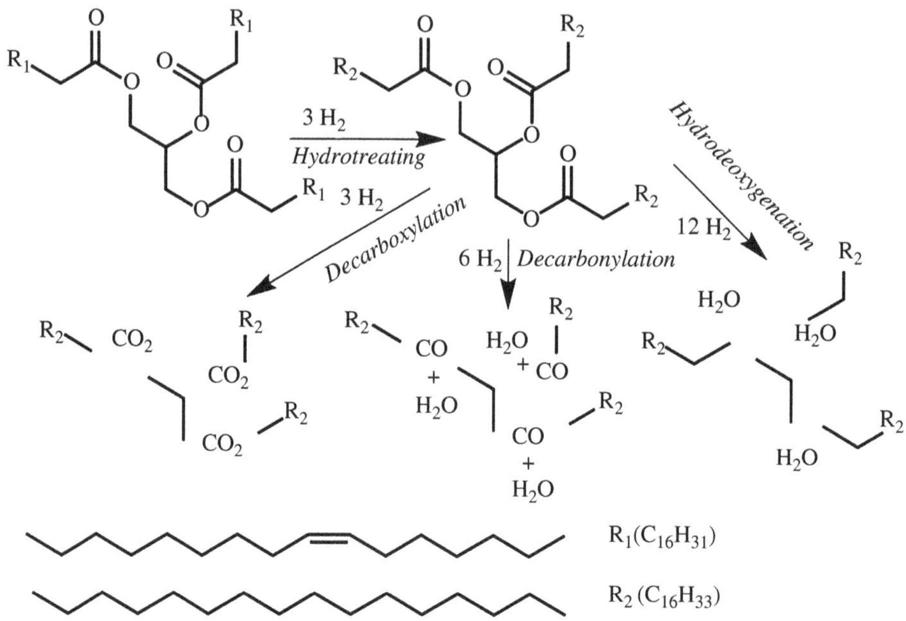

Figure 7.9 Conversion of triolein and tristearin into hydrocarbons via catalytic deoxygenation, producing green jet fuel and diesel.

oils can serve as a major source of the alkanes. Oleochemical feedstocks employed for biojet fuel include common vegetable oils such as soybean, oils from palm, palm kernel, jatropha, camelina, and waste cooking oils, algal oils, and animal fats.

An idealized example of deoxygenation is given in Figure 7.9 (reviewed in Diaz-Perez and Serrano-Ruiz 2020; Khan et al. 2019; Wei et al. 2019). H_2 under pressurized conditions (20–100 kPa) at moderately high temperatures (250–400 °C) adds across double bonds of TAG, yielding trisaturated TAG (e.g. conversion of triolein into tristearin; Figure 7.9). There are different deoxygenation pathways involving the trisaturated TAG that depend on the H_2 partial pressure, temperature, catalyst type and load, and residence time used. The reactions are exothermic and involve formation of saturated FFA (stearic acid in the case of Figure 7.9), aldehydes, ketones, and alcohols as intermediate products. Both decarboxylation and decarbonylation pathways convert FA groups into alkanes that are one carbon atom smaller in length (e.g. conversion of stearic [octadecanoic] acyl groups into heptadecane; Figure 7.9), with the carbonyl group being converted into CO_2 and CO, respectively. The glycerol backbone is converted into propane. Hydrodeoxygenation of TAG, which is more prominent than decarboxylation and decarbonylation at higher H_2 partial pressure, produces alkanes with the same number of carbon atoms as the fatty acyl groups (production of octadecane for the reactions depicted in Figure 7.9). In actuality, the reactions produce a mixture of products and can occur concurrently. Decarboxylation and decarbonylation of tristearin produce both heptadecanes and pentadecanes as major products, while hydrodeoxygenation yields octadecanes and hexadecanes. Several side reactions also occur during deoxygenation that yield by-product gases such as methane, including methanation ($CO_2 + 4\ H_2 \rightleftharpoons CH_4 + 2\ H_2O$) and the water gas shift reaction ($CO + H_2O \rightleftharpoons CO_2 + H_2$). To improve the properties of the deoxygenation product for use in jet fuels, catalytic cracking and isomerization is conducted, to reduce the alkane chain length (C_9–C_{15}) and introduce branching and cyclic compound formation. A fractionation process is then performed, with major fractions consisting of long-chain alkanes (applicable as diesel fuel), biojet fuel, naptha, and light gases such as CO_2, CO, and CH_4. Catalysts employed for deoxygenation are mainly of three types: noble metal-based (Pt, Pd, Ir, Rh, Ru, and Ag; supported on alumina or other metal oxides, zeolite, or activated carbon), sulfided catalysts, and non-sulfided catalysts. Further detail on the reaction mechanisms, catalysts, and other processing conditions is given in recent reviews (Diaz-Perez and Serrano-Ruiz 2020; Khan et al. 2019; Wei et al. 2019). A related approach is to employ catalytic thermolysis in the presence of near- or supercritical water (250–400 °C, 5–30 MPa) (Wei et al. 2019).

7.3.3 Surfactants

Seed oils serve as prominent building blocks in the synthesis of surfactants, which are derivatives that possess amphiphilicity, i.e. contain regions of polarity (hydrophilicity) and apolarity (lipophilicity) in their molecular structure. Amphiphilicity drives their partitioning to liquid-gas, liquid-liquid, and liquid-solid interfaces, where they lower the interfacial energy, thereby facilitating interfacial momentum, heat, and mass transfer. Surfactants are employed in many industrial sectors, with detergents and cleaners being the most major, followed by personal care products, foods, cosmetics, pharmaceuticals, agrochemicals, and paints and coatings. Their main role is in emulsification of oil in water, or vice versa. They also serve as wetting agents, solubilization agents, adjuvants (e.g. for the delivery of agrochemicals to plants) and modifiers of surface chemistry. Many surfactants also possess antimicrobial activity, through disrupting the function of cell walls and membranes of microorganisms. Thus, surfactants are in great demand during the current COVID-19 pandemic. The market size for surfactants was $27 billion and 15.6 million metric tons in 2012 (Markets and Markets 2012). Fatty acid soaps, produced via saponification of oils using alkali, have been used for centuries (Spitz 2009). During the twentieth century, biobased feedstocks were replaced by fossil fuels, particularly petroleum. However, during the current century, biobased surfactants have received increased interest, reaching current market share of ~25%, due to concerns relating to the impact of petrochemicals on increased atmospheric CO_2 levels, as described earlier. Further information on the biobased surfactants industrial sector is given in Hayes and Smith (2019).

Figure 7.10 depicts the possible integration of biobased surfactant synthesis with an oleochemical biorefinery, to utilize oilseed crops for the production of fuels, chemicals, materials, and foods. The major component of surfactants derived from oilseed TAG (or from phospholipids or the co-product soapstock) is the FA group, which serves as surfactants' lipophilic group. In a biorefinery setting, FA groups are mainly encountered in the form of FAME or FAEE due to their employment as biodiesel. Alternatively, FA groups can be reduced into fatty alcohols and fatty amines, to add versatility to the chemical synthesis of surfactants. Hydrophilic groups of surfactants can also be derived from biorefinery-based chemical intermediates, such as glycerol or its oligomers, an underutilized co-product from biodiesel (discussed earlier), sugars (derived from lignocellulosic polysaccharides, such as glucose and xylose), peptides and proteins, and the choline polar groups derived from phospholipids. In addition, glycerol and sugars are readily converted into biorefinery-based platform chemical

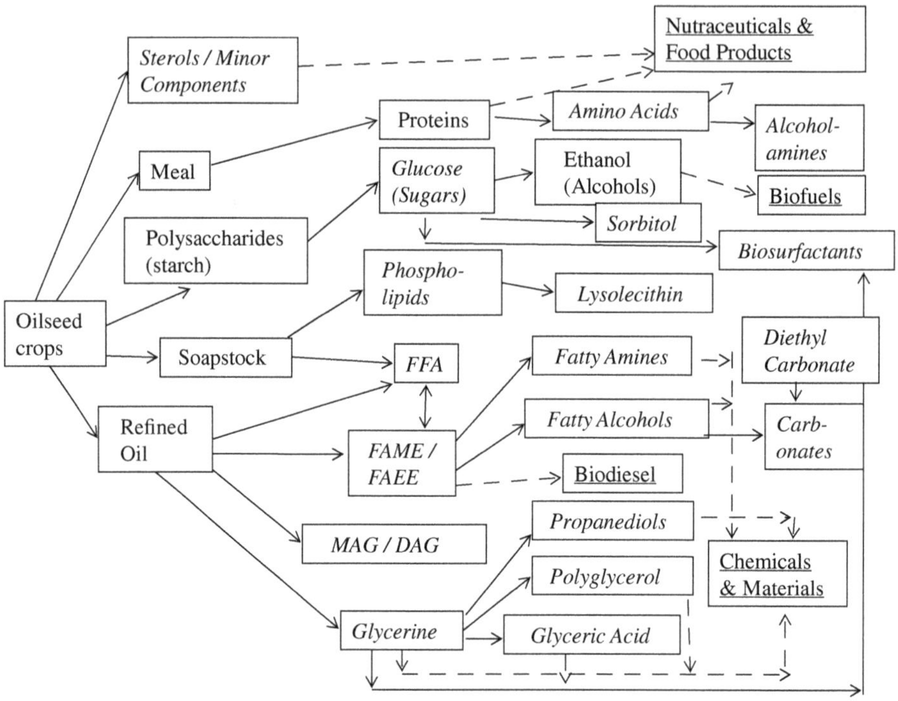

Figure 7.10 Biobased surfactant synthesis within the oleochemical refinery. *Source:* Hayes and Smith (2019). Reproduced with permission from AOCS Press.

intermediates that can be employed in surfactants, such as 1,3- and 1,2-propanediol. Alternatively, oils can serve as co-substrates in the fermentative production of surfactants by microorganisms, referred to as biosurfactants (e.g. glycolipids, such as sophorolipids [Roelants et al. 2019] and rhamnolipids [Suh et al. 2019] and lipoproteins such as surfactin [Geissler et al. 2019]). Fatty acids or alcohols can also be employed as building blocks in the synthesis of partially biobased surfactants, such as FA or fatty alcohol ethoxylates, where the ethoxylate (polyethylene oxide) groups are typically derived from petroleum-based ethylene (although the latter can also be derived from sugar cane and other biobased feedstocks). The European Union has recently adopted a classification system for "biobased" that is based on the content of renewable carbon, with *wholly* (≥95%) and *majority* (50–94%) *biobased* being the desired categories for manufacturers (Tropsch 2017). The surfactants described herein are minimally *majority* biobased.

The ideal fatty acyl architecture for non-food applications of surfactants (e.g. in detergents) consists of the absence of double bonds and a medium chain length. Therefore, high-lauric oils (i.e. enriched in lauric or dodecanoic acid) such as palm kernel and coconut oils are commonly employed. A recently developed competitor to the two oils is algal oil. Palm stearine (Table 7.1) is also a robust feedstock, enriched in palmitic (hexadecenoic) acid, therefore providing greater lipophilicity than the high-lauric oils. Biobased surfactant feedstocks are reviewed elsewhere (Hayes and Smith 2019).

A few prominent oleochemically derived biobased surfactants will now be discussed, and their molecular structures are given in Figure 7.11. Further information is available

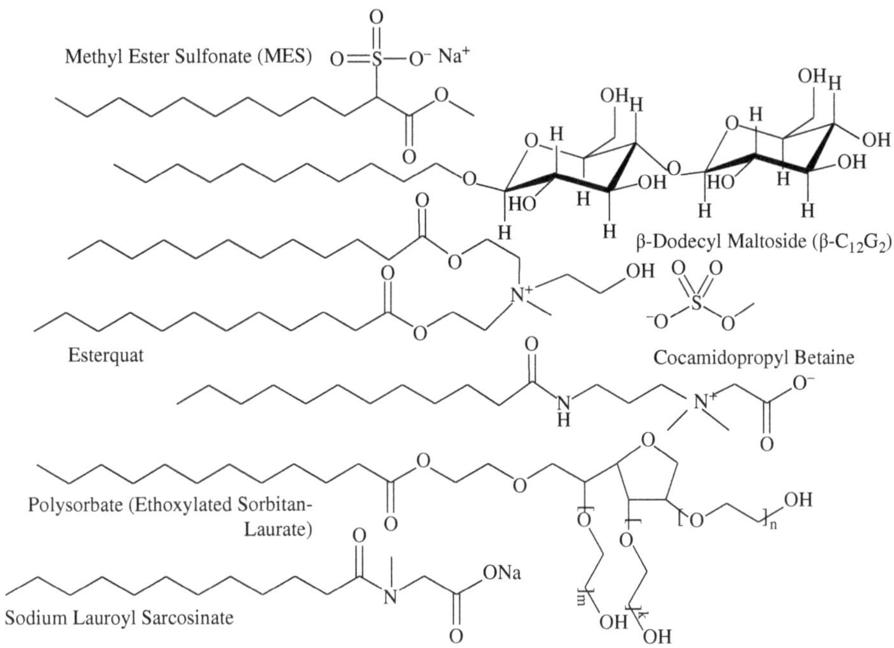

Figure 7.11 Biobased surfactants prepared from seed oils.

through a recently published book (Hayes et al. 2019), with its first chapter serving as an overview. *Anionic surfactants* are the most commonly employed surfactant class for detergents. Biobased lauryl alcohol can be readily incorporated into sodium lauryl (dodecyl) sulfate or sodium lauryl ether sulfate to increase the biobased content. Anionic surfactants receiving great attention, and employed in many Asian detergent products are methyl ester sulfonates (MES), commonly prepared using FAME derived from palm kernel or palm stearate oils (Figure 7.11) (Tobori and Kakui 2019). Biobased *nonionic surfactants* have numerous uses in detergents, personal care products, cosmetics, and pharmaceuticals due to their excellent biocompatibility, biodegradability, and low ecotoxicity. As described above, ethoxylates are very common biobased nonionic surfactants that are readily derived from vegetable oils or biobased fatty alcohols or amines (Smith 2019). Alkyl polyglycosides (APGs) are also abundantly employed, and are readily generated from cellulosic or hemicellulosic polysaccharides and fatty alcohols via acetal or ketal formation, or other synthetic routes (Figure 7.11) (Estrine et al. 2019). Although they are more expensive than ethoxylates, APGs possess much lower temperature-sensitivity and biobased content than ethoxylates. Sugar esters are also abundant surfactants employed in cosmetics and pharmaceutical products (Pyo et al. 2019). They are more biodegradable, but also more labile, than APGs due to the presence of ester bonds, which diminishes their application in detergents and personal care products. Sugar alcohol esters, particularly sorbitan esters and ethoxylated sorbitan esters (the latter often referred to as "polysorbates," Figure 7.11), are commonly employed surfactants in biochemical and pharmaceutical-related applications.

Esterquats (Figure 7.11), the most commonly encountered biobased *cationic surfactants*, are synthesized via transesterification of FAME with triethanolamine (or other alcohol amines) at ~250 °C for a few hours under vacuum pressure, to remove the methanol leaving group, and are then quaternized with methylenesulfate for a few hours at <100 °C (Mishra and Tyagi 2007; Overkempe et al. 2003). They are employed in fabric softeners due to their electrostatic attractive adsorption to negatively-charged fibers, and also in hair care products. Because of their high biodegradability and antimicrobial activity, they are also employed as biocides in pools.

Amphoterics are surfactants possessing both positive and negative charged groups. The most common class of amphoterics are phospholipids and their derivatives, which are obtained as co-products during the degumming of vegetable oils and are used in food and pharmaceutical products (Li et al. 2019). Betaines, amphoterics possessing both quaternary ammonium and carboxylate groups (Figure 7.11), are employed in detergents and personal care products (e.g. shampoos, dishwasher detergents, and liquid soaps) due to their good detergency, ability to foam, pH stability, and biodegradability (Clendennen and Boaz 2019).

Amino acid surfactants comprise a growing market sector (reviewed in Bordes and Holmberg 2015; Pinazo et al. 2016). Amino acids serve as useful and versatile polar groups of surfactants due to their high functionality. Moreover, amino acids can form surfactants via (i) ester and (ii) amide bond formation between their α-COOH group and fatty alcohols and amines, respectively, (iii) amide bonds between their α-amine groups and FA groups, and (iv) N-alkylation of α-amine groups. The latter is typically formed using FA chlorides; but, recently, a less expensive and greener approach has been developed that employs fatty alcohols and utilizes a Shvo catalyst (Yan et al. 2017). Reactions can occur also at ε-amine and carboxylate groups of amino acids, and, depending on the pH and pKa values of functional groups, protective group chemistry may be needed to control product selectivity. Surfactants synthesized using arginine undergo bond formation more readily at their α- rather than their ε-amine group, and the free ε-amine greatly enhance biological activity (Pinazo et al. 2019). N-acylated amino acids, such as N-lauryl sarcosinate (Figure 7.11) are used in hair and oral care products, carpet and hard floor cleaners, and corrosion inhibitors (Hayes and Smith 2019; Husmann 2008).

7.3.4 Lubricants

Lubricants are essential materials for the operation of machinery to reduce friction between moving parts. Seed oils and their derivatives have been employed in lubrication for several decades, but to a much lower extent than petroleum (reviewed in Chan et al. 2018; McNutt and He 2016; Panchal et al. 2017; Syahir et al. 2017). Lubricants represent a large industrial sector: approximately 40 million tons (Chan et al. 2018; Syahir et al. 2017). Oleochemical-based lubricants ("biolubricants") have several inherent advantages, as listed in Table 7.3: high viscosity index (i.e. lower change of viscosity with temperature), flash and boiling point temperatures, corrosion inhibition activity, biodegradability, low sulfur and phosphorous content, and often improved performance compared to mineral oil lubricants (Syahir et al. 2017). However, biolubricants are more expensive and possess poorer oxidative stability and cold temperature performance compared to petroleum-derived lubricants. Degradation caused by oxidation of bio lubricants can lead to a major increase of viscosity, which can hinder performance (Syahir et al. 2017). Biolubricants are employed for niche applications, such as those involving high-performance needs and particularly for machinery that comes in contact with the environment, such as chain saw oils, transformer

Table 7.3 Significant properties of bio-based lubricants for various applications.[a]

Properties	Tested lubricant	Remarks
All biolubricants • High biodegradability • Renewable • Good lubricity • Wide viscosity range • High viscosity index • Low volatility • High thermal stability • Good oxidation stability	VO; natural ester of various VOs	• Environmentally benign • Suitable for variety of applications • Suitable at high temperatures (chemically modified VO)
Engine oil • Low volatile organic compound emission • Good lubricity	Castor, palm, Karanja, and LEAR oils	• Reduced engine emissions (CO, hydrocarbons) • Improved engine performance
Hydraulic oil • Low compressibility • Fast air release rate	HOSO, LEAR oil	• Better pressure transmission • Less vibration and noise
Compressor oil • High thermal stability	ESBO	• Can tolerate the standard compressor discharge temperature (250 °C)
Chainsaw oil • Low volatility	VO	• Fewer harmful mist generation
Metal-Working Fluid • Low volatility • Good antirust capability • Good emulsifiability • Good lubricity	SBO; Karanja, neem, rice bran, coconut, and palm oils	• Fewer harmful mist generation • Longer tool life • Stable emulsions at high temperature
		• Reduced cutting temperature, surface roughness and tool flank wear • Longer tool life
Gear oil • Good lubricity • Higher weld load	Complex esters, PE tetraoleate ester	• Suitable additives
Insulation fluid • High water solubility • High dielectric constant	Corn, cottonseed, and LEAR oils, SBO	• Decreased the effect of humidity on insulation strength • Better insulation properties

Source: Syahir et al. (2017). Reproduced with permission.
[a] Abbreviations: ESBO: epoxidized soybean oil; HOSO: high-oleic sunflower oil; LEAR: low erucic acid rapeseed oil; SBO: soybean oil; VO: vegetable oil.

and marine fluids, asphalt release agents for road construction, and food processing equipment. Approximately 55% of lubricants come in contact with the environment during their service life or end-of-life, and the majority thereof (~95%) are derived from petroleum (Syahir et al. 2017). Several examples of biolubricant uses in lubrication and their inherent advantages and disadvantages are provided in Table 7.3 and reviewed elsewhere (Syahir et al. 2017). Several biolubricants are commercially available (Syahir et al. 2017).

The majority of biolubricant materials are vegetable oils and their simple derivatives, such as polyol esters (particularly trimethylolpropane and pentaerythritol esters), and epoxidized, hydroxylated acetylated derivatives (per Figure 7.3, **I→ II → III**, with subsequent acetylation of the OH groups of **III**) (McNutt and He 2016; Panchal et al. 2017; Syahir et al. 2017). Catalysts employed for the preparation of ester-based biolubricants, including hydrogenation thereof, have been recently reviewed (Hossain et al. 2018; Masudi and Muraza 2018; Panchal et al. 2017). Biolubricant materials are employed primarily as lubricant base oils, which represents 70–95% of most lubricant products (Chan et al. 2018). Soaps (e.g. lithium 12-hydroxystearate or stearate) are commonly used as biobased greases (Panchal et al. 2017).

An important class of lubricant materials in recent years have been estolides, which are oligomers of hydroxy fatty acids (oligoesters) (reviewed in Chen et al. 2020; McNutt and He 2016; Romsdahl et al. 2019). Estolides are readily formed from polycondensation of hydroxy fatty acids such as those described in Figure 7.1. Alternatively, they can be prepared from seed oils via acidic catalysis (e.g. *p*-toluene sulfonic acid or clays) under elevated temperature and pressure. Estolides can be further modified to improve performance via "end capping" of COOH or OH endgroups. Estolides possess high viscosity indices and low pour points. Their main deficiencies are oxidative stability and cost, the former of which can be addressed via additives. The main application areas have been as base oils for many lubricant, engine transmission, gear, compressor, and hydraulic oils, and greases.

7.3.5 Plasticizers

Plasticizers are a lucrative market at 13.2 million tons per year in 2018 (Bocqué et al. 2016) for which oleochemicals are anticipated to play an increasing role. Plasticizers are defined as "substances or materials incorporated into a plastic to increase its flexibility, workability or distensibility" (Wilkes et al. 2005). The list of polymeric materials that employ plasticizers is long, but includes rubber, polyethylene terephthalate (PET), polyvinylchloride (PVC), polyvinylalcohol (PVA), polymethyl methacrylate (PMMA), and biobased polymers such as polylactic acid (PLA) and starch (Bocqué et al. 2016; Wilkes et al. 2005). Governments and the general public have raised increased concerns with commonly used plasticizers, especially phthalates (e.g. diethylhexyl phthalate) due to their ability to migrate through plastics and be leached away, leading to bioacculuation in the environment (Bocqué et al. 2016). Phthalate plasticizers have been associated with ecotoxicity and harm to human health, especially children, leading to the banning of several phthalate plasticizers worldwide (Bocqué et al. 2016). Therefore, replacement of phthalates with biobased alternatives is desirable. The ideal biobased plasticizer will be nontoxic, miscible with the polymer, perform equally as well as conventional plasticizers, will not migrate out of the polymeric material, and will be inexpensive (Bocqué et al. 2016). Among the different plasticizer categories, vegetable oil-based plasticizers are most relevant to external and primary plasticizers, referring to those that are not integrated in the polymeric network when added and those that are soluble in the polymeric matrix, respectively.

The major seed oil-based plasticizers are the oils themselves (e.g. castor oil has served as a plasticizer for several decades), epoxidized oils (e.g. ESBO), dehydroxylated FA esters and TAG (obtained via ring-opening **II → III** in Figure 7.3), and acetylated versions of the latter (reviewed in Bocqué et al. 2016; Hassan et al. 2019). This list mimics that of lubricant materials described earlier. In addition, oils rich in erucic acid, namely high erucic acid rapeseed and crambe oils, are prominent lubricant materials (Piazza and Foglia 2001). Diacetylated MAG have also been employed (Bocqué et al. 2016). Glycerol can be used as a plasticizer for more hydrophilic polymers such as PLA (Bocqué et al. 2016). Estolides and polymerized or vulcanized vegetable oils have also been investigated as potentially valuable biobased plasticizers (Hassan et al. 2019; Stolp et al. 2019). Several vegetable oil-based plasticizers are commercially available (Table 7.4) (Bocqué et al. 2016).

Related to plasticizers are *slip agents*, additives for polymer films that reduce friction, improve mold release, reduce scratching and scuffing and improve surface quality. Fatty amides, particularly erucamide, are prominent slip agents (Peloso et al. 1998). Erucamide is derived from erucic acid via reaction with urea (Khalkar et al. 2013). Erucamide also has potential utility as an anti-friction additive for lubricants (Khalkar et al. 2013).

7.3.6 Cosmetics

The roles of seed oils and their derivatives in cosmetics and personal care products are numerous and complete coverage of this topic is beyond the scope of this chapter. However, a few applications will be briefly discussed. Vegetable oils are commonly used as emollients in cosmetic products (Botaneri 2020) and oleochemically derived surfactants (as discussed earlier) are commonly used as emulsifiers and can also impart antimicrobial activity (e.g. in the treatment of acne). Wax esters of long-chain fatty acids ($\geq C_{16}$) are also frequently employed in cosmetics, particularly for skin care. The ideal wax ester substance is jojoba oil, which is comprised entirely of wax esters. Transesterification of oils (including low-cost oils such as cooking oil or palm plant distillate) is readily achievable (Basri et al. 2013; Isbell et al. 1996; Lara-Hernandez et al. 2017). Castor and jojoba oils are also major components of lipstick (Labmuffin 2020). A novel development is the preparation of biosunscreens, via lipase-catalyzed interesterification (acidolysis) of TAG using monolignol derivatives such as ferulic acid, that serve as natural antioxidants (Laszlo et al. 2011; Yao and Sun 2020).

7.4 Conclusions and Future Prospects

Seed oils and their derivatives have many non-food applications as polymers (employed in paints and coatings, materials, and other application areas), biofuels, surfactants, lubricants, plasticizers, and cosmetics. These applications utilize both surplus vegetable oil feedstocks remaining after food production and industrial

Table 7.4 Industrial bio-plasticizers made from vegetable oils.[a]

Plasticizer trade name	Manufacturer	Feedstock	Biobased content, %	Vegetable oil derivative
Grinsted® SOFT-N-SAFE	Danisco	Castor oil	80	Acetylated TAG
Kalflex® 14A	Varterco	SBO	98	ESBO
Nexo® E01	Nexoleum	SBO	95	Methyl epoxy soyate
Resiflex® K50	Resypar	SBO	100	Amyl epoxy soyate
PLS Green® 9	Petrom	SBO	66	Nonyl epoxy soyate
DCS®	Jayant-Agro-Organics	Castor oil	100	Di-caprylsebacate
Radia®7295	Novance/Oleon	LEAR oil	78	Acetylated ester
Lankroflex® ED6	Akcros Chemicals	SBO	68	Octyl epoxy stearate
–	Gerflor	Tall and sunflower oils	–	Tall-oil fatty esters
Virkoflex®7010	Arkema	SBO	–	ESBO fatty acid esters
Hallgreen®	Hallstar	Linseed, castor, and SB oils	38–100	Fatty acid ester

Source: Bocqué et al. (2016). Reproduced with permission.
[a] Abbreviations: ESBO: epoxidized soybean oil; LEAR: low-erucic acid rapeseed; SBO: soybean oil.

oilseed crops such as jatropha and castor. The demand for oleochemicals will remain high due to increased interest of consumers in biobased products as sustainable alternatives to conventional fossil fuel-derived products, and will increase in prominence in the future as oil supplies are anticipated to be depleted several decades into the future. Innovations in plant genomics, heterogeneous catalysis and biocatalysis, biomolecular engineering, and artificial intelligence are anticipated to increase the economic viability of oleochemical products and widen the variety of products that can be derived.

References

Abdul Kapor, N.Z., Maniam, G.P., Ab Rahim, M.H. et al. (2017). Palm fatty acid distillate as a potential source for biodiesel production-a review. Journal of Cleaner Production 143: 1–9.

Ahmad, I. and Lie Ken Jie, M.S.F. (2008). Oleochemicals from isoricinoleic acid (*Wrightia tinctoria* seed oil). Industrial and Engineering Chemistry Research 47 (6): 2091–2095.

Alagi, P. and Hong, S.C. (2015). Vegetable oil-based polyols for sustainable polyurethanes. Macromolecular Research 23 (12): 1079–1086.

Anuar, M.R. and Abdullah, A.Z. (2016). Challenges in biodiesel industry with regards to feedstock, environmental, social and sustainability issues: a critical review. Renewable and Sustainable Energy Reviews 58: 208–223.

ASTM International (2020). *ASTM D6751: Standard Specification for Biodiesel Fuel Blend Stock (B100) for Middle Distillate Fuels*. West Conshohocken, PA: ASTM, International.

Barros, T.F.S., Arriel, N.H.C., Queiroz, M.F. et al. (2015). Fatty acid profiles of species of *Jatropha curcas* L., *Jatropha mollissima* (Pohl) Baill. and *Jatropha gossypiifolia* L. Industrial Crops and Products 73: 106–108.

Basri, M., Abd Rahman, R.N.Z.R., and Salleh, A.B. (2013). Specialty oleochemicals from palm oil via enzymatic syntheses. Journal of Oil Palm Research 25 (April): 22–35.

Bauer, S.T. (1946). The preparation of fatty acid chlorides. Oil & Soap 23 (1): 1–5.

Biermann, U., Metzger, J.O., and Meier, M.A.R. (2010). Acyclic triene metathesis oligo- and polymerization of high oleic sun flower oil. Macromolecular Chemistry and Physics 211 (8): 854–862.

Biermann, U., Bornscheuer, U., Meier, M.A.R. et al. (2011). Oils and fats as renewable raw materials in chemistry. Angewandte Chemie International Edition 50 (17): 3854–3871.

Bocqué, M., Voirin, C., Lapinte, V. et al. (2016). Petro-based and bio-based plasticizers: chemical structures to plasticizing properties. Journal of Polymer Science Part A: Polymer Chemistry. 54 (1): 11–33.

Bordes, R. and Holmberg, K. (2015). Amino acid-based surfactants – do they deserve more attention? Advances in Colloid and Interface Science 222: 79–91.

Botaneri The best emollients to use in natural skincare (2020). https://botaneri.com/the-best-emollients-to-use-in-natural-skincare (accessed May 15, 2020).

Busch, R., Kneuper, H.J., Weber, T. et al. Inventors; Google patents, assignee. Method for producing acid chlorides. US patent US6770783 B1. (2004).

Carre, C., Ecochard, Y., Caillol, S. et al. (2019). From the synthesis of biobased cyclic carbonate to polyhydroxyurethanes: a promising route towards renewable non-isocyanate polyurethanes. ChemSusChem 12 (15): 3410–3430.

Chan, C.-H., Tang, S.W., Mohd, N.K. et al. (2018). Tribological behavior of biolubricant base stocks and additives. Renewable and Sustainable Energy Reviews 93: 145–157.

Chen, G. (2016). Lesquerella (*Physaria* spp.). In: *Industrial Oil Crops* (eds. T.A. McKeon, D.G. Hayes, D.F. Hildebrand, et al.), 313–316. Amsterdam: Elsevier.

Chen, G. (2016). Pennycress (*Thlapsi* spp.). In: *Industrial Oilseeds* (eds. T.A. McKeon, D.G. Hayes, D.F. Hildebrand, et al.), 75–112. Amsterdam: Elsevier.

Chen, Y.-H., Chiang, T.-H., and Chen, J.-H. (2013). Properties of soapnut (Sapindus mukorossi) oil biodiesel and its blends with diesel. Biomass and Bioenergy 52: 15–21.

Chen, Y., Biresaw, G., Cermak, S.C. et al. (2020). Fatty acid estolides: a review. Journal of the American Oil Chemists' Society 97 (3): 231–241.

Christopher, L.P., Hemanathan, K., and Zambare, V.P. (2014). Enzymatic biodiesel: challenges and opportunities. Applied Energy 119: 497–520.

Clendennen, S.K. and Boaz, N.W. (2019). Betaine amphoteric surfactants — synthesis, properties, and applications. In: *Biobased Surfactants*, 2e (eds. D.G. Hayes, D.K.Y. Solaiman and R.D. Ashby), 447–469. Urbana, IL: AOCS Press.

Cruz, M., Pinho, S.C., Mota, R. et al. (2018). Enzymatic esterification of acid oil from soapstocks obtained in vegetable oil refining: effect of enzyme concentration. Renewable Energy 124: 165–171.

Cuperus, F.P., Boswinkel, G., and Derksen, J.T.P. (1996). The processing of new oilseed crops – an economic evaluation. Journal of the American Oil Chemists' Society 73 (12): 1635–1640.

Diaz-Perez, M.A. and Serrano-Ruiz, J.C. (2020). Catalytic production of jet fuels from biomass. Molecules 25 (4): 802.

Dong, Y., Matson, J.B., and Edgar, K.J. (2017). Olefin cross-metathesis in polymer and polysaccharide chemistry: a review. Biomacromolecules 18 (6): 1661–1676.

Egan, R.R. (1968). The preparation and properties of amines and cationic surfactants from fatty acids. Journal of the American Oil Chemists' Society 45 (7): 481–486.

Erhan, S.M. and Kleiman, R. (1990a). Meadowfoam oil factice and its performance in natural rubber mixes. Rubber World 203 (1): 33–36.

Erhan, S.M. and Kleiman, R. (1990b). Vulcanized meadowfoam oil. Journal of the American Oil Chemists' Society 67 (10): 670–674.

Erhan, S.M. and Kleiman, R. (1993). Factice from oil mixtures. Journal of the American Oil Chemists' Society 70 (3): 309–311.

Estevez, R., Aguado-Deblas, L., Bautista, F.M. et al. (2019). Biodiesel at the crossroads: a critical review. Catalysts 9 (12): 1033.

Estrine, B., Marinkovic, S., and Jérome, F. (2019). Synthesis of alkyl polyglycosides from glucose and xylose for biobased surfactants: synthesis, properties, and applications. In: *Biobased Surfactants*, 2e (eds. D.G. Hayes, D.K.Y. Solaiman and R.D. Ashby), 365–385. Urbana, IL: AOCS Press.

Fameau, A.-L. and Rogers, M.A. (2020). The curious case of 12-hydroxystearic acid – the Dr. Jekyll & Mr. Hyde of molecular gelators. Current Opinion in Colloid & Interface Science 45: 68–82.

Gandini, A., Lacerda, T.M., Carvalho, A.J.F. et al. (2016). Progress of polymers from renewable resources: furans, vegetable oils, and polysaccharides. Chemical Reviews (Washington, DC) 116 (3): 1637–1669.

Gebremariam, S.N. and Marchetti, J.M. (2018). Economics of biodiesel production: review. Energy Conversion and Management 168: 74–84.

Geissler, M., Heravi, K.M., Henkel, M. et al. (2019). Lipopeptide biosurfactants from bacillus species. In: *Biobased Surfactants*, 2e (eds. D.G. Hayes, D.K.Y. Solaiman and R.D. Ashby), 205–240. Urbana, IL: AOCS Press.

Giraldo, L., Camargo, G., Tirano, J. et al. (2010). Synthesis of fatty alcohols from oil palm using a catalyst of Ni-Cu supported onto zeolite. E-Journal of Chemistry 7 (4): 1138–1147.

Grand View Research Biodiesel market report. (2017). https://www.grandviewresearch.com/industry-analysis/biodiesel-market (accessed May 7, 2020).

Grinberg, S., Kolot, V., and Mills, D. (1994). New chemical derivatives based on Vernonia galamensis oil. Industrial Crops and Products 3 (1,2): 113–119.

Hassan, A.A., Abbas, A., Rasheed, T. et al. (2019). Development, influencing parameters and interactions of bioplasticizers: an environmentally friendlier alternative to petro industry-based sources. Science of the Total Environment 682: 394–404.

Hayes, D.G. and Dumont, M.-J. (2016). Polymeric products derived from industrial oils for paints, coatings, and other applications. In: *Industrial Oil Crops* (eds. T.A. McKeon, D.G. Hayes, D.F. Hildebrand, et al.), 43–73. Champaign, IL: AOCS Press.

Hayes, D.G. and Hildebrand, D.F. (2016). Perilla (*Perilla frutescens*). In: *Industrial Oilseeds* (eds. T.A. McKeon, D.G. Hayes, D.F. Hildebrand, et al.), 328–332. Amsterdam: Elsevier.

Hayes, D.G. and Smith, G.A. (2019). Biobased surfactants: overview and industrial state of the art. In: *Biobased Surfactants*, 2e (eds. D.G. Hayes, D.K.Y. Solaiman and R.D. Ashby), 3–38. Urbana, IL: AOCS Press.

Hayes, D.G., Solaiman, D.K.Y., and Ashby, R.D. (2019). *Biobased Surfactants: Synthesis, Properties, and Applications*, 2e. Champaign, IL: AOCS Press.

Henna, P.H., Kessler, M.R., and Larock, R.C. (2008). Fabrication and properties of vegetable-oil-based glass fiber composites by ring-opening metathesis polymerization. Macromolecular Materials and Engineering 293 (12): 979–990.

Hossain, M.A., Mohamed Iqbal, M.A., Julkapli, N.M. et al. (2018). Development of catalyst complexes for upgrading biomass into ester-based biolubricants for automotive applications: a review. RSC Advances 8 (10): 5559–5577.

Huang, K., Xia, J., Yang, X. et al. (2010). Properties and curing kinetics of C_21-based reactive polyamides as epoxy-curing agents derived from tung oil. Polymer Journal (Tokyo, Japan). 42 (1): 51–57.

Husmann, M. (2008). Surfactants derived from naturally-occurring amino acids. SOFW Journal 134 (3): 34–36, 38-40.

Isbell, T.A., Carlson, K.D., Abbott, T.P. et al. (1996). Isolation and characterization of wax esters in meadowfoam oil. Industrial Crops and Products 5 (3): 239–243.

Khalkar, S., Bhowmick, D., and Pratap, A. (2013). Synthesis and effect of fatty acid amides as friction modifiers in petroleum base stock. Journal of Oleo Science 62 (11): 901–904.

Khan, S., Kay Lup, A.N., Qureshi, K.M. et al. (2019). A review on deoxygenation of triglycerides for jet fuel range hydrocarbons. Journal of Analytical and Applied Pyrolysis 140: 1–24.

Knothe, G., Krahl, J., and van Gerpen, J. (2010). *The Biodiesel Handbook*, 2e. Champaign, IL: AOCS Press.

Koppireddi, S., Seo, J.-H., Jeon, E.-Y. et al. (2016). Combined biocatalytic and chemical transformations of oleic acid to ω-Hydroxynonanoic acid and α, ω-Nonanedioic acid. Advanced Synthesis and Catalysis 358 (19): 3084–3092.

Król, B. and Paszko, T. (2017). Harvest date as a factor affecting crop yield, oil content and fatty acid composition of the seeds of calendula (*Calendula officinalis* L.) cultivars. Industrial Crops and Products 97: 242–251.

Kunduru, K.R., Basu, A., Haim Zada, M. et al. (2015). Castor oil-based biodegradable polyesters. Biomacromolecules 16 (9): 2572–2587.

Labmuffin What's in a Lipstick? (2020). https://labmuffin.com/whats-in-a-lipstick (accessed May 15, 2020).

Lara-Hernandez, G., Flores-Cuautle, J.J.A., Hernandez-Aguilar, C. et al. (2017). Thermal properties of jojoba oil between 20°C and 45°C. International Journal of Thermophysics 38 (8): 1–9.

Laszlo, J.A., Yu, Y., Lutz, S. et al. (2011). Glycerol acyl-transfer kinetics of a circular permutated *Candida antarctica* lipase B. Journal of Molecular Catalysis B: Enzymatic 72 (3–4): 175–180.

Li, J., He, Y., Anankanbil, S. et al. (2019). Phospholipid-based surfactants. In: *Biobased Surfactants*, 2e (eds. D.G. Hayes, D.K.Y. Solaiman and R.D. Ashby), 243–286. Urbana, IL: AOCS Press.

Llevot, A. (2017). Sustainable synthetic approaches for the preparation of plant oil-based thermosets. *Journal of the American Oil Chemists' Society* 94 (2): 169–186.

Lligadas, G., Ronda, J.C., Galia, M. et al. (2013). Monomers and polymers from plant oils via click chemistry reactions. Journal of Polymer Science, Part A: Polymer Chemistry 51 (10): 2111–2124.

Lu, Y. and Larock, R.C. (2009). Novel polymeric materials from vegetable oils and vinyl monomers: preparation, properties, and applications. ChemSusChem 2 (2): 136–147.

Markets and Markets Market report on biobased surfactants. Pune, India. (2012).

Masudi, A. and Muraza, O. (2018). Vegetable oil to biolubricants: review on advanced porous catalysts. Energy & Fuels 32 (10): 10295–10310.

McKeon, T.A. (2016). Castor (*Ricinus communis* L.). In: *Industrial Oilseeds* (eds. T.A. McKeon, D.G. Hayes, D.F. Hildebrand, et al.), 75–112. Amsterdam: Elsevier.

McKeon, T.A. (2016). Epilogue. In: *Industrial Oilseeds* (eds. T.A. McKeon, D.G. Hayes, D.F. Hildebrand, et al.), 337–341. Amsterdam: Elsevier.

McKeon, T.A. (2016). Meadowfoam (*Limnanthes alba*). In: *Industrial Oilseeds* (eds. T.A. McKeon, D.G. Hayes, D.F. Hildebrand, et al.), 317–323. Amsterdam: Elsevier.

McNutt, J. and He, Q. (2016). Development of biolubricants from vegetable oils via chemical modification. Journal of Industrial and Engineering Chemistry (Amsterdam, Netherlands) 36: 1–12.

McVetty, P.B.E., Mietkiewska, E., Omonov, T. et al. (2016). *Brassica* spp. Oils. In: *Industrial Oilseeds* (eds. T.A. McKeon, D.G. Hayes, D.F. Hildebrand, et al.), 113–156. Amsterdam: Elsevier.

Melo, M.A.M.F., de Melo, M.A.R., Pontes, A.S.G.C. et al. (2014). Non-conventional oils for biodiesel production: a study of thermal and oxidative stability. Journal of Thermal Analysis and Calorimetry 117 (2): 845–849.

Mishra, S. and Tyagi, V.K. (2007). Ester quats: the novel class of cationic fabric softeners. Journal of Oleo Science 56 (6): 269–276.
Monteiro, M.R., Kugelmeier, C.L., Pinheiro, R.S. et al. (2018). Glycerol from biodiesel production: technological paths for sustainability. Renewable and Sustainable Energy Reviews 88: 109–122.
Montero de Espinosa, L. and Meier, M.A.R. (2011). Plant oils: the perfect renewable resource for polymer science? European Polymer Journal 47 (5): 837–852.
Montero de Espinosa, L. and Meier, M.A.R. (2012). Olefin metathesis of renewable platform chemicals. Topics in Organometallic Chemistry (Organometallics and Renewables) 39: 1–44.
Moreno, M., Gomez, M.V., Cebrian, C. et al. (2012). Sustainable and efficient methodology for CLA synthesis and identification. Green Chemistry 14 (9): 2584–2594.
Mubofu, E.B. (2016). Castor oil as a potential renewable resource for the production of functional materials. Sustainable Chemical Processes 4: 11/1–11/2.
Murphy, E.J. (2016). Camelina (*Camelina sativa*). In: *Industrial Oilseeds* (eds. T.A. McKeon, D.G. Hayes, D.F. Hildebrand, et al.), 207–242. Amsterdam: Elsevier.
Mutlu, H. and Meier, M.A.R. (2010). Castor oil as a renewable resource for the chemical industry. European Journal of Lipid Science and Technology 112 (1): 10–30.
Narine, S.S., Yue, J., and Kong, X.H. (2007). Production of polyols from canola oil and their chemical identification and physical properties. Journal of American Oil Chemistry Society 84 (2): 173–179.
Nautiyal, O.H. (2018). Castor oil and its derivatives with market growth, commercial perspective: review. Research & Reviews: Journal of Chemistry 7 (1): 1–13.
Navarro-Herrera, D., Aranaz, P., Eder-Azanza, L. et al. (2018). Borago officinalis seed oil (BSO), a natural source of omega-6 fatty acids, attenuates fat accumulation by activating peroxisomal beta-oxidation both in C. elegans and in diet-induced obese rats. Food & Function 9 (8): 4340–4351.
Nekhavhambe, E., Mukaya, H.E., and Nkazi, D.B. (2019). Development of castor oil–based polymers: a review. Journal of Advanced Manufacturing and Processing 1 (4): e10030.
Nielsen, P.M., Rancke-Madsen, A., Holm, H.C. et al. (2016). Production of biodiesel using liquid lipase formulations. Journal of the American Oil Chemists' Society 93 (7): 905–910.
Overkempe, C., Annerling, A., Van Ginkel, C.G. et al. (2003). Esterquats. In: *Novel Surfactants: Preparation, Application, and Biodegradability*, Surfactant Science Series, 2, Revised and Expanded, vol. 114 (ed. K. Holmberg), 347–384. New York: Marcel Dekker.
Panchal, T.M., Patel, A., Chauhan, D.D. et al. (2017). A methodological review on biolubricants from vegetable oil based resources. Renewable and Sustainable Energy Reviews 70: 65–70.
Peloso, C.W., O'Connor, M.J., Bigger, S.W. et al. (1998). Characterising the degradation of the polymer slip additive erucamide in the presence of inorganic antiblock agents. Polymer Degradation and Stability 62 (2): 285–290.
Piazza, G.J. and Foglia, T.A. (2001). Rapeseed oil for oleochemical usage. European Journal of Lipid Science and Technology 103 (7): 450–454.
Pinazo, A., Manresa, M.A., Marques, A.M. et al. (2016). Amino acid-based surfactants: new antimicrobial agents. Advances in Colloid and Interface Science 228: 17–39.
Pinazo, A., Pérez, L., del Carmen Morán, M. et al. (2019). Arginine-based surfactants: synthesis, aggregation properties, and applications. In: *Biobased Surfactants*, 2e (eds. D.G. Hayes, D.K.Y. Solaiman and R.D. Ashby), 413–445. Urbana, IL: AOCS Press.
Pyo, S.-H., Chen, J., Ye, R. et al. (2019). Sugar esters. In: *Biobased Surfactants*, 2e (eds. D.G. Hayes, D.K.Y. Solaiman and R.D. Ashby), 325–363. Urbana, IL: AOCS Press.

Resetco, C., Hendriks, B., Badi, N. et al. (2017). Thiol-ene chemistry for polymer coatings and surface modification – building in sustainability and performance. Materials Horizons 4 (6): 1041–1053.

Rezania, S., Oryani, B., Park, J. et al. (2019). Review on transesterification of non-edible sources for biodiesel production with a focus on economic aspects, fuel properties and by-product applications. Energy Conversion and Management 201: 112155.

Roelants, S., Solaiman, D.K.Y., Ashby, R.D. et al. (2019). Production and applications of sophorolipids☆. In: *Biobased Surfactants*, 2e (eds. D.G. Hayes, D.K.Y. Solaiman and R.D. Ashby), 65–119. Urbana, IL: AOCS Press.

Romsdahl, T., Shirani, A., Minto, R.E. et al. (2019). Nature-guided synthesis of advanced bio-lubricants. Scientific Reports 9 (1): 11711.

Rybak, A., Fokou, P.A., and Meier, M.A.R. (2008). Metathesis as a versatile tool in oleochemistry. European Journal of Lipid Science and Technology 110 (9): 797–804.

Sawpan, M.A. (2018). Polyurethanes from vegetable oils and applications: a review. Journal of Polymer Research 25 (8): 1–15.

Serson, W., Al-Amery, M., Patel, S. et al. (2016). Castor (*Ricinus communis* L.). In: *Industrial Oilseeds* (eds. T.A. McKeon, D.G. Hayes, D.F. Hildebrand, et al.), 278–287. Amsterdam: Elsevier.

Shan, R., Lu, L., Shi, Y. et al. (2018). Catalysts from renewable resources for biodiesel production. Energy Conversion and Management 178: 277–289.

Shockey, J., Rinehart, T., Chen, U. et al. (2016). Tung (*Vernicia fordii* and *Vernicia montana*). In: *Industrial Oil Crops* (eds. T.A. McKeon, D.G. Hayes, D.F. Hildebrand, et al.), 243–274. Amsterdam: Elsevier.

Singh, A. and Gaurav, K. (2018). Advancement in catalysts for transesterification in the production of biodiesel: a review. Journal of Biochemical Technology 7 (3): 1148–1158.

Singh, D., Sharma, D., Soni, S.L. et al. (2019). Chemical compositions, properties, and standards for different generation biodiesels: a review. Fuel 253: 60–71.

Singh, D., Sharma, D., Soni, S.L. et al. (2020). Review on feedstocks, production processes, and yield for different generations of biodiesel. Fuel 262: 116553.

Smith, G.A. (2019). Fatty acid, methyl ester, and vegetable oil ethoxylates. In: *Biobased Surfactants*, 2e (eds. D.G. Hayes, D.K.Y. Solaiman and R.D. Ashby), 287–301. Urbana. IL: AOCS Press.

Spitz, L. (ed.) (2009). The history of soaps and detergents. In: *Soap Manufacturing Technology*, 1–4. Urbana, IL: AOCS Press.

Stolp, L.J., Gronlund, P.J., and Kodali, D.R. (2019). Soybean oil fatty acid Ester Estolides as potential plasticizers. Journal of the American Oil Chemists' Society 96 (6): 727–738.

Suh, S.-J., Invally, K., and Ju, L.-K. (2019). Rhamnolipids: pathways, productivities, and potential. In: *Biobased Surfactants*, 2e (eds. D.G. Hayes, D.K.Y. Solaiman and R.D. Ashby), 169–203. Urbana. IL: AOCS Press.

Suryawanshi, Y., Sanap, P., and Wani, V. (2019). Advances in the synthesis of non-isocyanate polyurethanes. Polymer Bulletin (Heidelberg, Germany) 76 (6): 3233–3246.

Syahir, A.Z., Zulkifli, N.W.M., Masjuki, H.H. et al. (2017). A review on bio-based lubricants and their applications. Journal of Cleaner Production 168: 997–1016.

Sylwia, P. and Józef, K. (2016). Synthesis and bioactivity of (R)-Ricinoleic acid derivatives: a review. Current Medicinal Chemistry 23 (35): 4037–4056.

Tan, S.G. and Chow, W.S. (2010). Biobased epoxidized vegetable oils and its greener epoxy blends: a review. Polymer-Plastics Technology and Engineering 49 (15): 1581–1590.

Tobori, N. and Kakui, T. (2019). Methyl ester sulfonate. In: *Biobased Surfactants*, 2e (eds. D.G. Hayes, D.K.Y. Solaiman and R.D. Ashby), 303–324. Urbana. IL: AOCS Press.

Top, A.G.M. (2010). Production and utilization of palm fatty acid distillate (PFAD). Lipid Technology 22 (1): 11–13.

Tropsch, J.G. (2017). A journey to standardization of bio-based surfactants in Europe. Inform 28 (6): 20–22.

Tsoutsos, T.D., Tournaki, S., Paraíba, O. et al. (2016). The used cooking oil-to-biodiesel chain in Europe: assessment of best practices and environmental performance. Renewable and Sustainable Energy Reviews 54: 74–83.

USDA Oilseeds: world market and trade. (2019). https://downloads.usda.library.cornell.edu/usda-esmis/files/tx31qh68h/7d2797406/cj82kp34g/oilseeds.pdf (accessed December 6, 2019).

Van der Steen, M. and Stevens Christian, V. (2009). Undecylenic acid: a valuable and physiologically active renewable building block from castor oil. ChemSusChem 2 (8): 692–713.

Veldsink, J.W., Bouma, M.J., Schöön, N.H. et al. (1997). Heterogeneous hydrogenation of vegetable oils: a literature review. Catalysis Reviews 39 (3): 253–318.

Wang, Z., Yuan, L., and Tang, C. (2017). Sustainable elastomers from renewable biomass. Accounts of Chemical Research 50 (7): 1762–1773.

Wei, H., Liu, W., Chen, X. et al. (2019). Renewable bio-jet fuel production for aviation: a review. Fuel 254: 115599.

Wilkes, C., Summers, J., and Daniels, C. (2005). *PVC Handbook*. Cincinnati, OH: Hanser.

Winkler, M. and Meier, M.A.R. (2014). Highly efficient oxyfunctionalization of unsaturated fatty acid esters: an attractive route for the synthesis of polyamides from renewable resources. Green Chemistry 16 (4): 1784–1788.

Wood, L. Global polyurethane market (2019 to 2023) – identify growth segments for investment. (2019). https://www.businesswire.com/news/home/20200320005325/en/Global-Polyurethane-Market-2019-2023---Identify (accessed May 12, 2020).

World Bioenergy Association (2019). Global bioenergy statistics 2019. https://worldbioenergy.org/uploads/191129%20WBA%20GBS%202019_LQ.pdf (accessed May 7, 2020).

Xia, Y., Quirino, R.L., and Larock, R.C. (2013). Bio-based thermosetting polymers from vegetable oils. Journal of Renewable Materials 1 (1): 3–27.

Yan, T., Feringa, B.L., and Barta, K. (2017). Direct N-alkylation of unprotected amino acids with alcohols. Science Advances 3 (12): eaao6494. https://doi.org/10.1126/sciadv.aao6494.

Yao, N. and Sun, S. (2020). Hydrophilic glyceryl ferulates preparation catalyzed by free lipase B from *Candida antartica*. Journal of Oleo Science 69 (1): 43–53.

Yelchuri, V., Srikanth, K., Prasad, R.B.N. et al. (2019). Olefin metathesis of fatty acids and vegetable oils. Journal of Chemical Sciences (Berlin, Germany) 131 (5): 1–16.

Zhang, C., Garrison, T.F., Madbouly, S.A. et al. (2017). Recent advances in vegetable oil-based polymers and their composites. Progress in Polymer Science 71: 91–143.

Zhao, B., Gong, H., Li, H. et al. (2019). Fatty acid, triacylglycerol and unsaponifiable matters profiles and physicochemical properties of Chinese evening primrose oil. Journal of Oleo Science 68 (8): 719–728.

Zhu, L.H. (2016). Crambe (*Crambe abyssinica*). In: *Industrial Oilseeds* (eds. T.A. McKeon, D.G. Hayes, D.F. Hildebrand, et al.), 195–206. Amsterdam: Elsevier.

Zia, K.M., Noreen, A., Zuber, M. et al. (2016). Recent developments and future prospects on bio-based polyesters derived from renewable resources: a review. International Journal of Biological Macromolecules 82: 1028–1040.

Zion Market Research Biofuels market size will reach USD 218.7 billion by 2022. Sarasota, FL: Zion Market Research. (2018).

8
Occurrence and Determination of Contaminants in Edible Oils and Oilseeds

José L. Hidalgo-Ruiz, Roberto Romero-González, José Luis Martínez-Vidal and Antonia Garrido-Frenich
Department of Chemistry and Physics, Analytical Chemistry Area, Research Centre for Mediterranean Intensive Agrosystems and Agri-Food Biotechnology (CIAIMBITAL), Agrifood Campus of International Excellence ceiA3, University of Almería, Almería, Spain

8.1 Introduction

In 2018, the Rapid Alert System for Food and Feed (RASFF) has received a total of 3699 original notifications about food safety issues, and 1118 have been classified as an alert. This number of original alerts has been increasing since 2012, when 522 of them were received by the RASFF (Rapid Alert System for Food and Feed 2019). This means that in only seven years, the number of original alerts has increased more than seven fold, which gives us a perspective about the increasing concern about contaminants in food. Oils and seeds contribute significantly to these 3699 notifications: 25 were related to oils, while 667 were made for nuts and seeds, which comprise almost 20% of the total number of notifications (Figure 8.1) (Rapid Alert System for Food and Feed 2019).

The contamination of crops, and hence, their harvest, can come from several sources, either natural or anthropogenic. Natural sources are those that are not added willfully, as those substances that appear in production, storage, or transportation processes and these include mycotoxins. On the other hand, there are other contaminants that come from human activities. Within this group there are compounds that are intentionally added to the crops to protect them from infections, such as pesticides used to increase production yields, which can be detected in the samples as residues. There are also other substances that come from food processing processes such as materials that are in contact with food or environmental and/or ubiquitous contaminants, such as polycyclic aromatic hydrocarbons (PAHs), 3-monochloropropane-1,2-diol (3-MCPDs), mineral oils, or phthalates, among others.

Oil and Oilseed Processing: Opportunities and Challenges, First Edition.
Edited by Tomás Lafarga, Gloria Bobo, and Ingrid Aguiló-Aguayo.

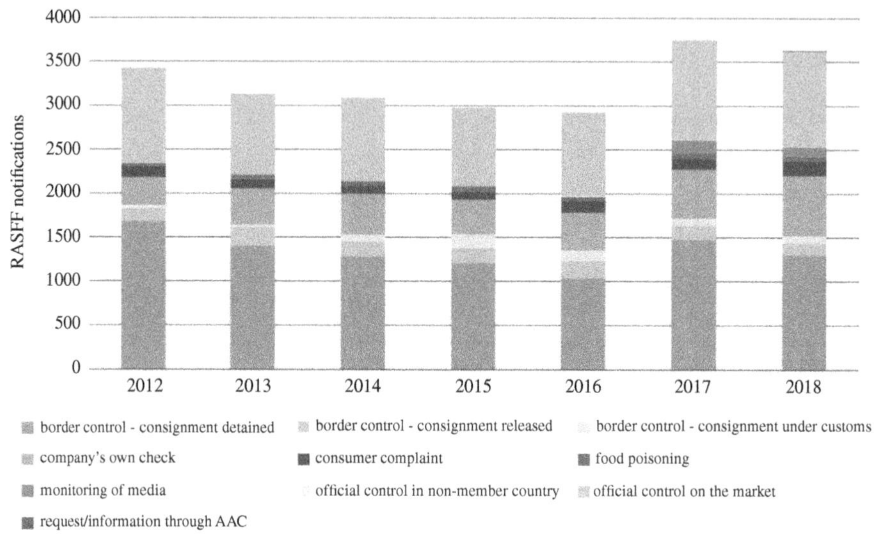

Figure 8.1 Evolution of total RASFF notifications between 2012 and 2018.

The appearance of mycotoxins in crops can happen either before or after harvest. Crops in general, including oils and seeds in this case, can be affected by mycotoxins produced by *Aspergillus* species, such as aflatoxins or *Fusarium* species, such as zearalenone and its derivatives (World Health Organization, and International Agency for Research on Cancer 2010). Oils and seeds can be easily contaminated with mycotoxins due to environmental conditions, such as high temperatures, relative humidity, and rainfall, as these favor fungal proliferation and as a direct consequence, mycotoxin production (Hidalgo-Ruiz et al. 2019a, b).

PAHs can be included in both natural and anthropogenic sources. They can appear during the degradation of vegetal matter carried out by plants and bacteria or formed when organic substances are burnt or exposed to high temperatures under low or no oxygen conditions, which occur in the treatment processes that oils are submitted to, such as refining (Abdel-Shafy and Mansour 2016).

Contaminants can also be produced during food processing similar to those developed during refining, i.e. 3-MCPDs and its glycidyl esters, and PAHs. 3-MCPD is formed in the reaction of hydrochloric acid with triacyclglycerols, phospholipids, and glycerol in the refining process in the case of edible oils, but also during thermal processes commonly used during the production of bakery products, malt-derived products, cooked/cured fish, or meat. In the latter, they are formed when lipids react with sodium chloride, which is naturally present or added to foods (Baer et al. 2009). Moreover, other food- processing steps like extractions or even packing can contaminate the product with mineral oils through leaks of lubricating oil or the direct contact of the product with inks, plastics, or paperboard (Brühl 2016). Also, mineral oil hydrocarbons (MOHs) can appear in edible oils when their seeds have been submitted to hard extraction processes such as centrifugation or solvent extraction. Indeed, the content of these compounds in oils, which are submitted to these processes like pomace oils, is considerably high (Gómez-Coca et al. 2016a). Something similar happens with the phthalates present in plastic materials.

They are fat-like and therefore they are easily released from plastic containers and commonly found in fat-containing foods as phthalates are not covalently bonded to the plastic (Rudel and Perovich 2009). Furthermore, oils and oilseeds are also commonly contaminated with pesticides due to the use of these compounds to increase the yields, or during the postharvest processes to maintain quality if the product is going to be stored for a long period (Amvrazi 2011).

Contamination often generates important economic losses due to the risk to the population and therefore, the profit of the product may be reduced. In order to control chemical contaminants present in the final commercial products, governments and international organizations have established maximum residue limits (MRLs) at toxicologically acceptable levels for all the contaminants present in oils or seeds as mycotoxins, PAHs, 3-MCPDs, glycidyl esters, MOHs, phthalates, or pesticides. Among these organizations, the European Union (EU), the Food and Drug Administration (FDA), Ministry of Health of the People's Republic of China, the CODEX Alimentarius Commission of the Food and Agriculture Organization of the United States (FAO) and the World Health Organization (WHO) have set different regulations related to the presence of these substances in oily matrices (Ma et al. 2016). The analysis and monitoring of chemical contaminants in oils and seeds is a challenge because of the complexity of the matrices. Lipid matrices usually require a deeper cleaning because of the high number of interferences that may be coextracted. Thus, many analytical methods employ clean-up procedures. Also, several analytical methods have been developed for the determination of all the contaminants indicated earlier, being chromatographic methods the most widely used.

In this chapter, chemical contaminants such as mycotoxins, PAHs, 3-MCPDs, mineral oils, phthalates, and pesticides present in edible oils and oilseeds will be studied. Furthermore, extraction, analytical techniques, and occurrence of them from 2010 will be reviewed in the target matrices.

8.2 Mycotoxins

Mycotoxins are secondary metabolites produced by certain species of fungi during their digestion processes (Adeyeye 2016). Some of these mycotoxins, such as aflatoxins, have been classified as group 1 human carcinogens by the World Health Organization (World Health Organization, and International Agency for Research on Cancer 2010). Among the mycotoxins that can cause any kind of illness, aflatoxins, citrinin, fumonisins, ochratoxin A, patulin, trichothecenes, zearalenone, or ergot alkaloids are the most important compounds (Bennett and Klich 2003).

8.2.1 Sources of Contamination

Aflatoxins usually affect crops including cereals, oilseeds, spices, and tree nuts (World Health Organization, and International Agency for Research on Cancer 2010). At certain environmental conditions, such as high temperatures, relative humidity, and rainfall, the contamination of the stored product occurs. These conditions favor fungal proliferation, which leads to the mycotoxin production that contaminates foods (Bahrami et al. 2015; Bhat and Reddy 2017; Hidalgo-Ruiz et al., 2019a, b). Because these compounds are easily produced, the FAO appraises that more than a quarter of all agricultural products are contaminated with them (Marin et al. 2013). To cite an example of the hazard of these compounds, in 2004, 125 people died and more than 200 fell ill in Kenya due to the consumption of contaminated corn (Ongoma 2013).

8.2.2 Legislation

The Joint Expert Committee on Food Additives (JECFA) evaluated the hazards derived from the consumption of these mycotoxins in food. Consequently, the European Commission has stablished a limit of 400 μg/kg of zearalenone in refined corn oil. Regarding the seeds, different criteria have been adopted by different countries around the world. Japan has set a limit of 10 μg/kg of aflatoxin B1 for mustard, rapeseed, soybean, and sunflower seeds, while 5 μg/kg is the limit for the same seeds in Russia. Also, a limit of 35 μg/kg for the sum of aflatoxins has been set for rapeseed, soybean, and sunflower seeds in Malaysia. Some countries have adopted the same limits for soybean and sunflower seeds. These limits are: 50 and 20 μg/kg for the sum of aflatoxins in soybean and sunflower seeds respectively in Brazil, Canada, Egypt, and Iran. Finally, a maximum of 5 μg/kg of aflatoxin B1 is allowed for soybean in China, while a maximum of 10 μg/kg of the same aflatoxin is permitted in Kenya (Bhat and Reddy 2017).

8.2.3 Analysis

8.2.3.1 Sample Treatment

Liquid–liquid extraction (LLE) combined with the QuEChERS (Quick, Easy, Cheap, Effective, Rugged, and Safe) method are widely used for the extraction of mycotoxins in edible oils and seeds. The most used solvents and salts are acetonitrile : water (4 : 1, v/v) with Na_2SO_4 and NaCl (Hidalgo-Ruiz et al. 2019a; Zhao et al. 2016), water and acetonitrile : formic (95 : 5, v/v) with $MgSO_4$ and NaCl (Eom et al. 2017), or acetonitrile alone with dispersive solid phase extraction (d-SPE) and $MgSO_4$ (Sharmili et al. 2016). Also, other extraction methods have been evaluated, such as magnetic solid phase extraction (MSPE), where magnetic Fe_3O_4 nanoparticles are dispersed into the sample solution to extract the analytes (Yu et al. 2019; Zhao et al. 2017), molecularly imprinted polymers SPE, that are tailor-made polymers with a predetermined selectivity toward a given analyte or a group of structurally related species (Wei et al. 2015), or gel permeation chromatography (GPC) that performs a separation based on the hydrodynamic behavior of the molecules using an ethyl acetate : cyclohexane solution (1 : 1, v/v) used by Qian et al. 2015.

Immunoaffinity chromatography is another widely used technique for the separation of mycotoxins in edible oils. For that, different solvents are used: acetonitrile : water (4 : 1, v/v) and a phosphate buffer saline solution to dissolve the matrix and pass it through the column (Marley et al. 2015), whereas methanol (Bao et al. 2012, 2013) or methanol : acetic acid (9 : 1 v/v) (Wei et al. 2015) were used for the elution of the compounds.

Finally, methods based on gas chromatography (GC) usually require a derivatization step like the silylation performed by Qian et al. (2015), using N,O-bis-trimethylsilyltrifluoroacetamide, containing 1% trimethylchlorosilane. Also, Yu et al. (2019) performed a derivatization by photochemical post-column derivatization. This technique is used to maintain the natural fluorescence of mycotoxins, avoiding the emission quenching in the aqueous mobile phase (see Table 8.1).

8.2.3.2 Determination

The determination of mycotoxins is usually performed by liquid chromatography (LC) with a simple detector like fluorescence detector (LC-FLD) used in the studies carried out by Wei et al. (2015), Yu et al. (2019), and the interlaboratory study carried out by 17 laboratories for the analysis of 4 aflatoxins in olive, peanut, and sesame oils (Bao et al. 2012, 2013). Also, sterigmastocystin was determined in samples of sunflower seeds by LC with ultraviolet (UV) detector by Marley et al. (2015). However, when a higher number of mycotoxins need to be determined, LC coupled to tandem

Table 8.1 Analytical methods for the determination of mycotoxins.

Matrix	Analytes	Sample extraction/ clean-up	Analysis	References
Vegetable oils	AFB1 and AFB2[a]	MSPE	HPLC-PCD-FLD	Yu et al. (2019)
Olive oils	AFB1, AFB2, AFG1, AFG2, α-ZOL, and ZEA	LLE – QuEChERS	UHPLC-QqQ-MS/MS	Hidalgo-Ruiz et al. (2019a)
Vegetable oils	11 mycotoxins	Liquid-liquid microextraction – QuEChERS	LC-QqQ-MS/MS	Eom et al. (2017)
Vegetable oils	FB1, ZEA, and OTA	LLE – MSPE	UHPLC-QqQ-MS/MS	Zhao et al. (2017)
Vegetable oils	ZEA and 5 derivatives	LLE – GPC – Derivatization	GC-QqQ-MS/MS	Qian et al. (2015)
Sunflower seeds	Sterigmatocystin	Immunoaffinity Chromatography	HPLC-UV/Vis LC-QqQ-MS/MS	Marley et al. (2015)
Vegetable oils	16 mycotoxins	QuEChERS	LC-QTRAP-MS/MS	Zhao et al. (2016)
Vegetable oils	7 mycotoxins	QuEChERS – SPE	LC-QqQ-MS/MS	Sharmili et al. (2016)
Vegetable oils	AFB1, AFB2, AFG1, and AFG2	LLE Immunoaffinity chromatography	LC FLD	Bao et al. (2012, 2013)
Peanut oil	AFB1 and AFM1	MISPE – Immunoaffinity chromatography	LC-FLD	Wei et al. (2015)

[a] AFB1: Aflatoxin B1; AFB2: Aflatoxin B2; AFG1: Aflatoxin G1; AFBG2: Aflatoxin G2; AFM1: Aflatoxin M1; FB1: Fumonixin B1; GC-QqQ-MS: Gas chromatography coupled to mass spectrometry with triple quadrupole; GPC: Gel permeation chromatography; HPLC-PCD-FLD: High performance liquid chromatography with post-column photochemical derivatization and fluorescence detector; HPLC-UV/Vis: High performance liquid chromatography with ultraviolet visible detector; LC-FLD: Liquid chromatography with fluorescence detector; LC-QTRAP-MS/MS: Liquid chromatography with ion trap coupled to triple quadrupole detector; LLE: Liquid-liquid extraction; MISPE: Molecularly imprinted solid phase extraction; MSPE: Magnetic solid phase extraction; OTA: Ochratoxin A; ZEA: Zearalenone α-ZOL: α-Zearalenol.

mode mass spectrometry (LC–MS/MS) is used. Within this category, the most used analyzer is triple quadrupole (QqQ); case of Hidalgo-Ruiz et al. (2019a), Eom et al. (2017), Y. Zhao et al. (2017), Marley et al. (2015), Sharmili et al. (2016). Also, QqQ analyzer is used by Qian et al. (2015) in a study carried out for the determination of zearalenone and five derivatives in 40 samples of vegetable oils.

Finally, H. Zhao et al. (2016) used LC coupled to an ion trap (QTRAP) for the determination of 16 mycotoxins in 25 samples of palm, corn, and sunflower oils. This analyzer offers a high sensitivity, accuracy, and an excellent mass resolution maintaining the same acquisition speed (see Table 8.1).

8.2.3.3 Occurrence

Almost 200 oil samples including olive, lampante, pomace, and refined oils were analyzed in the study carried out by Hidalgo-Ruiz et al. (2019a). The authors of that study observed that almost half of the analyzed samples were contaminated. Zearalenone was found at the highest concentration (25.6 μg/kg) but also other mycotoxins like aflatoxin B2 and aflatoxin G2 were present. Figure 8.2a shows the extracted ion chromatogram of a lampante oil sample contaminated with 25.6 μg/kg of zearalenone, while Figure 8.2b shows the extracted ion chromatogram of a crude olive pomace oil sample contaminated with 6.8 μg/kg of aflatoxin G2. Zearalenone was the most detected mycotoxin in other studies that analyzed oil samples, independently of the type of oil analyzed (Eom et al. 2017; Sharmili et al. 2016; Zhao et al. 2016, 2017). Specifically, Eom et al. analyzed nine edible oil samples, including soybean, corn, and rice bran oil and

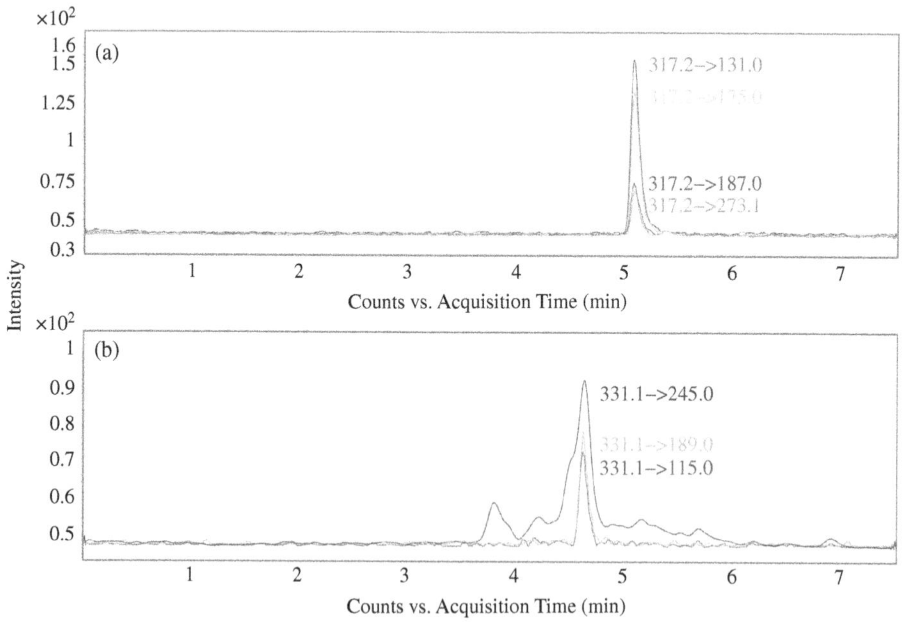

Figure 8.2 (a) Extracted ion chromatogram of a lampante olive oil sample contaminated with 25.6 μg/kg of zearalenone, and (b) extracted ion chromatogram of a crude olive pomace oil sample contaminated with 6.8 μg/kg of aflatoxin G2. *Source:* Hidalgo-Ruiz (2019a). Reprinted with permission from Elsevier.

detected zearalenone in six samples at a maximum concentration of 3.25 μg/kg (Eom et al. 2017). Sharmili et al. analyzed a total of 25 samples including palm, corn, and sunflower oil detecting zearalenone in all the palm oil samples analyzed at a maximum level of 69.1 μg/kg. They also found positives for aflatoxin B1, G1, and G2, while they did not detect aflatoxin B2, ochratoxin A, or deoxynivalenol (Sharmili et al. 2016). The highest number of mycotoxins analyzed in vegetable oils was performed by H. Zhao et al. analyzing 16 mycotoxins in six kinds of oils including sunflower, peanut, soybean, corn, linseed, and olive oil. Only six samples were positive among the 62 samples tested, detecting zearalenone (up to 42.5 μg/kg), aflatoxin B1 (up to 11.0 μg/kg), aflatoxin B2 (up to 4.6 μg/kg), aflatoxin G1 (up to 0.6 μg/kg),and α-zearalenol (up to 1.4 μg/kg) (Zhao et al. 2016). Finally, the highest concentration of zearalenone (111.0 μg/kg) was detected by Y. Zhao et al. in a sample of maize oil in a study carried out with three samples of maize, rapeseed, and soybean oil (Zhao et al. 2017).

8.3 Polycyclic Aromatic Hydrocarbons

PAHs are organic compounds, with two or more condensed benzene rings, produced by the incomplete combustion of organic material such as wood, petroleum products, coal, or food and they have been detected in different types of food (Lacoste 2014; Yao et al. 2015).

8.3.1 Sources of Contamination

There are three types of contamination sources: pyrogenic, petrogenic, and biological. In the first case, pyrogenic PAHs are formed when organic substances are burnt or exposed to high temperatures under low or no oxygen conditions. Secondly, petrogenic PAHs are derived from coal distillation as well as the cracking of petroleum. Also, they can appear during incomplete combustion of fuels, woods, or fuel oils. Finally, biological PAHs are those that come from natural sources; for example they can be synthetized by certain plants and bacteria or formed during the degradation of vegetative matter (Abdel-Shafy and Mansour 2016). PAHs are in the environment mainly due to the incomplete combustion of organic matter either coming from natural or anthropogenic sources, like vehicle exhaust, agricultural fires, or factories (Ravindra et al. 2008; Wang et al. 2013). PAHs can be found in the surroundings of the contamination source or they can be transported through air until they are deposited on the soil, where they can be absorbed by the crops (Cachada et al. 2012). Additionally, cooking of foods is a major source of PAHs as they are generated in situ (Zhao et al. 2012). Thus, edible oils should not have PAHs as they are extracted through cold processes, except pomace oils that are submitted to harder processes (Food and Agriculture Organization of the United Nations, and World Health Organization 2008).

8.3.2 Legislation

A limit of 2.0 μg/kg for benzo(a)pyrene and 10.0 μg/kg for the sum of benzo(a)pyrene, benz(a)anthracene, benzo(b)fluoranthene and chrysene is set for oils and fats by the European Commission (EC), whereas a maximum of 20.0 μg/kg for the sum of these

PAHs, has been set in coconut oil (European Commission 835 2011). The limit of benzo(a)pyrene is the lowest because it is considered the most carcinogenic PAH (United States Environmental Protection Agency 2000). For this reason, in 2003, the WHO set a unit risk of lung cancer of this PAH of 87×10^{-6} ng/m^3 for lifetime exposure (WHO 2003).

The US Agency for Toxic Substances and Disease Registry (ATSDR) and US Environmental Protection Agency (EPA) prepared a list of substances that are most commonly found and which are the most significant potential threat to human health due to their known or suspected toxicity and potential for human exposure. In that list, 16 PAHs were included, as they are frequently found in environmental monitoring samples. These PAHs are acenaphthene, acenaphthylene, anthracene, fluoranthene, fluorene, naphthalene, phenanthrene, pyrene, benz[a]anthracene, benzo[b]fluoranthene, benzo[k]fluoranthene, benzo[g,h,i]perylene, benzo[a]pyrene, chrysene, dibenz[a,h]anthracene, and indeno[1,2,3-cd]pyrene (Agency for Toxic Substances and Disease Registry 2017).

8.3.3 Analysis

PAHs are present in edible oils in concentrations of μg/kg (Sun et al. 2019). They are light sensitive, so the manipulation during the different steps of the analysis must be carried out within dark conditions. This will avoid the decomposition by photoirradiation and oxidation of the compounds (Plaza-Bolaños et al. 2010).

8.3.3.1 Sample Treatment

In recent years, the extraction and clean-up steps have been unified in animal and vegetable fats and oils, applying the ISO 15753 standard, where two LLEs are carried out (Spanish Association for Standardization and Certification [AENOR] 2012). To accomplish the extraction, a mixture of acetonitrile : acetone (3 : 2, v/v) is used (Amzad Hossain and Salehuddin 2012; Hao et al. 2016; Yousefi et al. 2018). Sun and Wu (2020) tested the solvent used in ISO 15753 and other combinations and they concluded that the mixture of acetonitrile : acetone (3 : 2 v/v) was the best option to extract PAHs from oils. They also tested acetonitrile : acetone (2 : 1, v/v) and noticed that the recovery was higher for three to four ring PAHs but lower for five to six ring PAHs.

Before ISO 15753, which was published in 2012, other methods were reported. For instance, a LLE with dimethylformamide : water (9 : 1, v/v) was developed by Camargo et al. (2011) for the analysis of PAHs in soybean oil. Also, a magnetic solid-phase extraction method with multiwalled carbon nanotubes was published in 2011 for the extraction of PAHs in edible oils (Zhao et al. 2011). Finally, the most recent method published uses a QuEChERS method, utilizing the same solvent as that in ISO 15753, but employing a new adsorbent material, named enhanced matrix removal to remove lipids that, according to the authors, exhibits better adsorption capacity for lipids in clean-up applications than C18, graphitized carbon black (GCB), Z-Sep, and Z-Sep+ (Sun and Wu 2020).

Moreover, a method in which no sample treatment is used was developed by Hollosi and Wenzl (2011). In this study, the olive oil is directly injected in the instrument after dilution with 30% of isopropanol. Nevertheless, the majority of the methods follow the ISO 15753 for the clean-up step. This consists in two SPE steps. Firstly, a C18 cartridge is used, and then, the extract is transferred to a Florisil cartridge (Yousefi et al. 2018; Zhao et al. 2018) (see Table 8.2).

Table 8.2 Analytical methods for the determination of PAHs.[a]

Matrix	Analytes	Sample extraction/ clean-up	Analysis	References
Edible oils	8 of the 16 EPA PAHs[b]	LLE	GC–MS	Amzad Hossain and Salehuddin (2012)
Edible oils	9 of the 16 EPA PAHs	LLE	HPLC-UV–Vis	Dost and Deli (2012)
WFO and VODD	16 EPA PAHs	QuEChERS – EMR	GC-QqQ-MS	Sun and Wu (2020)
Edible oils	8 of the 16 EPA PAHs	MSPE	GC–MS	Zhao et al. (2011)
Vegetable oils	4 of the 16 EPA PAHs	LLE – SPE	UHPLC-FLD	da Silva et al. (2017)
Soybean oils	13 of the 16 EPA PAHs	LLE – SPE	HPLC-FLD	Camargo et al. (2011)
Edible oils	13 of the 16 EPA PAHs	LLE – SPE	HPLC-FLD	Yousefi et al. (2018)
Edible oils	13 of the 16 EPA PAHs	LLE – SPE	HPLC-FLD	Molle et al. (2017)
Vegetable oils	16 EPA PAHs	LLE – SPE	GC–MS	Zhao et al. (2018)
Edible oils	16 EPA PAHs	LLE – SPE	GC–MS	Wu and Yu (2012)
Edible oils	16 EPA PAHs	LLE – Chromatography column	HPLC-UV	Hao et al. (2016)
Olive oil	16 EPA PAHs	Dilution – Direct injection	LC-QqQ-MS/ MS	Hollosi and Wenzl (2011)
Vegetable oils	16 EPA PAHs	LLE – SPE	GC–MS	Hua et al. (2016)

[a] EMR: Enhanced matrix removal; GC–MS: Gas chromatography coupled to mass spectrometry; GC-QqQ-MS: Gas chromatography coupled to mass spectrometry with triple quadrupole; HPLC-FLD: High performance liquid chromatography with fluorescence detector; LLE: Liquid-liquid extraction; MSPE: Matrix solid phase extraction; VODD: Vegetable oil deodorizer distillate; WFO: Waste frying oil.

[b] 16 EPA PAHs: Acenaphthene, Acenaphthylene, Anthracene, Benz[a]anthracene, Benzo[b]fluoranthene, Benzo[k]fluoranthene, Benzo[ghi]perylene, Benzo[a]pyrene, Chrysene, Dibenz[a,h]anthracene, Fluoranthene, Fluorene, Indeno[1,2,3-cd]pyrene, Naphthalene, Phenanthrene and Pyrene.

8.3.3.2 Determination

Regarding the analysis step, both LC and GC are commonly used. When GC is used, MS is coupled as detector in order to get a reliable analysis (Amzad Hossain and Salehuddin 2012; Hua et al. 2016; Wu and Yu 2012; Zhao et al. 2011, 2018), although only one of them used MS/MS (Sun and Wu 2020). Furthermore, the last study mentioned is the only one that used QqQ for the analysis of PAHs in waste frying oil (WFO) and vegetable oil deodorizer distillate (VODD) (Sun and Wu 2020). Whitin the MS analyzers, the single quadrupole is the most used analyzer. Hua et al. and X. Zhao et al. used it for the analysis of PAHs in soybean and rapeseed oils (Hua et al. 2016; Zhao et al. 2018), Q. Zhao et al. for the analysis of blend, peanut, olive, maize, rapeseed, sunflower, and soybean oils (Zhao et al. 2011) and Wu et al. determined the 16 EPA PAHs in peanut and olive oils (Wu and Yu 2012). On the other hand, Amzad Hossain et al. used an ion trap for the analysis of soybean, mustard, and coconut oils (Amzad Hossain and Salehuddin 2012).

LC can be coupled to a conventional detector as FLD (Camargo et al. 2011; da Silva et al. 2017; Molle et al. 2017; Yousefi et al. 2018) or UV–Vis (Dost and Deli 2012; Hao et al. 2016) for the analysis in edible oils. Only one study was found using LC coupled to MS/MS and QqQ as the analyzer (Hollosi and Wenzl 2011). This method highlights the use of anisole as dopant, utilizing assisted atmospheric pressure photo ionization. The photo ionization initiates the formation of dopant radical cations that react further with the analyte via several routes (McCulloch et al. 2017). Then the sample is analyzed by QqQ analyzer (see Table 8.2).

8.3.3.3 Occurrence

Regarding the real sample analyses, a difference was found between the heated and not heated oils, as it has been evaluated by Sun and Wu (2020) where WFO and VODD were measured. Very high concentrations were found for the 16 PAHs obtained in both categories, but in VODD, concentrations higher than 100 μg/kg were detected for the majority of the analytes, and the maximum concentration was detected for pyrene at 173.68 μg/kg. The difference between the two categories becomes 7.5-fold higher when the sum of all the 16 PAHs is done, obtaining 197.44 μg/kg in the case of WFO and 1482.25 μg/kg in the case of VODD. In addition, 38 out of 40 samples were found contaminated with benzo(a)pyrene in the study carried out by Yousefi et al. (2018) finding concentrations up to 74.89 μg/kg. Additionally, high concentrations of phenanthrene (58.80 μg/kg), anthracene (54.37 μg/kg), and pyrene (50.45 μg/kg) were detected in corn oil. Fluoranthene was also found as the most abundant PAH in edible oils by Dost and Deli (2012) detecting the highest concentration at 76.08 μg/kg in a sample of corn oil. In a different study, Amzad Hossain and Salehuddin (2012) analyzed 8 (naphthalene, anthracene, phenanthrene, fluorene, pyrene, chrysene, benzo(a)pyrene, and benzo(a)anthracene) of the 16 EPA PAHs in three types of edible oils; soybean, mustard, and coconut oils finding contamination of 7 out of the 8 analytes, except chrysene. Wu and Yu (2012) found high concentrations of acenaphthylene (457.12 μg/kg on average) and acenaphthene (222.63 μg/kg on average) in olive oil, finding contamination of all the 16 EPA PAHs in the four samples analyzed with ranges from 0.71 to 457.12 μg/kg on average.

Overall, positive samples were found in all the published studies. PAHs are ubiquitous contaminants that can contaminate oils and oilseeds either from natural sources or be produced during food processing. The analysis of these compounds is extremely necessary for the routine analysis of oils and oilseeds.

8.4 3-MCPD Esters and Glycidyl Esters

3-MCPD is a chloropropanol that can be formed during the refining process of edible oils. Also, 2-MCPD and their esters can appear. Glycidol is associated with them and usually forms monoesters with fatty acids during the refining processes (European Food Safety Authority 2013) (see Figure 8.3).

8.4.1 Sources of Contamination

In edible oils, 3-MCPD is formed during food processing due to the reaction of hydrochloric acid with triacyclglycerols, phospholipids, and glycerol. Furthermore, it can

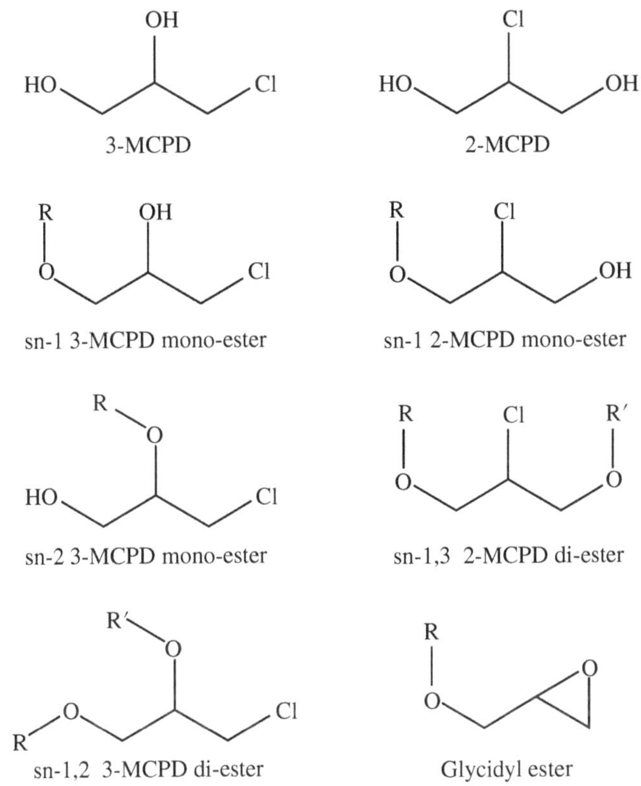

Figure 8.3 Molecules of 3-MCPD, 2-MCPD, their esters, and glycidyl ester.

appear in the thermal processes of other products such as baked goods, malt-derived products, or cooked/cured fish or meat. In the case of meat, it is formed when lipids and sodium chloride react, these components being naturally present or added to the food (Baer et al. 2009).

8.4.2 Legislation

Several criteria about the dairy intake of 3-MCPD and its esters have been adopted by different organizations. The European Food and Safety Authority (EFSA) has established 0.8 µg/kg per body weight/day for the sum of 3-MCPD and its esters (European Food Safety Authority 2016a), while the Joint Committee of FAO and WHO established a limit five times higher than the EFSA value (4.0 µg/kg) (Food and Agriculture Organization of the United Nations (FAO), and World Health Organization (WHO) 2007). The European Union has established a limit of 1000 µg/kg for glycidyl esters in edible oils and fats for adults, and 500 µg/kg if they are destined to baby food production or cereal-based foods for infants and young children. Also, a limit of 1250 µg/kg (2500 µg/kg for pomace oils) was established for the sum of 3-MCPD and 3-MCPD esters in vegetable oils and fats for adults, and 750 µg/kg if they are destined to baby food production or cereal-based foods for infants and young children (European Commission 1322, 2020).

8.4.3 Analysis

Determination of 2- and 3-MCPD esters is based on well-established sample preparation methods in lipid analysis, providing information about the profiles of 2- and 3-MCPD esters (Hori et al. 2012), with 3-MCPD more studied than 2-MCPD. Graziani et al. 2017, Haines et al. 2011, and Li et al. 2015 report the determination of a variable number of mono and diesters of the 2- and 3-MCPD esters.

8.4.3.1 Sample Treatment

Various analytical methods have been developed in recent years for the determination of 3-MCPD and its esters in several matrices, mainly in edible oils (Crews et al. 2013) or foods derived from them (Jędrkiewicz et al. 2017) distinguishing between direct and indirect approaches.

Indirect methods were developed first and they need some steps in order to transform 2- and 3-MCPD esters in free MCPD, performing a transesterification, neutralization, and subsequent quantification after derivatization with phenilboronic acid (Jędrkiewicz et al. 2017) or heptafluorobutyryl imidazole (Dubois et al. 2012). The analysis of glycidyl esters has been incorporated into these methodologies, due to the conversion of these substances into MCPD or bromopropanediol through the use of sodium bromide (Ermacora and Hrncirik 2013). However, a drawback of this approach is the complexity of the sample treatment and the time required to perform all these stages.

In turn, direct methods for the determination of glycidyl esters quantify the level of every species bearing different fatty acyl chains without chemical transformation (Cheng et al. 2017). LLE techniques are commonly used to extract 3-MCPD esters and glycidyl esters from edible oils, and different solvents are used. In fact, between all the revised studies, only two of them used the same, a mixture of cyclohexane : ethyl acetate (1 : 1, v/v) (Dubois et al. 2011, 2012; Weißhaar and Perz 2010). The rest of them use different solvents, as tert-butyl methyl ether : ethyl acetate (4 : 1, v/v) (Blumhorst et al. 2013), acetonitrile and *n*-heptane consecutively (Steenbergen et al. 2013), chloroform (Aniołowska and Kita 2015, 2016a, b, c), tetrahydrofuran (Ermacora and Hrncirik 2013; Steenbergen et al. 2013), acetonitrile (Masukawa et al. 2010, 2011), acetone (Becalski et al. 2012) or *n*-hexane (Hori et al. 2012). Other extraction techniques, such as GPC (Weißhaar and Perz 2010), using a mixture of 0.26 mM methanol–sodium acetate solution (MSA) : methylene chloride : acetonitrile (1 : 8 : 1, v/v/*v*) (Haines et al. 2011), or 20% ethyl acetate in methyl *tert*-butyl ether (MacMahon et al. 2013a, b) were used. An extraction making the oil dissolved in *n*-hexane go through a chromatography column (Song et al. 2015) was also published.

Regarding the clean-up step, most of the methods have followed the same 2-steps SPE method (Cheng et al. 2017), using first a C18 cartridge and then a silica cartridge, except in the case of Hori et al. (2012), who perform the first clean-up step using the silica cartridge and then the C18 one (see Table 8.2).

8.4.3.2 Determination

Both LC and GC are widely used for the determination of 3-MCPD esters and glycidyl esters, always coupled to MS. In the case of GC, all the studies used single quadrupole (Ermacora and Hrncirik 2013; Kuhlmann 2011, 2016; Steenbergen et al. 2013; Weißhaar and Perz 2010). LC–MS (Blumhorst et al. 2013; Shiro et al. 2011) and LC–MS/MS (Aniołowska and Kita 2016a; Becalski et al. 2012; MacMahon and Beekman 2019) are also widely used for the determination of 3-MCPD esters and glycidyl esters in edible oils. Two studies with time-of-flight analyzer (TOF) have been reported, using LC, for the analysis of glycidyl esters in vegetable oils (Dubois et al. 2011, 2012; Haines et al. 2011). Finally, a proton nuclear magnetic resonance determination method was developed by Song et al. (Song et al. 2015) (see Table 8.3).

Table 8.3 Analytical methods for the determination of 3-MCPDs and glycidyl esters.[a]

Matrix	Analytes	Sample preparation/ extraction	Analysis	References
Vegetable oils	5 GEs	LLE – 2-step SPE	LC–MS	Blumhorst et al. (2013), Shiro et al. (2011)
Extra virgin olive oils	5 GEs	LLE – 2-step SPE	GC–MS	Steenbergen et al. (2013)
Edible oils	Glycidol, 2- and 3-MCPD and 2 3-MCPD diesters	Alkaline transformation with Br^-	GC–MS	Kuhlmann (2016), Kuhlmann (2011)
Frying and refined oils	5 GEs	LLE – 2-step SPE	LC–MS/MS	Aniołowska and Kita (2015, 2016a, b, c),
Palm oil	14 GEs	GPC	GC–MS	Weißhaar and Perz (2010)
Vegetable oils	7 GEs and 20 3-MCPD mono- and diesters	Dilution with MSA/ methylene chloride/ acetonitrile (1 : 8 : 1)	LC-TOF-MS	Haines et al. (2011)
Vegetable oils	Glycidol and 2- and 3-MCPD	LLE – Derivatization	GC–MS	Ermacora and Hrncirik (2013)
Edible oils	6 GEs and 12 3-MCPD mono- and diesters	Dilution with ethyl acetate/methyl *tert*-butyl ether – 2-step SPE	LC–MS/MS	MacMahon et al. (2013a, b)
Edible oils	5 GEs	LLE – 2-step SPE	LC–MS	Masukawa et al. (2010, 2011)
Vegetable oils	7 GEs	LLE – GPC – SPE	LC-TOF-MS	Dubois et al. (2011, 2012)
Edible oils	5 GEs	LLE – 2-step SPE	LC–MS/MS	(Becalski et al. 2012)
Edible oils	5 GEs and 9 3-MCPD mono- and diesters	LLE – 2-step SPE	LC-TOF-MS	Hori et al. (2012)
Edible oils	Intact GEs	Chromatography column	^{1}H-NMR	Song et al. (2015)

[a] GC-FID: Gas Chromatography with Flame Ionization Detector; GC–MS: Gas Chromatography coupled to Mass Spectrometry; GEs: Glycidyl Esters; GPC: Gel Permeation Chromatography; ^{1}H-NMR: Proton Nuclear Magnetic Resonance; LC-TOF-MS/MS: Liquid Chromatography coupled to Mass Spectrometry in tandem with Time Of Flight Detector; LC–MS/MS: Liquid Chromatography coupled to Mass Spectrometry in tandem; LLE: Liquid–Liquid extraction; MCPD: Monochloropropanediol; MSA: 0.26 mM methanol–sodium acetate solution; SPE: Solid Phase Extraction.

8.4.3.3 Occurrence

Many types of oils have been analyzed, such as soybean, corn (Hori et al. 2012), olive (Steenbergen et al. 2013), sesame (Haines et al. 2011), rapeseed, sunflower (Kuhlmann 2016), peanut, almond, grapeseed (MacMahon et al. 2013b), walnut, and coconut oil (Kuhlmann 2011) and concentrations up to 28.0 mg/kg were found in palm oil and 28.8 mg/kg in rice oil (Shiro et al. 2011). Moreover, a collaborative study for the analysis of five glycidyl esters in edible oils has been developed and 17 laboratories around the world participated. In this study, several spiked samples were analyzed in 13 months to evaluate the degradation of the glycidyl esters and it was found that in the first three months, the amount of these compounds decreased by 13% while after 13 months, it had decreased by 24% for all five analytes (Blumhorst et al. 2013). Another interesting study is the one carried out by Aniołowska and Kita (2016a), where 20 refined vegetable oils (rapeseed, sunflower, and palm oils) from retail outlets were analyzed. It was found that refined palm oils showed a high contamination ratio in comparison with other oils, finding the highest contamination at 44.33 mg/kg. In another study, they evaluated the effect of frying on glycidyl esters content in palm oil (Aniołowska and Kita 2016c). They discovered that the content of glycidyl esters decreased when the amount of time and temperature was increased (to a maximum of 180 °C). These results confront the theory that glycidyl esters are formed in thermal processes (Baer et al. 2009). However, EFSA suggests that they appear with temperatures higher than 200 °C (European Food Safety Authority 2016b), while they do not increase until over 180 °C.

8.5 Mineral Oil

Mineral oils are either derived from or produced synthetically from coal, natural gas, or biomass. They are composed of MOHs, linear or branched (paraffins) (European Commission 2018). These mineral oils are used as lubricating oils for the machinery used in the food industry (Gómez-Coca et al. 2016b). Within the MOHs, two categories can be distinguished, depending if they are aliphatic (Mineral Oil Saturated Hydrocarbons, MOSH) (Figure 8.4) or aromatic (Mineral Oil Aromatic Hydrocarbons, MOAH) (Figure 8.5).

8.5.1 Sources of Contamination and Legislation

Direct contact of food or feed with materials containing MOHs is enough to contaminate the product. Also, MOHs can appear in edible oils when their seeds have been submitted to hard extraction processes, such as centrifugation or solvent extraction (Gómez-Coca et al. 2016a). Indeed, the content of these compounds in pomace oils is considerably higher. Finally, in the case of edible oils, fraud can be an important source of contamination, intentionally adding mineral oils to edible oils, which can be a very serious risk for human health.

The European Commission has set a maximum level of contamination for MOHs in edible oils. For crude and refined sunflower oils the limit is 50 mg/kg (Standing Committee on the Food Chain and Animal Health 2008).

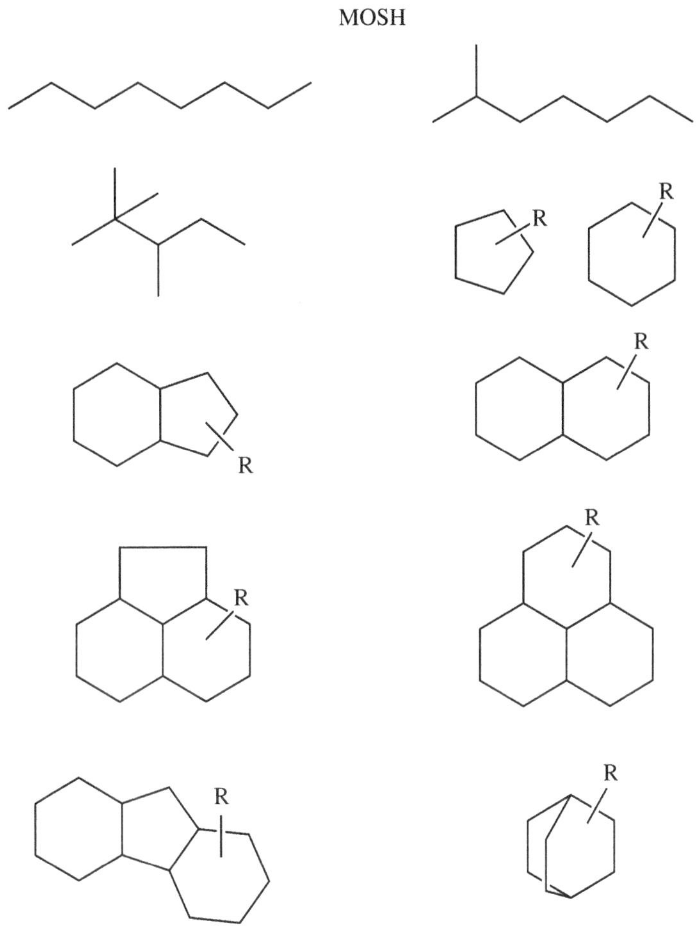

Figure 8.4 Examples of structures of MOSH.

8.5.2 Analysis

8.5.2.1 Sample Treatment

Due to the complexity of the matrices, a pre-treatment step is recommended before the analysis of MOHs. Sometimes only MOSH are analyzed. In these cases, an SPE using Ag-activated silica gel is enough to extract them (Li et al. 2016; Liu et al. 2017; Moret et al. 2011). However, the analysis of MOAH often needs a more sophisticated extraction, as there are compounds like olefins that interfere with the result. In this sense, several authors used the epoxidation reaction in order to eliminate the interferences that olefins produce (DIN EN 16995 2017; Gharbi et al. 2017; Zurfluh et al. 2014). Furthermore, Nestola et al. tried three reactions in order to eliminate those interferences: epoxidation, hydroboration, and bromohydrin formation reactions, finding the best results with epoxidation using 3-chloroperbenzoic acid (*m*CPBA) (Nestola and Schmidt 2017). In addition, Gharbi et al. 2017 used an Abencor extractor composed by a hammer crusher, a mixer, and a pulp centrifuge, which is used to smash and mix

MOAH

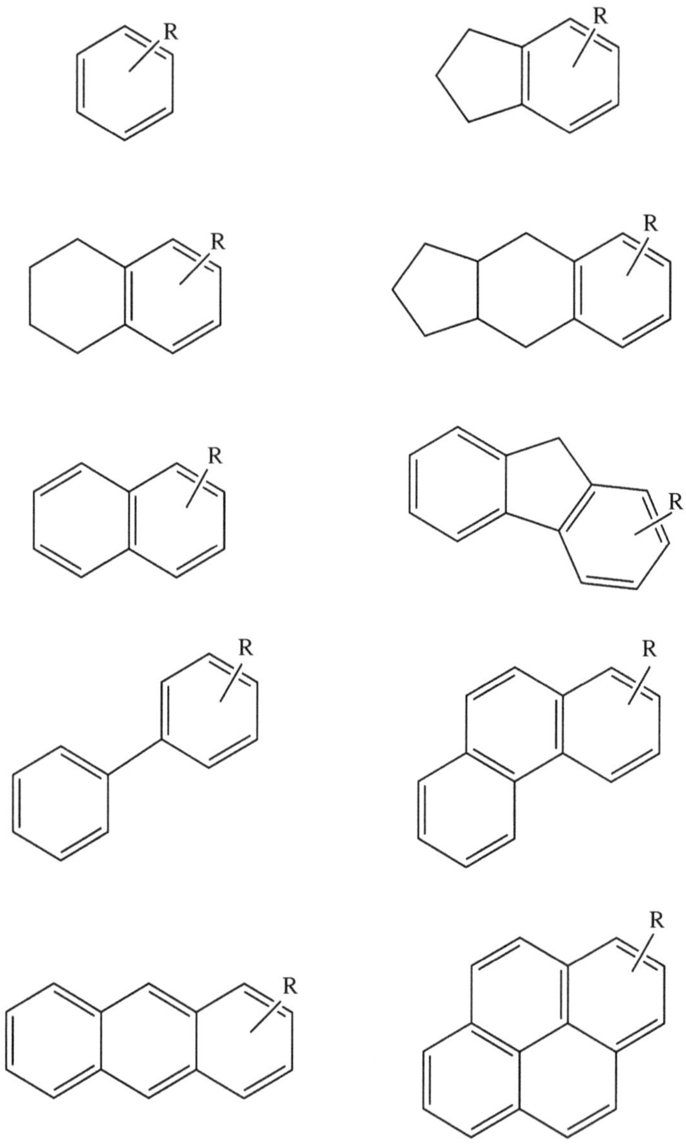

Figure 8.5 Examples of structures of MOAH.

the olives as it is the only study that analyses oilseeds. They also use microwave assisted extraction to speed-up the process and minimize the volume of organic solvents employed.

Overall, in most of the studies where these substances have been analyzed in oils, the separation has been carried out by LC and then analyzed by GC (DIN EN 16995 2017; Gharbi et al. 2017; Nestola and Schmidt 2017; Tranchida et al. 2011; Zoccali et al. 2016; Zurfluh et al. 2014) (see Table 8.4).

8.5.2.2 Determination

Few studies have performed the separation of the MOSH fraction from the MOAH one (Gharbi et al. 2017; Zoccali et al. 2016; Zurfluh et al. 2014). Most of these studies use LC prior to the injection of the sample into the GC. However, this high cost instrument can be skipped if a good epoxidation reaction is done (Nestola and Schmidt 2017) and very precise SPE cartridges are prepared, using 1% Ag-silica gel (Li et al. 2016; Liu et al. 2017), which are not commercially available. Finally, the most used detector is a flame ionization detector (FID) as it presents the advantage of cheapness against the high price of MS (Weber et al. 2018). Also, in the case of these compounds, a hump below the peaks of natural hydrocarbons has to be integrated, and compared to an internal standard, so the MS detection is not very useful. The injection method also plays an important role and different methods have been developed, such as splitless mode (Li et al. 2016), and large volume injection (Liu et al. 2017) when an LC equipment is not coupled to the GC to carry out the separation (see Table 8.4).

8.5.2.3 Occurrence

High concentrations of MOSH and MOAH are often found in oils and seeds. This fact does not mean that the sample is contaminated, as the seeds can have these compounds naturally (Gómez-Coca et al. 2016b). For example, Liu et al. (Liu et al. 2017) found concentrations below 60.9 mg/kg of MOSH in the majority of the samples, except in a blend oil (a mixture of rapeseed, soybean, peanut, corn, sunflower, fish,

Table 8.4 Analytical methods for the determination of mineral oil hydrocarbons (MOHs).[a]

Matrix	Analytes	Sample extraction/clean-up	Analysis	References
Edible oils	MOAH/PAH	Epoxidation	LC-GC-FID	Nestola and Schmidt (2017)
Vegetable oils	MOSH	Hexane, SPE-clean-up(Ag-activated silica gel)	LVI-GC-FID	Liu et al. (2017)
Vegetable oils	MOSH	Hexane	LC-GC-FID	Tranchida et al. (2011)
Virgin olive oil	MOSH/ MOAH/PAH	Epoxidation – Abencor extractor or MAE	LC-GC-FID	Gharbi et al. (2017)
Oils	MOSH/MOAH	Hexane, SPE-clean-up – Epoxidation	LC-GC-FID	DIN EN 16995 (2017)
Vegetable oils	MOSH	Hexane, SPE-clean-up(Ag-activated silica gel)	GC-FID	Li et al. (2016)
Vegetable oils	MOSH/MOAH	Hexane	LC-GC-FID/MS	Zoccali et al. (2016)
Vegetable oils	MOSH	Hexane, SPE-clean-up(Ag-activated silica gel)	GC-FID	Moret et al. (2011)
Edible oils	MOSH/MOAH	Hexane, SPE-clean-up – Epoxidation	LC-GC-FID	Zurfluh et al. (2014)

[a] GC-FID: Gas chromatography with flame ionization detector; GC-FID/MS: Gas chromatography with flame ionization detector coupled to mass spectrometry; LC-GC-FID: Liquid chromatography coupled to gas chromatography with flame ionization detector; LC-GC-FID/MS: Liquid chromatography coupled to gas chromatography with flame ionization detector and mass spectrometry; LVI-GC-FID: Large volume injection gas chromatography; MAE: Microwave assisted extraction; MOAH: Mineral oil aromatic hydrocarbons; MOSH: Mineral oil saturated hydrocarbons; PAH: Polycyclic aromatic hydrocarbons; SPE: Solid phase extraction.

sesame, linseed, and rice oils) that had 259.4 mg/kg. In the case of the extra virgin olive oil samples, the maximum concentration found was 30.3 mg/kg. Zoccali et al. (2016) found low levels of MOSH (below 21.8 mg/kg) and did not found MOAH in extra virgin olive oils, whereas levels up to 444.8 mg/kg of MOSH and 66.1 mg/kg of MOAH in olive pomace oils were found. Tranchida et al. (2011) analyzed 18 samples of extra virgin olive, olive, sunflower, grapeseed, corn, peanut, soybean, and pomace oils, and found concentrations ranging from 7.6 mg/kg in an extra virgin olive oil to 180.6 mg/kg in a pomace oil. Gharbi et al. (2017) analyzed five samples of extra virgin olive oil, detecting concentrations up to 38.0 mg/kg. Li et al. (2016) analyzed nine types of oils, and found that camellia oil contained the highest average contamination (78.49 mg/kg of maximum) compared with the other types of oils.

Overall, results summarized herein agree with the affirmation made by Gómez-Coca et al. (2016b), who indicated that hard extractions, such as centrifugation or solvent extractions used for olive pomace oils, extract more MOHs from the fruit, whereas extra virgin olive oils show the lowest levels of MOHs because they are not submitted to these processes.

8.6 Phthalates

Phthalates or phthalic acid esters are a group of chemical compounds used as plasticizers, and they are added to the plastics to increase their flexibility. The most common use is the conversion of polyvinylchloride (PVC), a rigid plastic, into a more flexible one. The most common phthalates used are bis (2-ethylhexyl) phthalate (DEHP), which represents approximately 50% of the world production of phthalates, dibutyl phthalate (DBP), butyl benzyl phthalate (BBP), diisodecyl phthalate (DIDP) and diisononyl phthalate (DINP). Phthalates are also used as solvents in perfumery and pesticides, in nail polish, adhesives, putties, paint pigments, lubricants, food packaging, plastic wrap, and toys (Lacoste 2014).

8.6.1 Sources of Contamination

These compounds are ubiquitous and atmospheric transport and deposition are the main sources of crops contamination of phthalates (Rakkestad et al. 2007). As there is no covalent bond between phthalates and the plastic, and they have a strong affinity for fat, they are easily released from its containers into fat-containing foods (Rudel and Perovich 2009). Phthalates do not persist in the outdoor environment because they can easily be degraded, either by photo, bio, or anaerobic degradation and in normal conditions, they do not persist more than 15 days in the environment before degradation (Xie et al. 2006).

8.6.2 Legislation

In 2011, the European Commission set a limit for the migration of five phthalates in food. The limit for DBP is 0.3 mg/kg and the limit for DEHP is 1.5 mg/kg. The sum of DINP and DIDP must not be higher than 9 mg/kg and finally, the limit for BBP is 30 mg/kg. The use of them in material containing fatty foods is forbidden (except infant formulae for DINP, DIDP and BBP), and DINP, DIDP, and BBP must be used only in single-use materials while DEHP and DBP can be used in reusable materials.

8.6.3 Analysis

Apart from the fact that usually foods are complex matrices to analyze, as discussed earlir, in this case, the ubiquity of the analytes makes their presence in any laboratory normal, so consequently, their determination is even harder due to background interference (Haji Harunarashid et al. 2017).

8.6.3.1 Sample Treatment

LLE is a very simple method that is used, utilizing different solvents such as chloroform : methanol (2 : 1, v/v) (Ostrovský et al. 2011), *n*-hexane (Guo et al. 2012), and acetonitrile (Xu et al. 2014). Head space solid phase microextraction (HS-SPME) was developed and used by Rios et al. (2010) and Amanzadeh et al. (2016). This last technique offers numerous benefits such as minimum solvent use, the integration of sampling and sample preparation steps, simple operation, low cost, and the possibility of an on-line analytical procedure (Amanzadeh et al. 2016). Rios et al. tested the performance of three fibers for the extraction (85 μm polyacrylate [PA], 100 μm-polydimethylsiloxane [PDMS], and divinylbenzene carboxen PDMS [DVB/CAR/PDMS]) of these compounds, finding the best results when DVB/CAR/PDMS was used (Rios et al. 2010). A great advantage of the HS-SPME extraction is that a clean-up step is not necessary. However, it is necessary when LLE is carried out. For this purpose, an innovative SPE method was developed with a nylon 6 nanofibers mat, which is characterized by small fiber diameters, controllable pore sizes, high porosities, and the consumption of solvent is also minimized. To perform the extraction, only 200 μL of acetone, 600 μL of water, and 20 μL of methanol were used (Cao et al. 2013; Xu et al. 2010) (see Table 8.5).

8.6.3.2 Determination

GC and LC are the most used techniques in liquid samples. In the case of GC, two detectors are used: FID (Amanzadeh et al. 2016; Ostrovský et al. 2011) and MS (Guo et al. 2012; Rios et al. 2010). Rios et al. (2010) used an ion trap analyzer based on the capture of charged particles after the desorption of the fiber. In the case of LC, the UV/Vis detector is commonly used (Cao et al. 2013; Xu et al. 2010) as well as MS (Xu et al. 2014) (See Table 8.5). The two methods that use UV/Vis detection, carry out the separation with the nylon 6 nanofibers mat, which means that a much better separation must be done prior to the analysis (see Table 8.5).

8.6.3.3 Occurrence

Rios et al. (2010) found high concentrations of four phthalates in olive oils stored in plastic bottles, with DEHP the phthalate detected at the highest concentration (840 μg/kg), followed by diisobutyl phthalate (DIBP), which was detected at 216 μg/kg, BBP (211 μg/kg) and DBP (175 μg/kg). Q. Xu et al. (2010) studied the migration of phthalates from plastic to cooking oil, maintaining the oils at 20, 40, and 60 °C for two months in plastic bottles. After the experiment, they found migration values up to 14% for DEHP. Ostrovský et al. (2011) analyzed DMP in four samples of vegetable oils and detected concentrations from 1.5 to 3.2 mg/kg. Guo et al. (2012) found DEHP at 79.0 μg/kg in a sample of cooking oil. Finally, Amanzadeh et al. (2016) did not found any of the studied samples (sunflower and olive oils) contaminated by these compounds either because their concentration was lower than their limit of quantification determined by this method or they were not contaminated by the analytes.

Table 8.5 Analytical methods for the determination of phthalates.[a]

Matrix	Analytes	Sample preparation/extraction	Analysis	References
Olive oil	DMP, DEP, DIBP, BBP, DEHP, DOP, etc.	Headspace solid-phase microextraction	Ion trap; MS	Rios et al. (2010)
Cooking oil	DMP, DEP, BBP, DBP, DEHP, DOP, etc.	Solid-phase extraction by nylon 6 nanofibers mat	HPLC-UV/VIS	Xu et al. (2010)
Edible oil	DMP, DEP, DBP, DEHP, etc.	Liquid-liquid extraction chloroform/metanol (2 : 1); NaCl	GC-FID	Ostrovský et al. (2011)
Cooking oil	DMP, DEP, DBP, DIBP, etc.	Liquid-liquid extraction with *n*-hexane	GC–MS	Guo et al. (2012)
Edible oil	DMP, DEP, DBP, DEHP, DOP	Solid-membrane extraction	HPLC-UV/VIS	Cao et al. (2013)
Oil	DMP, DEP, DBP, DCHP, DNOP, etc.	Liquid-liquid extraction with acetonitrile	HPLC-MS/MS	Xu et al. (2014)
Vegetable oil	DBP, DEHP, etc.	Graphene/polyvinyl chloride nanocomposite fiber headspace solid-phase microextraction	GC-FID	Amanzadeh et al. (2016)

[a] BBP: Butylbenzyl phthalate; DEHP: Bis (2-ethylhexyl) phthalate; DEP: Diethyl phthalate; DIBP: Diisobutyl phthalate; DMP: Dimethyl phthalate; DOP: Dioctyl phthalate; GC-FID: Gas chromatography with flame ionization detector; GC–MS: Gas chromatography coupled to mass spectrometry; HPLC: High performance liquid chromatography; HPLC-MS/MS: High performance liquid chromatography coupled to mass spectrometry in tandem; HPLC-UV/VIS: High performance liquid chromatography with ultraviolet-visible spectroscopy; MS: Mass spectrometry.

8.7 Pesticides

Pests are animals or plants that are considered harmful for humans or human concerns, including crops, livestock, and forestry (Britannica Online Encyclopedia 2019). Therefore, pesticides are chemical substances used to control pests (US Environmental Protection Agency 2019). Depending on where these substances are going to be used they can be classified as herbicides, insecticides, nematicides, molluscicides, piscicides, avicides, rodenticides, bactericides, insect repellents, animal repellents, antimicrobials, or fungicides (National Association of State Departments of Agriculture (NASDA) n.d.).

8.7.1 Sources of Contamination

According to the EPA report (Kiely et al. 2004), in 2001, more than five billion pounds of pesticides were used, and 76% of this amount was used in the agricultural sector, another 12.5% in the industry/commercial/government sector and finally, the remaining 11.5% was used in the home and garden sector. Transport of pesticides through the environment can also be a source of contamination. Volatile or semi-volatile compounds can be evaporated at high temperatures and condensed in zones where temperatures are lower (Bloomfield et al. 2006). This transportation can cause high levels of persistent bioaccumulative toxicants in areas far from the source of pollution (Blais et al. 2006). Parallel, they can be transported by other mobile particles, by migratory animals, or by hydrological flows (Blais et al. 2006). Finally, they can be added during the postharvest processes to maintain the quality of the crop if they are going to be stored for a long time or transported through a long distance (Amvrazi 2011).

8.7.2 Legislation

The use of pesticides in the European Union is regulated by Regulation No. 1107/2009 (European Commission 2009a) on Plant Protection Products in cooperation with other EU Regulations and Directives (e.g. the Regulation on MRLs in food; Regulation (EC) No. 396/2005 (European Commission 2010) and the Directive on sustainable use of pesticides, Directive 2009/128/EC (European Commission 2009b). This EU legislation sets MRLs of over 1100 pesticides for 315 fresh products. For any pesticide not specifically named in the legislation that is found in the matrix studied, a default MRL at 0.01 mg/kg is set. However, depending on the combination pesticide/matrix, this MRL could be up to 20 mg/kg (European Commission 2008).

The Codex Alimentarius has also set MRLs and the majority of the pesticides in oils and seeds are not higher than 0.8 mg/kg, except for some cases, as the limit of 5 mg/kg of chlormequat in rape seed, 70 mg/kg of fluorpyran in dill seed and 2800 mg/kg of propiconazole in orange oil (Joint FAO/WHO Expert Committee on Food Additives (JECFA) 2018).

8.7.3 Analysis

Currently, the determination of pesticides from fatty materials is a complex task. In fact, several different techniques have been developed to solve this problem.

8.7.3.1 Sample Treatment

In the majority of the cases, several steps are needed to extract these compounds from food. A solvent partitioning step with acetonitrile is used in the majority of the studies to extract and isolate the pesticides from the matrix (Deme et al. 2014; Moreno-González et al. 2014; Ramli et al. 2012; Ruiz-Medina et al. 2012; Tuzimski and Rejczak 2016). Then, low-temperature precipitation combined with QuEChERS using different salts is used. In the case of Anagnostopoulos and Miliadis (2013), primary secondary amine sorbent (PSA), GCB, and $MgSO_4$ are used. Deme et al. (2014) used PSA, activated charcoal and $MgSO_4$ and Sobhanzadeh et al. (2012) used PSA, C18, and GCB. SPE is widely used as a clean-up technique (Chung and Chen 2015; Han et al. 2016; Muhamad et al. 2012; Ramli et al. 2012; Tuzimski and Rejczak 2014, 2016). Muhamad et al. (2012) tested different SPE sorbents for the extraction of cypermethrin and λ-cyhalothrin in palm oil such as GCB, PSA, C18, silica, and florisil. The results exhibited that the combination GCB/PSA showed the best recoveries, while florisil showed the highest precision in terms of relative standard deviation (RSD) for cypermethrin and GCB for λ-cyhalothrin.

Other techniques that are applied include matrix solid phase dispersion (MSPD) (Liu et al. 2012), GPC for the determination of organophosphorus pesticides in camellia oil (Chung and Chen 2015), and other highly effective extraction techniques, like air-assisted liquid–liquid microextraction (AALLME), which was used for the determination of triazole pesticides in edible oils (Farajzadeh et al. 2015). AALLME uses a few microliters of an extraction solvent, such as dimethyl sulfoxide (DMSO), saving high amounts of solvent compared to a normal LLE (see Table 8.6).

Table 8.6 Analytical methods for the determination of pesticides.[a]

Matrix	Analytes	Sample preparation/ extraction and clean-up	Analysis	References
Camellia oil	Organophosphorus pesticides (15)	Solvent partitioning – MSPD	GC-FPD	Liu et al. (2012)
Palm oil	Pesticides from organophosphates, carbamates, triazines, and phenylureas groups (7)	Solvent partitioning – Low temperature precipitation; QuEChERS	LC-TOF-MS	Sobhanzadeh et al. (2012)
Palm oil	Diuron	Solvent partitioning – Low temperature precipitation; SPE	HPLC-UV	Ramli et al. (2012)
Palm oil	λ-cyhalothrin and cypermethrin	Solvent partitioning – Low temperature precipitation; SPE	GC-ECD	Muhamad et al. (2012)
Olive, sunflower, palm and rapeseed oils	Multiclass pesticides (44)	Solvent partitioning – QuEChERS	LC-QTRAP-MS/MS	Polgár et al. (2012)
Sunflower, rice bran and ground nut oils	Organochlorines, organophosphorus, and synthetic pyrethroids (35)	Solvent partitioning – Low temperature precipitation; QuEChERS	GC-QqQ-MS/MS	Deme et al. (2014)
Extra virgin olive, sunflower, maize, linseed and sesame oils	Carbamates (31)	Solvent partitioning – QuEChERS	UHPLC-QqQ-MS/MS	Moreno-González et al. (2014)
Sunflower, olive, grape and corn oils	Triazole pesticides (5)	AALLME	GC-FID	Farajzadeh et al. (2015)
Olive and grapeseed oils	Multiclass pesticides (21)	Solvent partitioning – SPE; QuEChERS	HPLC-DAD	Tuzimski and Rejczak (2016)
Olive oil and olives	Multiclass pesticides from 32 different groups	Solvent partitioning – Low temperature precipitation; QuEChERS	GC-QqQ-MS/MS LC-QqQ-MS/MS	Anagnostopoulos and Miliadis (2013)

Olive, sunflower oil and corn oils	Carbaryl	Solvent partitioning – QuEChERS	Automated Flow methodologies: Sequential Injection Analysis and Multicommutated Flow Injection Analysis	Ruiz-Medina et al. (2012)
Olive oil and peanuts	Organochlorines (33)	MSPD – GPC; SPE	GC–MS	Chung and Chen (2015)
Soybean and maize oils	Glufosinate	Derivatization – SPE	LC-QqQ-MS/MS	Han et al. (2016)
Sunflower seeds	Multiclass pesticides (10)	Solvent partitioning – SPE; QuEChERS	HPLC-DAD	Tuzimski and Rejczak (2014)
Flaxseeds	Multiclass pesticides (34)	Solvent partitioning – QuEChERS	GC-TOF-MS	Koesukwiwat et al. (2010)

[a] AALLME: Air-assisted liquid–liquid microextraction; GC-ECD: Gas chromatography with electron capture detector; GC-FPD: Gas chromatography with flame photometric detector; GC-TOF-MS: Gas Chromatography coupled to time of flight mass spectrometry; GC–MS: Gas chromatography coupled to mass spectrometry; GC-QqQ-MS/MS: Gas chromatography coupled to mass spectrometry in tandem with triple quadrupole detector; GPC: Gel permeation chromatography; HPLC-DAD: High performance liquid chromatography with diode array detector; HPLC-UV: High performance liquid chromatography ultraviolet; LC–MS: Liquid chromatography coupled to mass spectrometry; LC-QqQ-MS/MS: Liquid chromatography coupled to mass spectrometry in tandem with triple quadrupole detector; LC-QTRAP-MS/MS: Liquid chromatography coupled to mass spectrometry in tandem with ion trap-triple quadrupole detector; MSPD: Matrix solid phase dispersion; QuEChERS: Quick, easy, cheap, effective, rugged, and safe; SPE: Solid phase extraction; UHPLC-QqQ-MS/MS: Ultra performance liquid chromatography coupled to mass spectrometry in tandem with triple quadrupole detector.

8.7.3.2 Determination

The fact that an elevated number of pesticides need to be determined, makes the combination of chromatographic techniques (GC and LC) with MS or MS/MS very important. In this sense, a method for the determination of 33 organochlorines with GC–MS and single quadrupole in olive oil samples (Chung and Chen 2015) was developed. Other methods of MS/MS were applied, like the case of Anagnostopoulos and Miliadis (2013), who used both GC-QqQ-MS/MS and LC-QqQ-MS/MS for the determination of 32 groups of pesticides in 262 samples of olive oil and olives. Also a QqQ analyzer was coupled to GC for the determination of 35 organochlorines, organophosphorus, and synthetic pyrethroids in sunflower, rice bran, and ground nut oils (Deme et al. 2014). QqQ was also used in two studies with LC for the determination of carbamates (Moreno-González et al. 2014) and glufosinate (Han et al. 2016). In addition, the QTRAP analyzer coupled to LC was employed by Polgár et al. (2012) for the determination of 44 pesticides in vegetable oils. In addition, Sobhanzadeh et al. (2012) used a TOF analyzer coupled to LC for the determination of seven pesticides in palm oil. The same analyzer was used by Koesukwiwat et al. (2010) but coupled to GC for the determination of 34 pesticides in flaxseeds. Three different detectors were used apart from the MS cited ones. Flame photometric detection (FPD) was used by Y. Liu et al. (2012) for the determination of 15 organophosphorus pesticides in camellia oil. Muhamad et al. (2012) used an electron capture detector for the determination of λ-cyhalothrin and cypermethrin in samples of palm oil. Finally, Farajzadeh et al. (2015) used FID for the determination of five triazole pesticides in sunflower, olive, grape, and corn oils.

When high sensitivity is not crucial, LC with diode array detector (DAD) (Tuzimski and Rejczak 2014, 2016) can be used. Other detection techniques used are fluorescence spectrophotometry (Chen et al. 2015) and solid-phase spectroscopy (Ruiz-Medina et al. 2012) (see Table 8.6).

8.7.3.3 Occurrence

High concentrations of pesticides were found in samples of olive and olive oil in the study performed by Anagnostopoulos and Miliadis (2013). The authors of that study analyzed a total of 262 samples, of which 21 of olive oil and 17 of olives were found to be positive. The highest concentration was 458 μg/kg of omethoate followed by 184 μg/kg of dimethoate. Dimethoate and malathion were found at 1.5 and 3.5 μg/kg respectively in samples of palm oil in the study carried out by Sobhanzadeh et al. (2012). Deme et al. (2014) analyzed 35 pesticides in edible oils finding a maximum of 2.14 μg/kg of p,p′-DDD (Dichlorodiphenyldichloroethane). Hexaconazole was found at 22.0 μg/kg in a sample of grape oil analyzed by Farajzadeh et al. (2015). Tuzimski and Rejczak (2016) analyzed olive and grapeseed oils and found α-cypermethrin in extra virgin olive oil at concentrations ranging from 17.45 to 23.05 μg/kg, fenuron, dimethomorph, and propazine in grapeseed oil at concentrations ranging from 2.24 to 3.08, from 2.95 to 3.01, and from 0.46 to 1.12 μg/kg, respectively. Moreover, Muhamad et al. (2012) did not find λ-cyhalothrin and cypermethrin in any of the 30 palm oil samples analyzed.

8.8 Conclusions

The presence of most of the chemical contaminants in foods is controlled and regulated by different agencies around the world. However, there is still a lot of work to do in terms of researching on the different contaminants that can affect oils and seeds, as not all of them are regulated. For example, there is no MRL for aflatoxins in oils, which is surprising taking into account the tremendous hazard and health risk of these contaminants. When this happens, both institutions and laboratories must work in order to investigate the possible occurrence, toxicity, health effects and risks, and consequently elaborate regulations and analytical methods in order to control those substances.

Very often, the presence of contaminants in oils and seeds can be reduced by submitting food to processes like refining, although sometimes, these methods can enhance the presence of other contaminants like 3-MCPD. Regarding the determination of the contaminant studies, it must be noted that the robustness and reliability of the analysis is becoming higher and higher due to recent advances and the modern equipment that laboratories have nowadays. Furthermore, the automatization of the analysis of samples is also improving as it was seen in some studies where more than 200 samples were analyzed. Overall, research in food contaminants cannot be stopped as every day new dangerous substances and new hazards are discovered.

Acknowledgments

The authors gratefully acknowledge to the Spanish Ministry of Science, Innovation and Universities and FEDER (project ref. RTC-2017-6170-2) for ther financial support.

References

Abdel-Shafy, H.I. and Mansour, M.S.M. (2016). A review on polycyclic aromatic hydrocarbons: source, environmental impact, effect on human health and remediation. *Egyptian Journal of Petroleum* 25 (1): 107–123.

Adeyeye, S.A.O. (2016). Fungal mycotoxins in foods: a review. *Cogent Food and Agriculture* 2 (1): 1–11.

Agency for Toxic Substances and Disease Registry (2017). ATSDR's substance priority list. https://www.atsdr.cdc.gov/spl/index.html (accessed September 30, 2020).

Amanzadeh, H., Yamini, Y., Moradi, M. et al. (2016). Determination of phthalate esters in drinking water and edible vegetable oil samples by headspace solid phase microextraction using graphene/polyvinylchloride nanocomposite coated fiber coupled to gas chromatography-flame ionization detector. *Journal of Chromatography. A* 1465: 38–46.

Amvrazi, E.G. (2011). *Fate of Pesticide Residues on Raw Agricultural Crops After Postharvest Storage and Food Processing to Edible Portions*. London: IntechOpen.

Amzad Hossain, M. and Salehuddin, S.M. (2012). Polycyclic aromatic hydrocarbons (PAHs) in edible oils by gas chromatography coupled with mass spectroscopy. *Arabian Journal of Chemistry* 5 (3): 391–396.

Anagnostopoulos, C. and Miliadis, G.E. (2013). Development and validation of an easy multiresidue method for the determination of multiclass pesticide residues using GC-MS/MS and LC-MS/MS in olive oil and olives. *Talanta* 112: 1–10.

Aniołowska, M. and Kita, A. (2015). The effect of type of oil and degree of degradation on glycidyl esters content during the frying of French fries. *Journal of the American Oil Chemists' Society* 92 (11–12): 1621–1631.

Aniołowska, M.A. and Kita, A.M. (2016a). Monitoring of glycidyl fatty acid esters in refined vegetable oils from retail outlets by LC-MS. *Journal of the Science of Food and Agriculture* 96 (12): 4056–4061.

Aniołowska, M.A. and Kita, A.M. (2016b). The effect of raw materials on thermo-oxidative stability and glycidyl ester content of palm oil during frying. *Journal of the Science of Food and Agriculture* 96 (6): 2257–2264.

Aniołowska, M. and Kita, A. (2016c). The effect of frying on glicydyl esters content in palm oil. *Food Chemistry* 203: 95–103.

Baer, I., De Calle, B., and Taylor, P. (2009). 3-MCPD in food other than soy sauce or hydrolysed vegetable protein (HVP). *Analytical and Bioanalytical Chemistry* 396 (1): 443–467.

Bahrami, R., Shahbazi, Y., and Nikousefat, Z. (2015). Occurrence and seasonal variation of aflatoxin in dairy cow feed with estimation of aflatoxin M1 in milk from Iran. *Food and Agricultural Immunology* 27 (3): 388–400.

Bao, L., Liang, C., Trucksess, M.W. et al. (2012). Determination of aflatoxins B1, B2, G1, and G2 in olive oil, peanut oil, and sesame oil using immunoaffinity column cleanup, postcolumn derivatization, and liquid chromatography/fluorescence detection: collaborative study. *Journal of AOAC International* 95 (6): 1689–1700.

Bao, L., Liang, C., Trucksess, M.W. et al. (2013). Determination of aflatoxins B1, B2, G1, and G2 in olive oil, peanut oil, and sesame oil using immunoaffinity column cleanup, postcolumn derivatization, and liquid chromatography with fluorescence detection: first action 2013.05. *Journal of AOAC International* 96 (5): 1017–1018.

Becalski, A., Feng, S.Y., Lau, B.P.Y. et al. (2012). Glycidyl fatty acid esters in food by LC-MS/MS: method development. *Analytical and Bioanalytical Chemistry* 403 (10): 2933–2942.

Bennett, J.W. and Klich, M. (2003). Mycotoxins. *Clinical Microbiology Reviews* 16 (3): 497–516.

Bhat, R. and Reddy, K.R.N. (2017). Challenges and issues concerning mycotoxins contamination in oil seeds and their edible oils: updates from last decade. *Food Chemistry* 215: 425–437.

Blais, J.M., Charpentié, S., Pick, F. et al. (2006). Mercury, polybrominated diphenyl ether, organochlorine pesticide, and polychlorinated biphenyl concentrations in fish from lakes along an elevation transect in the French Pyrénées. *Ecotoxicology and Environmental Safety* 63 (1): 91–99.

Bloomfield, J.P., Williams, R.J., Gooddy, D.C. et al. (2006). Impacts of climate change on the fate and behaviour of pesticides in surface and groundwater: a UK perspective. *Science of the Total Environment* 369 (1–3): 163–177.

Blumhorst, M.R., Collison, M.W., Cantrill, R. et al. (2013). Collaborative study for the analysis of glycidyl fatty acid esters in edible oils using LC-MS. *Journal of the American Oil Chemists' Society* 90 (4): 493–500.

Britannica Online Encyclopedia (2019). Pest. 1–3.

Brühl, L. (2016). Occurrence, determination, and assessment of mineral oils in oilseeds and vegetable oils. *European Journal of Lipid Science and Technology* 118 (3): 361–372.

Cachada, A., Pato, P., Rocha-Santos, T. et al. (2012). Levels, sources and potential human health risks of organic pollutants in urban soils. *Science of the Total Environment* 430: 184–192.

Camargo, M.C.R., Antoniolli, P.R., and Vicente, E. (2011). HPLC-FLD simultaneous determination of 13 polycyclic aromatic hydrocarbons: validation of an analytical procedure for soybean oils. *Journal of the Brazilian Chemical Society* 22 (7): 1354–1361.

Cao, Y., Yin, X.-Y., Zhou, F.-Q. et al. (2013). Determination of phthalates esters in edible oil by high performance liquid chromatography with solid-membrane extraction based on electrospun nylon 6 nanofibrous membrane. *Chinese Journal of Analytical Chemistry*, (Chinese version) 41: 1837.

Chen, M., Zhao, Z., Chen, Y. et al. (2015). Determination of carbendazim and metiram pesticides residues in reapeseed and peanut oils by fluorescence spectrophotometry. *Measurement* 73: 313–317.

Cheng, W.W., Liu, G.Q., Wang, L.Q. et al. (2017). Glycidyl fatty acid esters in refined edible oils: a review on formation, occurrence, analysis, and elimination methods. *Comprehensive Reviews in Food Science and Food Safety* 16 (2): 263–281.

Chung, S. and Chen, B. (2015). Development of a multiresidue method for the analysis of 33 organochlorine pesticide residues in fatty and high water content foods. *Chromatographia* 78: 565–577.

Crews, C., Chiodini, A., Granvogl, M. et al. (2013). Analytical approaches for MCPD esters and glycidyl esters in food and biological samples: a review and future perspectives. *Food Additives & Contaminants – Part A Chemistry, Analysis, Control, Exposure, Risk & Assessment* 30 (1): 11–45.

Deme, P., Azmeera, T., Prabhavathi Devi, B.L.A. et al. (2014). An improved dispersive solid-phase extraction clean-up method for the gas chromatography-negative chemical ionisation tandem mass spectrometric determination of multiclass pesticide residues in edible oils. *Food Chemistry* 142: 144–151.

DIN EN 16995 (2017). Foodstuffs – vegetable oils and foodstuff on basis of vegetable oils – determination of mineral oil saturated hydrocarbons (MOSH) and mineral oil aromatic hydrocarbons (MOAH) with on-line HPLC-GC-FID analysis. Berlin: German Institute for Standardisation.

Dost, K. and Deli, C. (2012). Determination of polycyclic aromatic hydrocarbons in edible oils and barbecued food by HPLC/UV-Vis detection. *Food Chemistry* 133 (1): 193–199.

Dubois, M., Tarres, A., Goldmann, T. et al. (2011). Determination of seven glycidyl esters in edible oils by gel permeation chromatography extraction and liquid chromatography coupled to mass spectrometry detection. *Journal of Agricultural and Food Chemistry* 59 (23): 12291–12301.

Dubois, M., Tarres, A., Goldmann, T. et al. (2012). Comparison of indirect and direct quantification of esters of monochloropropanediol in vegetable oil. *Journal of Chromatography A* 1236: 189–201.

Eom, T., Cho, H.-D., Kim, J. et al. (2017). Multiclass mycotoxin analysis in edible oils using a simple solvent extraction method and liquid chromatography with tandem mass spectrometry. *Food Additives and Contaminants Part A* 34 (11): 2011–2022.

Ermacora, A. and Hrncirik, K. (2013). A novel method for simultaneous monitoring of 2-MCPD, 3-MCPD and glycidyl esters in oils and fats. *Journal of the American Oil Chemists' Society* 90 (1): 1–8.

European Commission (2008). New rules on pesticide residues in food. Brussels: Directorate-General for Health and Consumer Protection.

European Commission (2009a). Regulation (EC) No. 1107/2009 of the European Parliament and of the Council of 21 October 2009 concerning the placing of plant protection products on the market and repealing Council Directives 79/117/EEC and 91/414/EEC. *Official Journal of the European Union* 309 (1): 1–50.

European Commission (2009b). Directive 2009/128/EC of the European Parliament and of the council of 21 October 2009. *Official Journal of the European Union* 309: 71–86.

European Commission (2010). Regulation (EC) No 396/2005 of the European Parliament and of the Council of 23 February 2005 on maximum residue levels of pesticides in or on food and feed of plant and animal origin and amending Council Directive 91/414/EEC. *Official Journal of the European Union.*

European Commission (2018). Commission Recommendation (EU) 2017/84 of 16 January 2017 on the monitoring of mineral oil hydrocarbons in food and in materials and articles intended to come into contact with food. *Official Journal of the European Union* 10 (6): 95–96.

European Commission 1322 (2020). Commission Regulation (EU) 2020/1322 of 23 September 2020 amending Regulation (EC) No 1881/2006 as regards maximum levels of 3-monochloropropanediol (3-MCPD), 3-MCPD fatty acid esters and glycidyl fatty acid esters in certain foods. *Official Journal of the European Union* 310: 2–5.

European Commission 835 (2011). Commission Regulation (EU) No 835/2011 of 19 August 2011 amending Regulation (EC) No 1881/2006 as regards maximum levels for polycyclic aromatic hydrocarbons in foodstuffs. *Official Journal of the European Union* 215: 4–8.

European Food Safety Authority (2013). Analysis of occurrence of 3-monochloropropane-1, 2-diol (3-MCPD) in food in Europe in the years 2009-2011 and preliminary exposure assessment. *EFSA Journal* 11 (9): 3381.

European Food Safety Authority (2016a). Risks for human health related to the presence of 3- and 2-monochloropropanediol (MCPD), and their fatty acid esters, and glycidyl fatty acid esters in food. *EFSA Journal* 14 (5): e04426.

European Food Safety Authority (2016b). Process contaminants in vegetable oils and foods. *EFSA Journal*: 1–3.

Farajzadeh, M.A., Feriduni, B., and Mogaddam, M.R.A. (2015). Determination of triazole pesticide residues in edible oils using air-assisted liquid-liquid microextraction followed by gas chromatography with flame ionization detection. *Journal of Separation Science* 38 (6): 1002–1009.

Food and Agriculture Organization of the United Nations (FAO), and World Health Organization (WHO) (2007). Safety evaluation of certain contaminants in food. Rome and Geneva: FAO/WHO.

Food and Agriculture Organization of the United Nations, and World Health Organization (2008). Proposed draft code of practice for the prevention and reduction of contamination of food with polycyclic aromatic hydrocarbons (PAH) from smoking and direct drying processes. Rome and Geneva: FAO/WHO.

Gharbi, I., Moret, S., Chaari, O. et al. (2017). Evaluation of hydrocarbon contaminants in olives and virgin olive oils from Tunisia. *Food Control* 75: 160–166.

Gómez-Coca, R.B., Pérez-Camino, M.C., and Moreda, W. (2016a). Saturated hydrocarbon content in olive fruits and crude olive pomace oils. *Food Additives and Contaminants Part A* 33 (3): 391–402.

Gómez-Coca, R.B., Cert, R., Pérez-Camino, M.C. et al. (2016b). Determination of saturated aliphatic hydrocarbons in vegetable oils. *Grasas y Aceites* 67 (2): e127.

Graziani, G., Gaspari, A., Chianese, D. et al. (2017). Direct determination of 3-chloropropanol esters in edible vegetable oils using high resolution mass spectrometry (HRMS-Orbitrap). *Food Additives and Contaminants Part A* 34 (11): 1893–1903.

Guo, Y., Zhang, Z., Liu, L. et al. (2012). Occurrence and profiles of phthalates in foodstuffs from China and their implications for human exposure. *Journal of Agricultural and Food Chemistry* 60 (27): 6913–6919.

Haines, T.D., Adlaf, K.J., Pierceall, R.M. et al. (2011). Direct determination of MCPD fatty acid esters and glycidyl fatty acid esters in vegetable oils by LC–TOFMS. *Journal of the American Oil Chemists' Society* 88: 1–14.

Haji Harunarashid, N.Z.I., Lim, L.H., and Harunsani, M.H. (2017). Phthalate sample preparation methods and analysis in food and food packaging: a review. *Food Analytical Methods* 10 (12): 3790–3814.

Han, Y., Song, L., Zhao, P. et al. (2016). Residue determination of glufosinate in plant origin foods using modified Quick Polar Pesticides (QuPPe) method and liquid chromatography coupled with tandem mass spectrometry. *Food Chemistry* 197: 730–736.

Hao, X., Li, J., and Yao, Z. (2016). Changes in PAHs levels in edible oils during deep-frying process. *Food Control* 66 (835): 233–240.

Hidalgo-Ruiz, J.L., Romero-González, R., Martínez-Vidal, J.L. et al. (2019a). A rapid method for determination of mycotoxins in edible vegetable oils by ultra-high performance liquid chromatography-tandem mass spectrometry. *Food Chemistry* 288: 22–28.

Hidalgo-Ruiz, J.L., Romero-González, R., Martínez Vidal, J.L. et al. (2019b). Determination of mycotoxins in nuts by ultra high-performance liquid chromatography-tandem mass spectrometry: looking for a representative matrix. *Journal of Food Composition and Analysis* 82: 103228.

Hollosi, L. and Wenzl, T. (2011). Development and optimisation of a dopant assisted liquid chromatographic-atmospheric pressure photo ionisation-tandem mass spectrometric method for the determination of 15+1 EU priority PAHs in edible oils. *Journal of Chromatography A* 1218 (1): 23–31.

Hori, K., Koriyama, N., Omori, H. et al. (2012). Simultaneous determination of 3-MCPD fatty acid esters and glycidol fatty acid esters in edible oils using liquid chromatography time-of-flight mass spectrometry. *Food Science and Technology* 48 (2): 204–208.

Hua, H., Zhao, X., Wu, S. et al. (2016). Impact of refining on the levels of 4-hydroxy-trans-alkenals, parent and oxygenated polycyclic aromatic hydrocarbons in soybean and rapeseed oils. *Food Control* 67: 82–89.

Jędrkiewicz, R., Głowacz-Różyńska, A., Gromadzka, J. et al. (2017). Novel fast analytical method for indirect determination of MCPD fatty acid esters in edible oils and fats based on simultaneous extraction and derivatization. *Analytical and Bioanalytical Chemistry* 409 (17): 4267–4278.

Joint FAO/WHO Expert Committee on Food Additives (JECFA) (2018). Report of the 50th session of the Codex Committee on pesticide residues. Rome and Geneva: FAO/WHO

Kiely, T., Donaldson, D., and Grube, A. (2004). Pesticide industry sales and usage. 2000 and 2001 market estimates. Washington, DC: US Environment Protection Agency.

Koesukwiwat, U., Lehotay, S.J., Mastovska, K. et al. (2010). Extension of the QuEChERS method for pesticide residues in cereals to flaxseeds, peanuts, and doughs. *Journal of Agricultural and Food Chemistry* 58 (10): 5950–5958.

Kuhlmann, J. (2011). Determination of bound 2,3-epoxy-1-propanol (glycidol) and bound monochloropropanediol (MCPD) in refined oils. *European Journal of Lipid Science and Technology* 113 (3): 335–344.

Kuhlmann, J. (2016). Analysis and occurrence of dichloropropanol fatty acid esters and related process-induced contaminants in edible oils and fats. *European Journal of Lipid Science and Technology* 118 (3): 382–395.

Lacoste, F. (2014). Undesirable substances in vegetable oils: anything to declare? *OCL* 21 (1): A103.

Li, H., Chen, D., Miao, H. et al. (2015). Direct determination of fatty acid esters of 3-chloro-1, 2-propanediol in edible vegetable oils by isotope dilution – ultra high performance liquid chromatography – triple quadrupole mass spectrometry. *Journal of Chromatography A* 1410: 99–109.

Li, B., Wu, Y., Liu, L. et al. (2016). Determination of mineral oil-saturated hydrocarbons (MOSH) in vegetable oils by large scale off-line SPE combined with GC-FID. *Journal of the American Oil Chemists' Society* 94 (2): 215–223.

Liu, Y., Shen, D., and Tang, F. (2012). Multiresidue determination of organophorous pesticides in camellia oil by matrix solid-phase dispersion followed by GC-FPD. *Bulletin of Environmental Contamination and Toxicology* 89 (5): 1057–1061.

Liu, L., Huang, H., Wu, Y. et al. (2017). Offline solid-phase extraction large-volume injection-gas chromatography for the analysis of mineral oil-saturated hydrocarbons in commercial vegetable oils. *Journal of Oleo Science* 66 (9): 981–990.

Ma, F., Wu, R., Li, P. et al. (2016). Analytical approaches for measuring pesticides, mycotoxins and heavy metals in vegetable oils: a review. *European Journal of Lipid Science and Technology.* 118 (3): 339–352.

MacMahon, S. and Beekman, J. (2019). 3-Chloro-1,2-propanediol (3-MCPD), 2-chloro-1,3-propanediol (2-MCPD) and glycidyl esters in infant formula: a review. *Current Opinion in Food Science* 30: 67–72.

MacMahon, S., Mazzola, E., Begley, T.H. et al. (2013a). Analysis of processing contaminants in edible oils. Part 1. Liquid chromatography-tandem mass spectrometry method for the direct detection of 3-monochloropropanediol monoesters and glycidyl esters. *Journal of Agricultural and Food Chemistry* 61 (20): 4737–4747.

MacMahon, S., Begley, T.H., and Diachenko, G.W. (2013b). Occurrence of 3-MCPD and glycidyl esters in edible oils in the United States. *Food Additives and Contaminants: Part A Chemistry, Analysis, Control, Exposure, and Risk Assessment* 30 (12): 2081–2092.

Marin, S., Ramos, A.J., Cano-Sancho, G. et al. (2013). Mycotoxins: occurrence, toxicology, and exposure assessment. *Food and Chemical Toxicology* 60: 218–237.

Marley, E., Brown, P., Mackie, J. et al. (2015). Analysis of sterigmatocystin in cereals, animal feed, seeds, beer and cheese by immunoaffinity column clean-up and HPLC and LC-MS/MS quantification. *Food Additives and Contaminants: Part A Chemistry, Analysis, Control, Exposure, and Risk Assessment* 32 (12): 2131–2137.

Masukawa, Y., Shiro, H., Nakamura, S. et al. (2010). A new analytical method for the quantification of glycidol fatty acid esters in edible oils. *Journal of Oleo Science* 59 (2): 81–88.

Masukawa, Y., Shiro, H., Kondo, N. et al. (2011). Generalized method to quantify glycidol fatty acid esters in edible oils. *Journal of the American Oil Chemists' Society* 88 (1): 15–21.

McCulloch, R., Alvaro, A., Astudillo, A.M. et al. (2017). A novel atmospheric pressure photoionization – mass spectrometry (APPI-MS) method for the detection of polychlorinated dibenzo P- dioxins and dibenzofuran homologues in real environmental samples collected within the vicinity of industrial incinerators. *Internation Journal of Mass Spectrometry* 421: 135–143.

Molle, D.R.D., Abballe, C., Gomes, F.M.L. et al. (2017). Polycyclic aromatic hydrocarbons in canola, sunflower and corn oils and estimated daily intake. *Food Control* 81: 96–100.

Moreno-González, D., Huertas-Pérez, J.F., García-Campaña, A.M. et al. (2014). Determination of carbamates in edible vegetable oils by ultra-high performance liquid chromatography-tandem mass spectrometry using a new clean-up based on zirconia for QuEChERS methodology. *Talanta* 128: 299–304.

Moret, S., Barp, L., Grob, K. et al. (2011). Optimised off-line SPE-GC-FID method for the determination of mineral oil saturated hydrocarbons (MOSH) in vegetable oils. *Food Chemistry* 129 (4): 1898–1903.

Muhamad, H., Zainudin, B.H., and Abu Bakar, N.K. (2012). Comparative study of different clean-up techniques for the determination of λ-cyhalothrin and cypermethrin in palm oil matrices by gas chromatography with electron capture detection. *Food Chemistry* 134 (4): 2489–2496.

National Association of State Departments of Agriculture (NASDA) (n.d.). Pest management. 1–18. https://s3.amazonaws.com/nasda2/media/Reports/Core-Chpt-1_Pest-Management.pdf?mtime=20171025135710 (accessed September 30, 2020).

Nestola, M. and Schmidt, T.C. (2017). Determination of mineral oil aromatic hydrocarbons in edible oils and fats by online liquid chromatography–gas chromatography–flame ionization detection – evaluation of automated removal strategies for biogenic olefins. *Journal of Chromatography A* 1505: 69–76.

Ongoma, V. (2013). A review of the effects of climate change on occurrence of aflatoxin and its impacts on food security in semi-arid areas of Kenya. *International Journal of Agricultural Science Research* 2 (11): 307–311.

Ostrovský, I., Čabala, R., Kubinec, R. et al. (2011). Determination of phthalate sum in fatty food by gas chromatography. *Food Chemistry* 124 (1): 392–395.

Plaza-Bolaños, P., Frenich, A.G., and Vidal, J.L.M. (2010). Polycyclic aromatic hydrocarbons in food and beverages. Analytical methods and trends. *Journal of Chromatography A* 1217 (41): 6303–6326.

Polgár, L., Kmellár, B., García-Reyes, J.F. et al. (2012). Comprehensive evaluation of the clean-up step in QuEChERS procedure for the multi-residue determination of pesticides in different vegetable oils using LC-MS/MS. *Analytical Methods* 4 (4): 1142–1148.

Qian, M., Zhang, H., Wu, L. et al. (2015). Simultaneous determination of zearalenone and its derivatives in edible vegetable oil by gel permeation chromatography and gas chromatography-triple quadrupole mass spectrometry. *Food Chemistry* 166: 23–28.

Rakkestad, K.E., Dye, C.J., Yttri, K.E. et al. (2007). Phthalate levels in Norwegian indoor air related to particle size fraction. *Journal of Environmental Monitoring* 9 (12): 1419–1425.

Ramli, M.I., Zakaria, Z., Sahid, I. et al. (2012). Determination of herbicide diuron levels in palm oil matrices using HPLC-UV. *Sains Malaysiana* 41 (11): 1451–1459.

Rapid Alert System for Food and Feed (2019). The Rapid Alert System for Food and Feed 2018 annual report. Luxembourg: Publications Office of the European Union.

Ravindra, K., Sokhi, R., and Van Grieken, R. (2008). Atmospheric polycyclic aromatic hydrocarbons: source attribution, emission factors and regulation. *Atmosperic Environment* 42 (13): 2895–2921.

Rios, J.J., Morales, A., and Márquez-Ruiz, G. (2010). Headspace solid-phase microextraction of oil matrices heated at high temperature and phthalate esters determination by gas chromatography multistage mass spectrometry. *Talanta* 80 (5): 2076–2082.

Rudel, R.A. and Perovich, L.J. (2009). Endocrine disrupting chemicals in indoor and outdoor air. *Atmospheric Environment* 43 (1): 170–181.

Ruiz-Medina, A., Llorent-Martínez, E.J., Fernández-de Córdova, M.L. et al. (2012). Automated optosensor for the determination of carbaryl residues in vegetable edible oils and table olive extracts. *Journal of Food Composition and Analysis* 26 (1–2): 66–71.

Sharmili, K., Jinap, S., and Sukor, R. (2016). Development, optimization and validation of QuEChERS based liquid chromatography tandem mass spectrometry method for determination of multimycotoxin in vegetable oil. *Food Control* 70: 152–160.

Shiro, H., Kondo, N., Kibune, N. et al. (2011). Direct method for quantification of glycidol fatty acid esters in edible oils. *European Journal of Lipid Science and Technology* 113: 356–360.

da Silva, S.A., Sampaio, G.R., and Torres, E.A.F.d.S. (2017). Optimization and validation of a method using UHPLC-fluorescence for the analysis of polycyclic aromatic hydrocarbons in cold-pressed vegetable oils. *Food Chemistry* 221: 809–814.

Sobhanzadeh, E., Bakar, N.K.A., Abas, M.R.B. et al. (2012). A simple and efficient multiresidue method based on QuEChERS for pesticides determination in palm oil by liquid chromatography time-of-flight mass spectrometry. *Environmental Monitoring and Assessment* 184 (9): 5821–5828.

Song, Z., Wang, Y., Li, G. et al. (2015). A novel 1H NMR spectroscopic method for determination of glycidyl fatty acid esters coexisting with acylglycerols. *European Journal of Lipid Science and Technology* 117 (7): 918–925.

Spanish Association for Standardization and Certification (AENOR) (2012). UNE-EN ISO 15753. Animal and vegetable fats and oils. Determination of polycyclic aromatic hydrocarbons. Madrid: AENOR.

Standing Committee on the Food Chain and Animal Health (2008). Summary minutes of the meeting of the Standing Committee on the food chain and animal health. June 2008, 1–5. Brussels: European Commission.

Steenbergen, H., Hrnčiřík, K., Ermacora, A. et al. (2013). Direct analysis of intact glycidyl fatty acid esters in edible oils using gas chromatography-mass spectrometry. *Journal of Chromatography A* 1313: 202–211.

Sun, Y. and Wu, S. (2020). Analysis of PAHs in oily systems using modified QuEChERS with EMR-lipid clean-up followed by GC-QqQ-MS. *Food Control* 109: 106950.

Sun, Y., Wu, S., and Gong, G. (2019). Trends of research on polycyclic aromatic hydrocarbons in food: a 20-year perspective from 1997 to 2017. *Trends in Food Science and Technology* 83: 86–98.

Tranchida, P.Q., Zoccali, M., Purcaro, G. et al. (2011). A rapid multidimensional liquid-gas chromatography method for the analysis of mineral oil saturated hydrocarbons in vegetable oils. *Journal of Chromatography A* 1218 (42): 7476–7480.

Tuzimski, T. and Rejczak, T. (2014). Determination of pesticides in sunflower seeds by high-performance liquid chromatography coupled with a diode array detector. *Journal of AOAC International* 97 (4): 1012–1020.

Tuzimski, T. and Rejczak, T. (2016). Application of HPLC-DAD after SPE/QuEChERS with ZrO-based sorbent in d-SPE clean-up step for pesticide analysis in edible oils. *Food Chemistry* 190: 71–79.

United States Environmental Protection Agency (2000). Deposition of air pollutants to the Great Waters. Research Triangle Park, NC: US EPA.

US Environmental Protection Agency (2019). What is a pesticide. https://www.epa.gov/ingredients-used-pesticide-products/basic-information-about-pesticide-ingredients (accessed September 28, 2020).

Wang, Z., Ren, P., Sun, Y. et al. (2013). Gas/particle partitioning of polycyclic aromatic hydrocarbons in coastal atmosphere of the north Yellow Sea, China. *Environmental Science and Pollution Research International* 20 (8): 5753–5763.

Weber, S., Schrag, K., Mildau, G. et al. (2018). Analytical methods for the determination of mineral oil saturated hydrocarbons (MOSH) and mineral oil aromatic hydrocarbons (MOAH): a short review. *Analytical Chemistry Insights* 13: 117739011877775.

Wei, S., Liu, Y., Yan, Z. et al. (2015). Molecularly imprinted solid phase extraction coupled to high performance liquid chromatography for determination of aflatoxin M1 and B1 in foods and feeds. *RSC Advances* 5 (27): 20951–20960.

Weißhaar, R. and Perz, R. (2010). Fatty acid esters of glycidol in refined fats and oils. *European Journal of Lipid Science and Technology* 112 (2): 158–165.

World Health Organization (2003). Polynuclear aromatic hydrocarbons in drinking-water: background document for development of WHO guidelines for drinking-water quality Geneva: WHO.

World Health Organization, and International Agency for Research on Cancer (2010). *Some Non-Heterocyclic Polycyclic Aromatic Hydrocarbons Some Related Exposures*, IARC Monographs on the Evaluation of Carcinogenic Risks to Humans, vol. 92. Geneva; Lyon: WHO; IARC.

Wu, S. and Yu, W. (2012). Liquid-liquid extraction of polycyclic aromatic hydrocarbons in four different edible oils from China. *Food Chemistry* 134 (1): 597–601.

Xie, Z., Lakaschus, S., Ebinghaus, R. et al. (2006). Atmospheric concentrations and air-sea exchanges of nonylphenol, tertiary octylphenol and nonylphenol monoethoxylate in the North Sea. *Environmetal Pollution* 142 (1): 170–180.

Xu, Q., Yin, X., Wang, M. et al. (2010). Analysis of phthalate migration from plastic containers to packaged cooking oil and mineral water. *Journal of Agricultural and Food Chemistry* 58 (21): 11311–11317.

Xu, D., Deng, X., Fang, E. et al. (2014). Determination of 23 phthalic acid esters in food by liquid chromatography tandem mass spectrometry. *Journal of Chromatography A* 1324: 49–56.

Yao, Z., Li, J., Wu, B. et al. (2015). Characteristics of PAHs from deep-frying and frying cooking fumes. *Environmental Science and Pollution Research* 22 (20): 16110–16120.

Yousefi, M., Shemshadi, G., Khorshidian, N. et al. (2018). Polycyclic aromatic hydrocarbons (PAHs) content of edible vegetable oils in Iran: A risk assessment study. *Food and Chemical Toxicology* 118 (May): 480–489.

Yu, L., Ma, F., Zhang, L. et al. (2019). Determination of aflatoxin B1 and B2 in vegetable oils using Fe3O4/rGO magnetic solid phase extraction coupled with high-performance liquid. *Toxins (Basel)* 11: 621.

Zhao, Q., Wei, F., Luo, Y.B. et al. (2011). Rapid magnetic solid-phase extraction based on magnetic multiwalled carbon nanotubes for the determination of polycyclic aromatic hydrocarbons in edible oils. *Journal of Agricultural and Food Chemistry* 59 (24): 12794–12800.

Zhao, Z.Y., Chu, Y.L., and Gu, J.D. (2012). Distribution and sources of polycyclic aromatic hydrocarbons in sediments of the Mai Po Inner Deep Bay Ramsar Site in Hong Kong. *Ecotoxicology* 21 (6): 1743–1752.

Zhao, H., Chen, X., Shen, C. et al. (2016). Determination of 16 mycotoxins in vegetable oils using a QuEChERS method combined with high-performance liquid chromatography-tandem mass spectrometry. *Food Additives and Contaminants. Part A Chemistry, Analysis, Control, Exposure and Risk Assessment* 34 (2): 255–264.

Zhao, Y., Wan, L.H., Bai, X.L. et al. (2017). Quantification of mycotoxins in vegetable oil by UPLC-MS/MS after magnetic solid-phase extraction. *Food Additives and Contaminants. Part A Chemistry, Analysis, Control, Exposure and Risk Assessment* 34 (7): 1201–1210.

Zhao, X., Gong, G., and Wu, S. (2018). Effect of storage time and temperature on parent and oxygenated polycyclic aromatic hydrocarbons in crude and refined vegetable oils. *Food Chemistry* 239: 781–788.

Zoccali, M., Barp, L., Beccaria, M. et al. (2016). Improvement of mineral oil saturated and aromatic hydrocarbons determination in edible oil by liquid-liquid-gas chromatography with dual detection. *Journal of Separation Science* 39 (3): 623–631.

Zurfluh, M., Biedermann, M., and Grob, K. (2014). Enrichment for reducing the detection limits for the analysis of mineral oil in fatty foods. *Journal fur Verbraucherschutz und Lebensmittelsicherheit* 9 (1): 61–69.

9

By-Products from Oilseed Processing and Their Potential Applications

María Janeth Rodríguez-Roque[1], Rogelio Sánchez-Vega[2], Ramona Pérez-Leal[1], Mayra Cristina Soto-Caballero[1], Nora Aideé Salas-Salazar[1] and María Antonia Flores-Córdova[1]

[1] *Faculty of Agrotechnological Sciences, Autonomous University of Chihuahua, Chihuahua, México*
[2] *Faculty of Zootechnics and Ecology, Autonomous University of Chihuahua, Chihuahua, México*

9.1 Introduction

Oilseeds are seeds from different plant species containing a high level of oil and other food fats, as well as a range of proteins and carbohydrates in smaller quantities (Rodrigues et al. 2012). According with the Food and Agriculture Organization of United Nations (FAO) (FAO 2020), the most produced oilseeds worldwide are soybean, sunflower, and rapeseed. Canola, coconut, cotton, groundnut, linseed palm fruit, peanut, sesame, and pulses are also considered as interesting oilseeds in the market. Pulses are defined as the dried edible seeds of cultivated legumes, such as beans, peas, and lentils.

The main application of oilseeds is oil extraction by means of different technologies that are selected on the basis of the initial commodity and the desired final product. The extracted oils are usually used as edible oils for human consumption. Once the oil is extracted, there remain several by-products, such as shells, leaves, petals, stalks, pollen, heads, and pith.

There is a need to reassess these by-products both to reduce the amount of food waste and to add value to the production process. According with the European Commission, around ninety million tons of food waste is generated in the European Union per year (European Commission, Report T 2010). The environmental impact of food waste is a great problem but at the same time its reuse is a challenge. However, we should be aware that these by-products are important source of compounds that can be ingredients in a wide range of new products. Using these by-products makes it possible to contribute to the sustainability of food chain and, at the same time, help to reduce the environmental impact of this waste.

Oil and Oilseed Processing: Opportunities and Challenges, First Edition.
Edited by Tomás Lafarga, Gloria Bobo, and Ingrid Aguiló-Aguayo.

Oilseed by-products, are primarily used as ingredients for feeding animals and pets. However, since oilseed by-products contain important amounts of carbohydrates, proteins, fiber, minerals, vitamins, phenolic compounds, lignans, and tocopherols, it could be interesting to explore other areas that allow us to take advantage of their high nutritional content.

In this sense, oilseed by-products can be feedstock for producing enriched foods, such as flour, milk, sauce, concentrate and isolates, margarine, meat analogues, shortening, among others. They can be a source of bioactive ingredients such as carbohydrates, proteins, fiber, minerals, vitamins, phenolic compounds, lignans, and tocopherols. Thus, oilseed by-products can be used as ingredients for industrial products (i.e. personal care products, bakery, stock-feeds/supplements for ruminant animals), for producing food proteins (in form of eggs, meat, or milk), or creating innovative products containing oilseed by-products with applications in the food, agro-, pharmaceutical, and cosmetic industries. Adding value to oilseed by-products is achieved by resorting both skills and knowledge for properly combining these by-products with raw materials, ingredients, and/or foodstuffs. In this way, undesirable interactions that decrease the nutritional value of these products could be avoided. Additionally, it is necessary to look for environmentally friendly technologies that may provide sustainable solutions in the revalorization of oilseed by-products.

This chapter provides a review of the main by-products obtained from oilseed processing, their origin, characteristics, and composition. Moreover, the nutritional composition and functional properties are also considered. The final section of this chapter comprises the current applications in the valorization of oilseed by-products, as well as the future trends.

9.2 Oilseed by-Products: Origin, Characteristics, and Composition

Once oilseeds have been harvested, they are incorporated into the processing phase, which according to the FAO mainly consists of the following steps:

- Raw material preparation
- Conditioning
- Extraction
- Clarification
- Oil packaging

As shown in Figure 9.1, oilseeds processing generates several by-products which in this chapter are divided in two groups: first, those come from unprocessed oilseeds (shells, leaves, petals, stalks, pollen, heads and pith), and second, those come from processed oilseeds (cake). This section describes the origin, characteristics, and composition of oilseed by-products.

9.2.1 By-Products from Unprocessed Oilseeds

Oilseeds must be prepared before processing for removing shells, leaves, petals, stalks, pollen, heads, and pith, with these becoming the first by-products generated during oilseed processing. An overview of the composition of oilseeds by-products and their main uses is presented in Table 9.1.

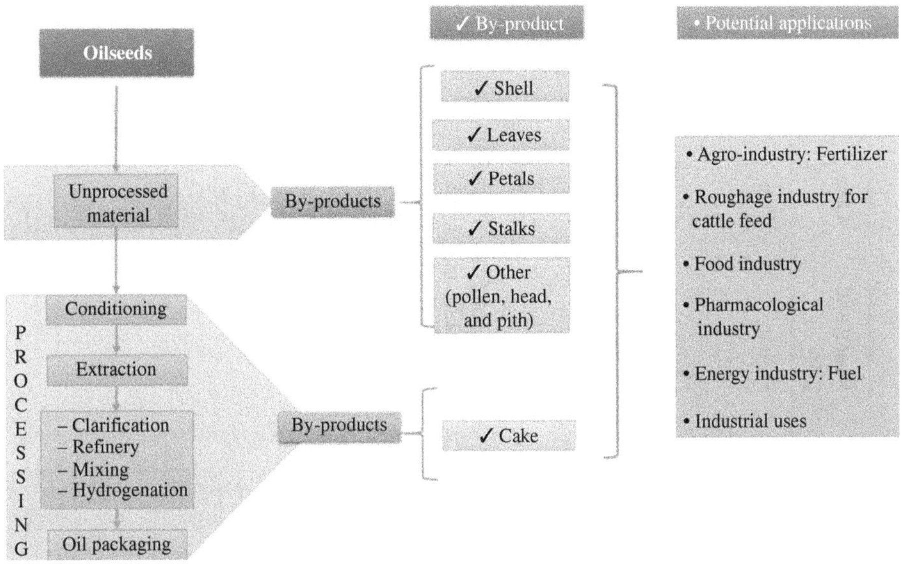

Figure 9.1 By-products generated during oilseed processing and their potential applications.

9.2.1.1 Shells, Hulls or Husk

Shells, hulls, or husk, represent approximately 5–30% of total oilseed weight, they contain around 38% fiber, 12% protein plus other constituents such as carbohydrates, minerals, and lipids, depending on the source oilseed.

There is a great interest in revalorizing shells since their protein represents a quarter of that contained in the cake after oil extraction. Several applications have been found for shells from oilseeds (Nagaraj 2009), of which the main are:

- Fertilizer: Sells also can be pulverized and used to enrich compost since they contain around 1.24% of potash.
- Fuel: They can be used as fuel either directly or combined with other compounds (e.g. dung).
- Roughage for cattle feed: Shells are mixed with molasses and mineral supplements.
- Litter for poultry houses
- Industrial applications: high grade activated charcoal, abrasive for polishing steel and aluminum, insulation for farm buildings, compressed blocks, filler for plastic wood, a base for veneer boards or bottle caps.
- Pellet: A short time ago, shells were used fresh, but nowadays there is a new product derived from shells called pellet with the potential for being used as ingredient in human or animal diets. In this sense, it has been reported that production of the pellet from soy shells is cheaper than that of oilseed.

Groundnut possess two shell types; the external shell is a thick woody shell while the inner is a thick skin called "testa." The color of testa in groundnut can be red, brown, purple or white. The composition of testa consist of carbohydrates (48–52%), crude fiber (21–35%), protein (11–13%), oil (0.5–2%), reducing sugar (1–1.2%), ash (2–5%).

Safflower shells represent around 45–50% of the seeds, and contain around 70% cellulose, 40% fiber, 21% lignin, 1% ash, and protein. The high percentage of hull

Table 9.1 Composition of oilseeds' by-products and their main uses.

	Shells	Leaves	Petals	Stalks	Pollen	Heads and pith	Cake
Composition	Fiber	Minerals	Fatty acids	Carbohydrates, lignin, and cellulose	Minerals	Protein	Fiber
	Protein Carbohydrates Minerals Lipids	Protein Carbohydrates Carotenoids Fiber	Minerals		Amino acids	Lignin Micronutrients lipids	Proteins Carbohydrates Oil Fatty acids
Main uses	Organic fertilizer	Aromatic herbs	Dye	Biomass: paper industry and hard boards	Pharmacological industry	Animal feeding	Organic fertilizer
	Bio-Fuel: bioethanol	Pharmaceutical use	Color and flavor food		Food industry	Pharmaceutical use (pains and infection)	Animal feeding
	Roughage		Cosmetic use				Human food as edible flour, extruded snacks, flours, fermented products.
	Industrial applications		Organic fertilizer				Protein isolated and concentrates
	Pellet						Fuel
	Food industry: protein concentrate and isolated, supplement, confectionary and bakery products						Baked products (i.e. bread, biscuit, bun, rusk, and cake)
							Texturized soy protein

must reduce the nutritional value of this oilseed and depending on their composition can be used for feeding different animal types. For this, it is graded in two groups:

1. Low fiber content – Contains 42% crude protein and 16% crude fiber. It is used for feeding monogastric animals.
2. High fiber content – Contains 20% crude protein and 30% crude fiber. It is used for feeding ruminants.

Sunflower hulls comprise between 20 and 30% of the seeds, containing great amounts of fiber and low quantities of fat and ash. The applications of this by-products are: fuel, production of ethanol and furfurol, substrate for microorganisms (i.e. for growing yeasts), organic fertilizer (rich source of phosphorus and potassium), and in lining plywood.

Although **sesame** is a small seed (4 mm long, 2 mm wide, and 1 mm thick), it contains hull (testa), endosperm, cotyledons, and embryo. Hulls represent around 10–20% of the seed, providing features such as color (white to brown) and bitterness, but also important amounts of oxalate and non-digestible fiber. Overall, white varieties contain higher hull quantity. Hulls are difficult to remove from sesame oilseed, but it can be replaced by soaking in water overnight and by rubbing between rough surfaces. To obtain a high-quality cake it is necessary to remove the hull from sesame seed, prior to oil extraction. By this way, a protein concentrate rich in amino acids containing sulfur can be obtained. Sesame flour can be also produced from cake, and it can be used as supplement with high amino acid content, or as ingredient in confectionary and bakery products.

On the other hand, **rapeseed hull** is mainly composed of carbohydrates such as cellulose, hemicellulose, and lignin with a low quantity of antinutritional carbohydrates (raffinose and stachyose) (Shahidi et al. 1990). This by-product can be used for producing bio-fuel, as well as a substrate for microorganism growth, and as an ingredient in the food industry due to the low level of antinutrients.

Around 8% of hull is contained in **soybean** seed, containing 9% crude protein, 1% crude fat (oil), 86% carbohydrates and 4.5% ash. The high protein content of this by-product makes it suitable to be applied as an ingredient in the food industry. Bio-fuel production is another application that can be considered for revalorizing soybean hull.

9.2.1.2 Leaves, Petals, and Stalks

As with shells, *leaves*, *petals*, and *stalks* are obtained from oilseeds, prior to processing. Overall, these by-products can be used for the animal feed (Rodrigues et al. 2012), food, and pharmaceutical industries.

- Rapeseed leaves – animal feed
- Safflower leaves – pharmaceutical effects and food industry.
- Sesame leaves – pharmaceutical effects.
- Safflower stalks – industrial use as source of lignocellulosic biomass
- Sunflower stalks – fuel, industrial use, fertilizer, and substrate for microorganisms.
- Safflower petals – colorant for food industry, industrial and cosmetic uses.
- Safflower pollen – food industry.
- Sunflower head – pharmaceutical effects and food industry.
- Sunflower pith – pharmaceutical effects.

Sesame leaves have a great amount of minerals. They are similar in flavor to fennel and licorice and have a strong aroma. Thus, they can be used as an aromatic herb. Sesame leaves are also cooked for preparing folk medicine – mainly in Asiatic countries.

Safflower leaves, stalks, and petals are by-products with high nutritional value and palatable, similar to that of oats and alfalfa. **The leaves** contain an important amount of vitamin A, thiamine, riboflavin, vitamin C, minerals (iron, phosphorus, and calcium), and carotene, as well as protein (28%), carbohydrates (44%), and fiber (9%). Safflower leaves have a pharmaceutical effect in reducing myocardial ischemia, thrombosis, inflammation, and cancer, but also have undesirable effects on male and female fertility and are used as an abortive when consumed in tea (Delshad et al. 2018). They can be directly applied to swellings and fractures. In the food industry, young, thin leaves are cooked and used as a vegetable side-dish. **Safflower stalks** are generated in large quantities – around one ton per hectare, which correspond to 7×10^5 tons/annum of lignocellulosic biomass used in the industry of paper and hard boards. The content of lignocellulose in stalks is distributed as follows: lignin (4%), alpha cellulose (14.4%), beta cellulose (14.1%), and gamma cellulose (17.7%) (Nagaraj 2009). **Safflower petals** are generated at around 100 kg/ha. This by-product's importance is its medicinal/therapeutic use, as well as the high content of natural pigments. Safflower petals are red and orange and contain carthamin and carthamidin, the proportion and concentration of which depends on the petal color (Rudometova et al. 2001). Carthamin is a water insoluble red dye that can be solubilized into alkaline medium, its proportion within florets is around 0.3–0.6%. On the other hand, carthamidin is a water-soluble dye contained in yellow petals with a proportion of 28–36%. Petals from safflower contain important amounts of low molecular weight fatty acids, such as decanoic, dodecanoic, and myristic acids. In addition they also contain palmitic, stearic, oleic, linoleic, and linolenic fatty acids. Medium-chain fatty acids, such as gamma linoleic acid possess antimicrobial effects and also prevent cracking of skin, skin aging, and eczema. The mineral composition of safflower petals is calcium, magnesium, iron, copper, zinc, and manganese at 530, 287, 7.3, 0.8, 1.0, and 0.8 mg/100 g, respectively (Nagaraj 2009). Petals are used for color and flavor foods such as soups, rice, bread, sauces, yogurt, and pickles. After drying, they are also used as a saffron substitute. Other industrial uses of safflower petals are as a dye for clothes, carpets, cotton, and silk and as ointment for religious ceremonies (Zhaomu and Lijie 2001). Safflower petals are used in the cosmetics industry in lipsticks and cosmetic rouge. When the safflower plant is charred and made soot, other types of cosmetic can be made. One of these is a traditional Egyptian cosmetic, *kohl*, has been produced by this method. After color extraction, petals can be further valorized using an anaerobic fermentation technique where a microbial consortium is used to obtain a bio-enriched product with high protein content. This bio-enriched by-product is used for poultry feed.

Sunflower petals are an important source of yellow and purple dyes with applications in the food industry, as well as industrial uses. **Sunflower stalks** are considered as residue, however, they are a rich source of phosphorus and potassium and their ash can fertilize the soil. In some cases, sunflower stalks have been combined with *Trichoderma viridae* to achieve fast stalk decomposition. They can also be mixed with farmyard manure to help meet the nutritional requirement of the crops (Nagaraj 2009).

9.2.1.3 Other by-Products from Some Oilseeds

Pollen is a valued ingredient, which is easily obtained and contains a high nutrient content. Around 4% of pollen is collected from the safflower corolla, it contains amino acids and microelements.

Heads and pith from sunflowers can be used for feeding animals. Heads contain around 12.5% crude protein, 12% lignin, 11% ash, 2% lipids, pectin (150–250 g/kg), and inorganic micronutrients (183 mg/kg). The head of a sunflower can be consumed as an infusion for alieving chest pain, while pectin has a wide range of application in the food industry as a thickener. Sunflower pith can be used to reduce pains and infection, as well as to treat burns.

9.2.2 By-Products from Oilseed Processing

There are two ways to obtain this by-product, the first when oilseed is processed by mechanical methods (i.e. pressing or milling) and the second by extraction with organic solvents (i.e. hexane, xylene, and toluene). The composition of this by-product depends on the processing type. Cakes obtained by mechanical methods are rich in fiber, oils and fatty acids but they have a lower protein levels than those obtained by solvent extraction. On the other hand, cakes obtained through solvents are rich in proteins but low in fats. The cake expelled by mechanical processing contains more than 10% of residual oil while those coming from solvent extraction around 1% (Schmidt and Mejia 2014). Due to their high protein content, cakes are usually used for manufacturing protein rich foods for both human and animal consumption Table 9.1.

Groundnut cake is obtained after oil extraction. It contains around 6% of oil but after extraction of residual oil with solvents, this percentage diminishes. It also contains important amounts of proteins and carbohydrates. It can be used as organic fertilizer due to the presence of phosphoric acid (1–4%) and potash (1–2%), as well as animal feed (poultry, calves, and ducklings), human food as flour (alone or in combination with wheat) and as a diet for diabetic patients. An edible flour with low fat and high protein content can be obtained from groundnut cake by decorticating the pods and roasting/decuticling/ degermening the kernels.

Safflower cake from solvent extraction is a rich source of proteins (up to 60%) but it also contains high amounts of crude fiber and a strong bitter flavor, which is reduced by applying pre-treatments. Dehulling is one of the main pretreatments applied to safflower for reducing the crude fiber content and thus, making it suitable for human consumption. Safflower cake composition depends on the oilseed processing type. In this sense, it has been reported significant differences in the content of protein, oil, crude fiber, N-free extract, and minerals in rotary pressed cake versus that solvent extracted. Overall, this by-product is used for animal feeding but an amino acid supplementation with lysine and methionine must be considered to improve its nutritional quality because its protein is low in amino acids. Another safflower cake application is as manure, but in comparison with groundnut hull its decomposition is slower.

Sunflower cake possesses a high fiber content that is usually used as feedstuffs when it has been defatted. Its composition depends on both method of extraction and hull content. Decorticated cake contains higher protein and fat content compared with partially or undehulled cake, reaching up to 42% of protein and 10% of fat. A large amount of carbohydrates and fiber were observed in undehulled cake, corresponding to 32 and 29%, respectively (Nagaraj 2009). The color varies from gray to black. When sunflower cake has not been dehulled or only partially dehulled, it is used for feeding ruminant animals. On the other hand, when cake is completely dehulled, it is used for feeding swine or poultry. Sunflower cake has been also utilized as protein isolates in snacks, biscuits and similar items as sunflower cake flour. However, protein isolates have shown an important improvement in the protein content of these foodstuffs (by 29%) in comparison with that obtained in flour (by 3–4%).

Rapeseed cakes are rich in proteins (up to 52%), which is the major constituent, but also contain sugars, oil, and crude fiber. They are mainly used as fuel, fertilizer, or as livestock feed. To improve the quality of rapeseed cake, it is important to take off the hulls prior to oilseed processing. **Sesame cake** consists of oil (13% by applying pressure, and 1–3% by solvent extraction), proteins (35% in whole cakes, and 46% in dehulled cake), carbohydrates (up to 35%), minerals (5–10%), and oxalic acid (4% and 0.5% in whole and dehulled cakes, respectively). For its nutritional composition, sesame cake is a valued feedstock for producing human food, while low quality cakes can be utilized as organic fertilizer.

On the other hand, **soybean cake** is used for protein supplementation for all kind of animals, for this reason is the most widely utilized oilseed by-product. Its composition consists of around 44–50% of crude proteins and only a little proportion of oil (1%) since this oilseed is mainly processed by solvent extraction. Soybeans cakes are good protein supplements, as well as meat and fish protein substitutes. They can be added to food for improving their properties (i.e. emulsification, water absorption and adhesion) and nutritional content (i.e. children and adults consuming soy protein flour improved their nutritional status). This by-product, can be used as an ingredient in baked products (i.e. bread, biscuit, bun, rusk, and cake), texturized soy protein, protein isolates and concentrates, extruded snacks, flours, and fermented products. Similarly, edible defatted soy flour can be obtained from soybean cake that comes from solvent extraction. However, operations such as cleaning, cracking, drying, conditioning, and flaking must be carried out. The edible defatted soy flour is utilized for manufacturing isolated and protein concentrates from soy. It is important to note that soy flour possesss a similar composition as non-fat milk power and in overall, this product can be used as milk substitute for people with galactosemia and lactose intolerance (Rodríguez-Roque et al. 2013).

9.3 Nutritional Composition and Functional Properties of Oilseed by-Products

The nutritional composition of oilseed by-products depends on both oilseed type and the processing conditions. Similarly, their functional properties are related to the composition, structural, and physicochemical features. Overall, oilseed by-products contain carbohydrates, proteins, fiber, minerals, vitamins, phenolic compounds, lignans, tocopherols, and antinutritional compounds (i.e. glucosinolates, phytic and oxalic acids, erucic and brassidic acids, stachyose, and raffinose).

9.3.1 Carbohydrates

They are contained in oilseed meals by around 48%. The composition of this constituent depends on the oilseed type. In rapeseed meals, carbohydrates are contained in lower proportion compared with proteins, mainly as amyloid compounds, cellulose (2.8–7%), pectin (6.5–14.5%), arabinogalactan (3.5%), arabinan (6.9%), amyloid (15.5%) (Rodrigues et al. 2012: Naczk and Shahidi 1990), fructose (0.15–2.1%), glucose (0.2–0.96%), raffinose (0.3–1.2%), stachyose (1.5–2.4%), and sucrose (1.7–7.5%) (Nagaraj 2009).

Safflower de-oiled hulls contain the following carbohydrates: cellulose (35%), lignin (24%), pentosans (21.5%), soluble sugars (4.6%), raffinose (0.1%), and sucrose (0.2%).

Carbohydrates from dehulled and deffated cake from sunflower are raffinose (3.2%), sucrose (2.3%), trehalose (0.8%), and glucose (0.6%). Reducing sugars from sunflower hulls reaches up to 23% as follows: xylose (14%), arabinose (7%), and galactose (2%).

Regarding carbohydrates, it has been reported some digestive problems in both humans and animals due to the lack of enzymes for hydrolyzing derivatives of sucrose, such as stachyose and raffinose, leading to their fermentation in the large intestine and producing flatulence (Shahidi et al. 1990). Raffinose is contained in great amounts in sunflower.

Carbohydrates from sesame represent around 21–25%. In deffated flour, the composition of carbohydrates is: glucose (3.6%), fructose (3.4%), raffinose (0.6%), galactose (0.4%), stachyose (0.4%), sucrose (0.17%), and sesamosa (0.14%). In comparison with soybean and peanut, sesame possess lower quantities of antinutrional carbohydrates (stachyose and raffinose), which cause flatulence and bad taste.

In soybean defatted flour, the content of carbohydrates is in the range of 13–15%, containing 7–8% sucrose, 5–6% stachyose, and 1% raffinose. The hull possesses 40% cellulose but overall, the starch content in all soybean by-products is less than 5%.

9.3.2 Proteins

These compounds are complex polymers containing up to 20 amino acids linked with peptide bonds (Rodrigues et al. 2012). The physicochemical features of proteins, their biological and functional properties are determined by the order in which amino acids are connected, as well as the chain size. Food proteins are contained in oilseed by-products, mainly in the cake, with the following features: non toxicity, easy digestion, nutritive, and functional (Damodaran 2000).

Around 70% of the proteins in rapeseed meals are globulines, however, the meal also contains albumins and in a lower quantity olesines (Krause and Schwenke 2001). Globulines and albumines are soluble in a saline solution and in water, respectively, while oleosines are structural proteins from lipidic constituents (Ghodsvali et al. 2005). The amino acids composition in rapeseed proteins are aspartic acid, isoleucine, leucine, glutamic acid, leucine, phenylalanine, proline, and tryptophan (Yoshie-Stark et al. 2006).

Globulins are the main sunflower cake proteins with 55–60% of content, followed by albumins (17–23%), glutelins (11–17%) and prolamines (1–4%). The essential amino acid concentration reaches 279 g/100 g and contains in descendant order: arginine, leucine, valine, phenylalanine, isoleucine, threonine, lysine, histidine, cysteine+cystine, methionine, and tryptophan. Non-essential amino acids are, in descendant order: glutamic acid, aspartic acid, glycine, proline, serine, alanine, and tyrosine and they corresponded to 328 g/100 g (Nagaraj 2009).

Proteins contained in sesame are the most nutritious among oilseed proteins, consisting in: globulins (67%, alpha globulin corresponded to 70% and beta-globulin 25%), albumins (9%), glutelins (7%), and prolamins (1%). Essential amino acids from sesame protein are: arginine (547 mg/g), leucine and phenylalanine (461 mg/g each one), isoleucine (261 mg/g), valine (244 mg/g), threonine (194 mg/g), methionine (160 mg/g), histidine (121 mg/g), and tryptophan (91 mg/g). However, sesame protein is deficient in amino acids such as lysine.

Soybean is the oilseed with the highest protein content. The average content of soybean proteins is 47% where lysine is the main amino acid, while methionine and cysteine are limited. The amino acid composition consists in: glutamic acid (11.8%), leucine (7.8%), arginine (7.3%), lysine (6.4%), proline (5.5%), serine

(5.5%), phenylalanine (5%), isoleucine (4.6%), valine (4.6%), glycine (4.3%), threonine (3.9%), tyrosine (3.8%), histidine (2.6%), cysteine (1.4%), tryptophan (1.4%) and methionine (1.1%).

9.3.3 Fiber

Fiber is the edible part of the plant which is resistant to both enzymatic digestion and absorption in the human small intestine with complete or partial fermentation. Dietary fiber includes cellulose, hemicellulose, pectic substances, gums, mucilages, and lignins (Dhingra et al. 2012). Fiber is divided in soluble and insoluble, the first group (water-soluble fiber) is composed of pectins, gums, and mucilages. In the second group (non-water soluble fiber) are cellulose, hemicellulose, and lignin.

Oilseed by-products contain important amounts of fiber. The fiber content in oilseed cake in sunflower is around 15–29%, in rapeseed 12%, in safflower 31%, in sesame 3–6%, and soybean 3%. These by-products can be revalorized in the food industry by means of fiber supplementation in cookies, crackers, cereal-based products, snacks, beverages, imitation cheeses, sauces, and meat (Chau and Huang 2003).

9.3.4 Minerals

The mineral content of rapeseed meal is very variable but it contains in descending order: calcium (6.6%), potassium (1.2%), phosphorus (1%), sulfur (0.9%), magnesium (0.5%), iron (0.1%), zinc (<1%), manganese (<1%), copper (<1%) and selenium (<1%) (Nagaraj 2009; Shahidi et al. 1990). The content of minerals and trace elements (zinc, manganese, copper, and selenium) can be increased by applying pretreatments such as dehulling and deffating. The sulfur content is related to the presence of glucosinolates, which are considered as antinutrients. A large portion of phosphorus is contained as phytic acid, interfering with the absorption of calcium, zinc, and iron.

Overall, sunflower is considered as a rich source of minerals. It contains nitrogen (3%), potassium (2%), calcium (0.6%), magnesium (0.4%) and phosphorus (0.1%), while micronutrients are iron, zinc, manganese, cuprum, and iron with 360, 78, 47 and 30 μg/g, respectively. Iron is the main mineral from sesame cake with a concentration of 19.3 mg/g. The mineral composition of soybean flour is around 6%, which mainly consist of phosphorus (546 mg/100 g), magnesium (236 mg/100 g), calcium (226 mg/100 g), sodium (27.9 mg/100 g), and cuprum 2.4 mg/100 g.

9.3.5 Vitamins

Rapeseed cake contains the following vitamins: choline, niacin, vitamin E, pantothenic acid, pyridoxine, riboflavin, thiamine, folic acid, and biotin (Nagaraj 2009; Shahidi et al. 1990). Sunflower possesses important amounts of B complex vitamins which are water soluble: nicotinic acid, thiamine, pantothenic acid, and riboflavin corresponding to 318, 38, 45 and 4 mg/kg, respectively. Sesame cake is rich in niacin (5.3 mg/g), thiamine (1 mg/g), and rivoflavin (0.4 mg/g). The vitamin composition of soybean cake is similar than that of whole soybean, which is a rich source of vitamins A, vitamin B (thiamine, riboflavin, niacin, and folic acid).

9.3.6 Phenolic Compounds

Phenolic compounds contribute to the bitter flavor of some oilseed producing a dark colored complex when they are oxidized. In rapeseed, p-hydroxybenzoic acid, trans-cinnamic acid, and sinapic acid are the most abundant; while p-coumaric, ferulic, caffeic, and chlorogenic acids are contained in less quantity.

In some cases, phenolic compounds are considered as antinutrients, such as those found in safflower which contains no more than 100 mg/100 g. Drying, autoclaving, solvents, and alkali solutions are used as pre-treatments for reducing the phenolic content of this oilseed. In the case of sunflower, chlorogenic acid is considered an important antinutrient due to its producing a dark color in the cake by an oxidative process. Sunflower possess between 1.4 and 4.0% of chlorogenic acid; while the concentration of total phenolic compound ranged from 80 and 87 mg/100 g in fresh whole seed cake, and 25 mg/100 g in hulls. Oxidation of this phenolic compound can be avoided by applying an aqueous extraction, washing with acidified acetone (1 : 9 and 0.5 M of HCl) from the seed.

9.3.7 Lignans

These compounds are contained in sesame oil, but also can be found as a remainder quantity in by-products (i.e cake). Lipid-soluble lignans (sesamin and sesamolin) and water-soluble lignans glycosides (sesamol triglucoside and sesaminol diglucoside) possess high antioxidant activity and prevent oil oxidation and provide a long shelf-ife stability (Kim et al. 2006). Sesamin is contained in 0.5–1.1% of raw sesame oil, and around 0.2–0.6% of sesamolin and sesamol in trace amounts. On the other hand, sesamol (3,4-methylene dioxyphenol) is found in roasted sesame oil (Mahendra Kumar and Singh 2015).

These phenolic constituents reduce the risk of cardiovascular disease, cancer, coronary heart disease, stroke, atherosclerosis, and disease associated with oxidative stress (Shahidi et al. 2006).

9.3.8 Tocopherols

They are natural lipid antioxidants with the role of preventing oxidation of polyunsaturated fatty acids, lipoproteins, and biological material. They reduce the risk of cancer, cataracts, cell damage, ischemia, and perfusion. Sunflower possess both alpha and beta tocopherols, corresponding to 464.1 and 15.2 μg/g oil, respectively.

9.4 Antinutritional Compounds

Some compounds from oilseed by-products have an undesirable effect on the nutritional composition by reducing the availability and digestibility of other constituents, or in the worst case affecting animal or human health. For this reason, it is important to remove the antinutritional or toxic compounds before revalorizing oilseed by-products, mainly when designing human food.

9.4.1 Glucosinolates

Glucosinolates are compounds with biological activity found in the *Brassicaceae* family of plants (oilseeds as rapeseed and mustard, but also broccoli, cabbage, cauliflower, and horseradish) (Bheemreddy and Jeffery 2006). Although several studies have shown the beneficial effect of these compounds in regulating stress response, anti-inflammatory effect and antioxidant activity, they are considered as antinutrients in oilseed by-products used for feeding livestock. Diets with high glucosinolate content produce gastrointestinal irritation, goiter, anemia, and hepatic and renal lesions in livestock. It also reduces the feed intake and growth (Bischoff 2016). This compound is hydrolysated by the enzyme myrosinases, producing nitriles, thiocyanates, isothiocyanates, and oxazolidine-2-thiones which possess a diverse biological activity. These degradation products produce damage in thyroid, the vital organs, and metabolism (Halkier and Gershenzon 2006; Tripathi and Mishra 2007).

Around 90 glucosinolates have been isolates from plants and only seven have been identified in rapeseed, with 3-butenyl, 4-pentnenyl and 2-hydroxy-3 butenyl glucosinolates the most important (Nagaraj 2009). Glucosinolates reduce the nutritional level of rapeseeds and their by-products. The presence of glucosinolates and fiber make the digestibility of proteins difficult. For this reason, rapeseed meal containing great amounts of glucosinolates is usually used in animal feeding.

9.4.2 Phytic Acid or Phytate

Phytic acid or phytate is a compound contained in plants for phosphorus storage. It possesses a strong chelation activity binding to mono- and divalent metal ions and forming complexes. In this way, the compound interacts with the basic residues of proteins, as well as with essential elements such as calcium, iron, magnesium, manganese, copper, and zinc, forming complexes and thus affecting their metabolic activity, availability, and absorption. In the case of sesame, phosphorus is unavailable due to the presence of phytic acid. Phytates are contained in the range of 2.0–5 mg/100 g in sesame and around 1–2% in soybeans (Nagaraj 2009; Rodrigues et al. 2012).

9.4.3 Oxalic Acid

Oxalic acid reduces the availability of elements such as calcium and produces a bitter taste due to the formation of calcium oxalates. Safflower seed contains around 10 mg/100 g of oxalic acid. Both phytate and oxalate antinutrients are contained in hull and cake from sesame. Treatments with H_2O_2 at pH 5.0, as well as dehulling sesame seeds in saline solution reduce total free oxalates. Roasting had no effect on diminishing the concentration of oxalic acid.

9.4.4 Erucic and Brassidic Acids

Erucic acid (*cis*-13-docosenoic acid) and brassidic acid (*trans*-13-docosenoic acid) are omega-9 monounsaturated fatty acids. This compound represents a risk to human health when consumed in high doses due to its toxicity affecting the growth, functioning, and morphology of the heart and skeletal muscles. In fact, it has been established in the *Codex Alimentarius* that food and oils with erucic acid content equal to or higher than 5% of the total fraction of fatty acid, are not suitable for human consumption (Vianni and Braz-Filho 1996).

9.4.5 Carbohydrates with Antinutritional Properties

Carbohydrates classed as antinutrientes are linked with oligosaccharide derivatives of sucrose, these are stachyose and raffinose. These compounds cause digestive problems in both humans and animals related with flatulence, and they also produce bad taste. The problem comes from the fact that the human digestive system does not possess the enzyme α-galactosidase, which hydrolyzes these carbohydrates, thus, conducting the anaerobic fermentation with gas production in the large intestine (Shahidi et al. 1990). The main oilseed by-products containing these compounds are rapeseed, sesame, and soybean (mainly in the hulls and cake).

9.4.6 Other Antinutritional Factors

Other antinutritional factors in soybeans are protease and trypsin inhibitors, hemagglutinins, and lectins. These antinutrients can be diminished by treatments (i.e. dry). It has been reported that crude fiber must be removed from cake by taking off the hulls (prior to oil extraction). The enzyme linasi, which release a poisonous acid called "prussic acid," must be inactivated from linseed by heat processing for around 10 minutes.

9.5 Current Applications in the Valorization of Oilseed by-Products

In recent years, there has been an increase in the production of oilseed to satisfy the demand for fuel. This has provoked a rise in the generation of oilseed by-products that have a high potential for being valorized (Rodrigues et al. 2012). It is necessary to apply procedures such as extraction, concentration, and isolation of target compounds with sustainable techniques but at the same time capable for reducing undesirable compounds or antinutrients (i.e. glucosinolates, phytates, oxalic acid, erucic acid, among others).

9.5.1 Vegetable Proteins Source

Overall, most of the oilseed by-products possess an important amount of proteins. For this reason, they are considered as an alternative to animal proteins in human nutrition. Additionally, they can be considered as ingredients for developing functional foods or as source of bioactive constituents for pharmaceutical or cosmetics applications (Živanović et al. 2011). Rapeseed cake possesses the required amino acid composition for the human diet, thus, it is an adequate ingredient for developing balanced and nutritious foods with higher functional protein properties compared to soybean, wheat, and casein (Ghodsvali et al. 2005).

Protein isolates and protein concentrate are the main products obtained after valorization of oilseed by-products for food applications. When the protein level reaches

65% on a dry base it is considered as protein concentrate, while it is called protein isolate when it reaches 90% (Moure et al. 2006). The first step for obtaining protein concentrates is the extraction of non-protein material in the oilseed by-product (i.e. carbohydrates) with alcoholic solutions (60–80%) and acidic water (pH 4.5 for minimizing the fraction of solubilized protein) (Grandison and Lewis 1996). By this way is obtained a protein concentrate rich in fiber and proteins.

Protein isolates are obtained as follows: first, the protein must be solubilized in saline solution where the pH is far from the isoelectric point. Second, the protein concentration is carried out by ultra- or diafiltration membranes or through protein precipitation (pH close to isoelectric point). This method involves both protein solubilization and precipitation which depends on the protein features (Moure et al. 2001). Finally, proteins are centrifuged and dried in a conventional way or by lyophilization for obtaining a powder.

Additionally, consideration should be given to the protein digestibility as well as the presence of antinutrients to avoid undesirable interactions that may affect the stability of foods produced with these products. Rapeseed and canola extracts contain a low content of phytate when proteins were solubilized in a solution with a pH between 9.5 and 12 and a ratio of 5–10% w/v (solid/solution) (Ghodsvali et al. 2005; Pedroche et al. 2004). Overall, interactions between proteins and phytates depend on two factors: pH and concentration of ions. At pH 12, protein and phytic acid are negatively charged, the phytate solubility is reduced and the protein–phytate interaction only occurs in the presence of multivalent ions, forming protein–mineral–phytate complexes. The formation of these complexes is reduced by removing ions and adding a chelator such as EDTA (Rodrigues et al. 2012). On the contrary, proteins and phytic acid are positively and negatively charged, respectively, forming the complex phytate-protein, however, a high concentration of calcium can dissociate this complex.

9.5.2 Natural Antioxidants and Preservatives

As previously stated, oilseed by-products possess a great variety of both lipophilic and hydrophilic phytochemicals contained in seeds, leaves, and shells. The most important phytochemicals from oilseed by-products that are reported in the literature are phenolic compounds, tocopherols, and vitamins (Bodoira et al. 2017). However, there is great need for much more research regarding carotenoids and pigments.

At the same time, the utilization of antioxidants as natural additives has becoming popular due to their ability to improve the quality of food products. They also can preserve the microbiological stability of food and extend the shelf-life. Food supplemented with oilseed by-products can be considered as functional food due to their increased quality and nutritional value (Koprivnjak et al. 2008).

Lignans contained in oilseed by-products, such as those coming from sesame, can be used as both antioxidants and antimicrobial preservative for food and in the cosmetics and pharmaceutical industries. For instance, it has been used as an effective stabilizer for prolonging the shelf life of oils and oil-based products (Mahendra Kumar and Singh 2015).

9.5.3 Organic Fertilizer

Nowadays, agricultural management techniques are focused on environmental sustainability. Organic fertilizers increase the soil's organic matter and provide a continuous release of nutrients to the crops (Willer and Kilcher 2011). Moreover, they improve the physicochemical features and the biological functions of soil. They also enhance the production of antioxidant metabolites in plants (Dumas et al. 2003).

In this sense, some oilseed by-products, mainly those coming from raw material (shells, leaves, petals, stalks, pollen), can be used as organic fertilizers. They possess important amounts of proteins and carbohydrates, but also minerals (calcium, copper, iron, magnesium, manganese, phosphorus, potassium, sulfur, and zinc). However, future studies must be carried out for elucidating the influence of oilseed by-products on the soil features and nutritional composition of crops.

9.5.4 Livestock Diets

Oilseed by-products, mainly residual cakes can be used for animal feeding. They are an alternative protein source in poultry, pigs, and aquatic animals, among others. When oilseed by-products possess lower protein digestibility and solubility, it is suggested to use these products for feeding ruminant animals due to their possession of a complex digestive system (Moure et al. 2006). This is the case of rapeseed cake.

Previous studies have shown variable effects on growth and weight performance due to poor digestibility of proteins (Gn et al. 2011; Jaworski et al. 2014). Also, the presence of antinutrients (i.e. glucosinolates, phytates, oxalates, erucic acid, brassidic acid, stachyose, and raffinose) affect protein and mineral absorption in animals by decreasing digestibility, metabolism, and palatability and have a toxic influence on organs (e.g. the liver) (Duodu et al. 2018; Panhwar 2005).

The nutritional value of oilseed by-products can be increased by inactivating toxins and decreasing the compounds with antinutritional features through processing techniques, such as fermentation, boiling, heating, autoclaving, treatment with NaOH (Duodu et al. 2018). However, future studies are necessary to optimize these processing techniques and elucidate their influence on the decrease of these compounds.

9.5.5 Renewable Energy

Energy consumption around the world is increasing and the main source of energy is hydrocarbon fuel combustion, comprising around 80%. This fact has led to shortages of fuel supply as well as an increase in greenhouse emissions (Montenegro-Camacho et al. 2016). For this reason, alternative sources and sustainable/efficient methods for obtaining fuel must be found.

Conversion of waste by-products (biomass) into energy reduce the gas emission by: (i) reducing the dependence on conventional power plants, (ii) greenhouse gas emission (i.e. methane) from landfills is reduced, and (iii) waste-to-energy plants are efficient and allow the revalorization of waste by-products, such as those coming from oilseeds. Overall, the main constituents of biomass are cellulose, hemicellulose, lignin, and minerals (White et al. 2011).

Diverse energy forms can be obtained from biomass, depending on the processing and raw material features. Thermochemical conversion, such as pyrolysis and hydrothermal liquefaction are suitable technologies for producing bio-oil (liquid fuel). Flash pyrolysis consists in the rapid decomposition of organic compounds, in the absence of oxygen, to produce gases, liquids, and char. In hydrothermal liquefaction, the biomass is introduced in water with high temperature and pressure in the presence or not of catalyst.

Bio-oil can be produced from organic waste; thus oilseed by-products represent a good option for its valorization. Bio-oils are a promising renewable energy source and an environmentally friendly fuel as compared with those coming from petroleum or chemical feedstock. The decomposition of organic material is carried out by pyrolysis, which is a thermal method that rapidly converts the biomass from by-products into chemical energy, in the form of bio-oil and bio-char (Yang et al. 2014). The product of pyrolysis is a liquid composed by two fractions: hydrophobic fraction (tar) and hydrosoluble fraction (bio-oil).

The industrial applications of bio-oils are the following (Fini et al. 2010; Radlein 2002; Strenziok et al. 2001; Xiu and Shahbazi 2012):

- Heat generator – fuel in boiler, burner, and furnace.
- Power generator – fuel in diesel engines and turbines
- Production of chemicals and resins (agrochemicals, fertilizers, acids, emission control agents), adhesives (asphalt bio-binder), anhydro-sugars (levoglucosan, which is used as surfactant, biodegradable polymer, as well as in pharmaceutical industry), liquid smoke and wood flavors.

However, it should be taken into consideration the disadvantages of bio-oil, such as its high viscosity, high water and ash content, low heating value (high oxygen content), high instability and corrosiveness (acidity), as well as high upgrading cost (hydrotreating, hydrocracking, supercritical fluids extraction, solvent addition/esterification, emulsification, steam reforming, and chemical extraction).

Another application of oilseed by-products is the production of bioethanol by alcoholic fermentation of sugars. Bioethanol is the main biofuel source in the world at this time. It is produced by microbial fermentation of carbohydrates and starch from agricultural waste, forestry waste, livestock, and manures. The first step is hydrolyzing carbohydrate through chemical or enzymatic procedures. The second step consists of fermentation using different types of microorganisms. During the fermentation progress, hydrolyzed carbohydrates are converted into alcohol (Pandey et al. 2019). Similar to bio-oils, bioethanol is widely used in the pharmaceutical and chemical industries.

9.6 Future Trends

A large number of by-products are generated during oilseed processing. In the past, these by-products were considered as a waste, its disposal being one of the main problems in processing plants. As a result, there was an increased demand for valorizing oilseed by-products and for looking for new alternatives of extraction.

It has been demonstrated that oilseed by-products (shells, leaves, petals, stalks, cake, pollen, head, pith) possess rich amounts of proteins, carbohydrates, fiber, bioactive constituents, antioxidants, vitamins, and minerals, which are feasible for industrial applications (i.e chemical, energy, food, personal care, and pharmacological industries). For this reason, the development of technologies focused on sustainable processes for all these industries is compulsory. Revalorization of oilseed by-products by means of efficient, sustainable, and profitable technologies is nowadays the greatest challenge.

The new trends in oilseed by-products revalorization are:

- Extraction and recovery of valuable bio-molecules from oilseed-waste through efficient, sustainable, and innovative technologies.

- Improvement of functional properties of substances contained in oilseed by-products.
- Transformation of bio-molecules into high added-value products for food-, pharmaceutical-, agro-, and cosmetic industries.
- Fortified food with increased valued products such as the functional peptides and essential amino acids.
- Generation of environmentally friendly fuel through highly efficient and sustainable processes.

Future studies must be carried out to elucidate the feasibility of these new trends and procedures, its impact in the revalorization of oilseed by-products, their advantages and disadvantages, as well as the stability of obtained products in terms of nutritional, physicochemical, organoleptic, and microbiological features.

References

Bheemreddy, R.M. and Jeffery, E.H. (2006). Glucosinolates. In: *Nutritional Oncology*, 2e (eds. D. Herber, G.L. Blackburn, V.L.W. Go, et al.), 583–596. San Diego, CA: Elsevier.

Bischoff, K.L. (2016). Glucosinolates. In: *Nutraceuticals Efficacy, Safety, and Toxicity*, 1e (ed. R.C. Gupta), 551–554. San Diego, CA: Springer.

Bodoira, R.M./., Penci, M.C., Ribotta, P.D. et al. (2017). Chia (*Salvia hispanica* L.) oil stability: study of the effect of natural antioxidants. *LWT - Food Science and Technology* 75: 107–113. http://dx.doi.org/10.1016/j.lwt.2016.08.031.

Chau, C.F. and Huang, Y.L. (2003). Comparison of the chemical composition and physicochemical properties of different fibers prepared from the peel of citrus sinensis L. Cv. Liucheng. *Journal of Agricultural and Food Chemistry* 51 (9): 2615–2618.

Damodaran, S. (2000). Aminoácidos, péptidos y proteínas. In: *Química de los Alimentos*, 2e (ed. O.R. Fennema), 383–511. Zaragoza, Spain: Acribia.

Delshad, E., Yousefi, M., Sasannezhad, P. et al. (2018). Medical uses of Carthamus tinctorius L. (Safflower): a comprehensive review from traditional medicine to modern medicine. *Electronic Physician* 10 (4): 6672–6681.

Dhingra, D., Michael, M., Rajput, H. et al. (2012). Dietary fibre in foods: a review. *Journal of Food Science and Technology* 49 (3): 255–266.

Dumas, Y., Dadomo, M., Di Lucca, G. et al. (2003). Effects of environmental factors and agricultural techniques on antioxidant content of tomatoes. *Journal of the Science of Food and Agriculture* 83 (5): 369–382.

Duodu, C.P., Adjei-Boateng, D., Edziyie, R.E. et al. (2018). Processing techniques of selected oilseed by-products of potential use in animal feed: effects on proximate nutrient composition, amino acid profile and antinutrients. *Animal Nutrition* 4 (4): 442–451.

Duodu, C.P., Adjei-Boateng, D., Edziyie, R.E. et al. (2018). Processing techniques of selected oilseed by-products of potential use in animal feed: effects on proximate nutrient composition, amino acid profile and antinutrients. *Animal Nutrition* 4 (4): 442–451. https://doi.org/10.1016/j.aninu.2018.05.007.

European Commission Preparatory study on food waste across EU 27. (2010). http://ec.europa.eu/environment/eussd/pdf/bio_foodwaste_report.pdf (accessed September 30, 2020).

FAO Oilseeds, oils & meals monthly price and policy update. No. 126 (Internet]. (2020) (cited 2020 Feb 5). http://www.fao.org/3/ca7623en/ca7623en.pdf

Fini, E.H., Yang, S., Xiu, S. et al. Characterization and application of manure-based biobinder in asphalt industry. In: 89th Annual Meeting Transportation Research Board (ed. The Transportation Research Board), 14. Washington, DC: The National Academies of Sciences. (2010).

Ghodsvali, A., Khodaparast, M.H.H., Vosoughi, M. et al. (2005). Preparation of canola protein materials using membrane technology and evaluation of meals functional properties. *Food Research International* 38 (2): 223–231.

Gn, E., Oladimeji, O., Bo, E. et al. (2011). Biochemical characteristics of rats fed on Shea nuts (Butyrospermum Parkii) meal. *Journal of Basic and Applied Sciences Research* 1 (3): 124–130.

Grandison, A.S. and Lewis, M.J. (eds.) (1996). *Separation Processes in the Food and Biotechnology Industries. Principles and Applications*, 1e. Cambridge, UK: Woodhead Publishing 296 p.

Halkier, B.A. and Gershenzon, J. (2006). Biology and biochemistry of glucosinolates. *Annual Review of Plant Biology* 57 (1): 303–333.

Jaworski, N.W., Shoulders, J., González-Vega, J.C. et al. (2014). Effects of using copra meal, palm kernel expellers, or palm kernel meal in diets for weanling pigs. *The Professional Animal Scientists* 30 (2): 243–251. http://dx.doi.org/10.15232/S1080-7446(15]30108-X.

Kim, K.S., Park, S.H., and Choung, M.G. (2006). Nondestructive determination of lignans and lignan glycosides in sesame seeds by near infrared reflectance spectroscopy. *Journal of Agricultural and Food Chemistry* 54 (13): 4544–4550.

Koprivnjak, O., Škevin, D., Valić, S. et al. (2008). The antioxidant capacity and oxidative stability of virgin olive oil enriched with phospholipids. *Food Chemistry* 111 (1): 121–126.

Krause, J.P. and Schwenke, K.D. (2001). Behaviour of a protein isolate from rapeseed (Brassica napus) and its main protein components – globulin and albumin – at air/solution and solid interfaces, and in emulsions. *Colloids and Surfaces. B, Biointerfaces* 21 (1–3): 29–36.

Mahendra Kumar, C. and Singh, S.A. (2015). Bioactive lignans from sesame (Sesamum indicum L.): evaluation of their antioxidant and antibacterial effects for food applications. *Journal of Food Science and Technology* 52 (5): 2934–2941.

Montenegro-Camacho, Y.S., Bensaid, S., Ruggeri, B. et al. (2016). Valorisation of by-products/waste of agro-food industry by the pyrolysis process. *Journal of Advanced Catalysis Science & Technology* 3 (1): 1–11.

Moure, A., Sineiro, J., and Domínguez, H. (2001). Extraction and functionality of membrane-concentrated protein from defatted Rosa rubiginosa seeds. *Food Chemistry* 74 (3): 327–339.

Moure, A., Sineiro, J., Domínguez, H. et al. (2006). Functionality of oilseed protein products: a review. *Food Research International* 39 (9): 945–963.

Naczk, M. and Shahidi, F. (1990). Carbohydrates of canola and rapeseed. In: *Canola and Rapeseed* (ed. F. Shahidi), 211–220. Boston, MA: Springer.

Nagaraj, G. (ed.) (2009). *Oilseeds: Properties, Processing, Products and Procedures*, 1e. New Delhi: New India Publishing 590 p.

Pandey, A., Larroche, C., Dussap, C.G. et al. (eds.) (2019). *Biofuels: Alternative Feedstocks and Conversion Processes for the Production of Liquid and Gaseous Biofuels*, 2e. London: Academic Press 867 p.

Panhwar, F. (2005). Anti-nutritional factors in oil seeds as aflatoxin in ground nut. *Edition ChemLin* (February): 1–8.

Pedroche, P.J., Yust, M.M., Megías, C. et al. (2004). Utilisation of rapeseed protein isolates for production of peptides with angiotensin I-converting enzyme (ACE)-inhibitory activity. *Grasas y Aceites* 55 (4): 354–358.

Radlein, D. (2002). Study of levoglucosan production – a review. In: *Fast Pyrolysis of Biomass: A Handbook 2*, 1e (ed. A.V. Bridgwater), 205–241. Newbury, UK: CPL Press.

Rodrigues, I.M., Coelho, J.F.J., and Carvalho, M.G.V.S. (2012). Isolation and valorisation of vegetable proteins from oilseed plants: methods, limitations and potential. *Journal of Food Engineering* 109 (3): 337–346. https://doi.org/10.1016/j.jfoodeng.2011.10.027.

Rodríguez-Roque, M.J., Rojas-Graü, M.A., Elez-Martínez, P. et al. (2013). Soymilk phenolic compounds, isoflavones and antioxidant activity as affected by in vitro; gastrointestinal digestion. *Food Chemistry* 136 (1):206–212. http://dx.doi.org/10.1016/j.foodchem.2012.07.115.

Rudometova, N.V., Pasovskij, A.P., and Blohina, E. (2001). Method of isolation and identification of carthamin in safflower: application's perspectives in Russian food products. In: *Proceedings of the Fifth International Safflower Conference* (eds. J.W. Bergman, H.H. Mündel, J.L. Jensen, et al.), 309–313. Fargo, ND: North Dakota State University.

Schmidt, O.G. and Mejia, D.. Oilseeds: post-harvest operations. International Development Research Centre, Canada (IDRC) and FAO (Internet). IDRC-FAO. (2014). http://www.fao.org/in-action/inpho/crop-compendium/oilseeds/en (accessed September 30, 2020).

Shahidi, F. (ed.) (1990). *Canola and Rapeseed Production, Chemistry, Nutrition and Processing Technology*, 1e. New York;: Springer 354 p.

Shahidi, F., Liyana-Pathirana, C.M., and Wall, D.S. (2006). Antioxidant activity of white and black sesame seeds and their hull fractions. *Food Chemistry* 99 (3): 478–483.

Strenziok, R., Hansen, U., and Kunster, H. (2001). Combustion of bio-oil in a gas turbine. In: *Progress in Thermochemical Biomass Conversion*, 1e (ed. A.V. Bridgwater), 1452–1458. Oxford: Blackwell Science.

Tripathi, M.K. and Mishra, A.S. (2007). Glucosinolates in animal nutrition: a review. *Animal Feed Science and Technology* 132 (1–2): 1–27.

Vianni, R. and Braz-Filho, R. (1996). Ácidos Graxos Naturais: Importância e Ocorrência em Alimentos. *Quimica Nova* 19 (4): 400–407.

White, J.E., Catallo, W.J., and Legendre, B.L. (2011). Biomass pyrolysis kinetics: a comparative critical review with relevant agricultural residue case studies. *Journal of Analytical and Applied Pyrolysis* 91 (1): 1–33. http://dx.doi.org/10.1016/j.jaap.2011.01.004.

Willer, H. and Kilcher, L. (2011). *The World of Organic Agriculture: Statistics and Emerging Trends*. Frick, Switzerland: Forschungsinstitut für Biologischen Landbau.

Xiu, S. and Shahbazi, A. (2012). Bio-oil production and upgrading research: a review. *Renewable and Sustainable Energy Reviews* 16 (7): 4406–4414. http://dx.doi.org/10.1016/j.rser.2012.04.028.

Yang, S.I., Wu, M.S., and Wu, C.Y. (2014). Application of biomass fast pyrolysis part I: pyrolysis characteristics and products. *Energy* 66: 162–171.

Yoshie-Stark, Y., Wada, Y., Schott, M. et al. (2006). Functional and bioactive properties of rapeseed protein concentrates and sensory analysis of food application with rapeseed protein concentrates. *LWT - Food Science and Technology* 39 (5): 503–512.

Zhaomu, W. and Lijie, D. (2001). Current situation and prospects of safflower products development in China. In: *Proceedings of the Fifth International Safflower Conference* (eds. J.W. Bergman, H.H. Mündel, J.L. Jensen, et al.), 315–319. Fargo, ND: North Dakota State University.

Živanović, I., Vaštag, Ž., Popović, S. et al. (2011). Hydrolysis of hull-less pumpkin oil cake protein isolate by pepsin. *World Academy of Science, Engineering and Technology* 75 (3): 67–71.

10 Proteins and Peptides Derived from Rapeseed: Techno-Functional and Bioactive Properties

Maria Hayes
Food BioSciences Department, Teagasc Food Research Centre, Ashtown, Dublin, Ireland

10.1 Introduction

Rapeseed and Canola are often used interchangeably. However, Canola refers to a rapeseed variety that was introduced to Canada in 1974 and is low in both erucic acid and glucosinolates. It is derived from the species *Brassica napus* (summer rape) and *Brassica campestris* (Turnip rape), (Alashi et al. 2013). Rapeseed proteins can be classified according to their solubility as globulins, albumins, and oleosins (Uppstrom 1995). Salt soluble globulins make up 70% of rapeseed proteins, prolamins, which exist exclusively as oleosin in rapeseed make up 20% and 10–15% consists of water-soluble albumins. Rapeseed proteins can also be characterized according to their corresponding sedimentation coefficient values in Svedberg units (S). This value indicates the speed of sedimentation of a macromolecule in a centrifuge. The protein fractions for rapeseed are 12 S, 7 S, and a split 2 S, 1.7 S or 1.8 S. Cruciferin and napin are the two major rapeseed protein storage proteins and make up 20 and 60% of the total protein content of rapeseeds (Höglund et al. 1992).

Rapeseed proteins are ordinarily extracted using an alkali process. However, denaturation can occur and this renders the alkali-extracted proteins unsuitable for food ingredient use. Furthermore, alkali-extracted proteins are poorly soluble at neutral pH values and therefore can have poor techno-functional characteristics. Despite this, value can be added to rapeseed protein meal if it can be used for human food or functional food purposes due to its nutritional and health benefits. Rapeseed proteins contain all of the essential amino acids, and have a profile similar to that of soybean protein isolates and comply with the amino acid nutrition requirements set out by Food and Agriculture Organization of the United Nations (FAO), World Health Organization (WHO), and the United Nations University (UNU). Furthermore, toxic and anti-nutritional elements including glucosinolates and sinapinic acid can be

Oil and Oilseed Processing: Opportunities and Challenges, First Edition.
Edited by Tomás Lafarga, Gloria Bobo, and Ingrid Aguiló-Aguayo.

removed from rapeseed protein meal using hydrolysis or fermentation, which results in the generation of a bioactive peptide containing hydrolysates with potential preventative healthcare benefits for consumers.

The aim of this chapter is to collate information concerning rapeseed proteins and isolates, how they are extracted by industry, and to describe their use as techno-functional ingredients such as emulsifiers and their potential for use as functional foods.

10.2 Summary of Existing Rapeseed Meal Protein Extraction Processes

A number of protein-extraction processes already exist concerning rapeseed meal (RM) proteins. Most rapeseed protein isolates are prepared by alkaline extraction followed by isoelectric precipitation (Cosgrove 1980). These processes are preceded by the oil extraction process that usually involves crushing and pressing at high temperatures to produce a high quality oil product which can impact on downstream protein production yields and quality (Dekker 2018). This process can be replaced by cold-press techniques to produce the oil but lower yields of oil are obtained in this process. Solvents are then applied to the rapeseed cake (alkali and acids) and are often combined with high temperatures to concentrate and isolate the protein but this can denature proteins and lead to poor functional attributes of the resulting protein extracts including solubility, low yields, and a dark-colored protein extract due to the presence of phenolic compounds and glucosinolates. Alternative RM protein extraction strategies are therefore required to optimize the potential of RM and the use of aqueous extraction combined with filtration methods have been examined as potential alternatives. For example, proteins were extracted previously from RM at pH 12 followed by isoelectric protein precipitation. This involves adjusting the pH value at which a specific protein has no net electrical charge and thus minimum solubility (the pI value of the protein). pI values are protein specific and not all RM proteins will be precipitated and to recover soluble proteins membrane technology can be applied to concentrate proteins using a combination of molecular weight cut off (MWCO) and ultrafiltration (UF). Dia-filtration can be used to remove impurities and protein isolates can be concentrated using spray-drying (Tzeng et al. 1990). However, phenolic compounds are still present in this extract and can be oxidized to quinones that result in a dark colored protein and poor organoleptic properties (Dekker 2018; Tzeng et al. 1990).

Improvements were made to this process to enhance the yield of rapeseed protein isolate and reduce phenolic content. Sodium sulphide (an antioxidant) can be added to the extraction solution to prevent oxidation of the phenolic compounds. An additional membrane processing and mild heat treatment and precipitation of protein with polyvinyl pyrrolidine was used (Xu and Diosady 2002). Heat broke down the protein-phenolic complex and phenolic fractions were removed using filtration. The protein isolates that resulted had lower levels of phenolics and protein yields were not affected. The negative effects of this process include denaturation of proteins, which results in poor functional attributes of the end proteins. This is a result of the high pH values used to precipitate the proteins. pH values of 12 can result in protein racemization and lysinoalanine formation, both of which have negative impact on the functional attributes of proteins (Rodrigues et al. 2012).

The Canadian company Burcon NutraSciences Corporation produce several RM protein isolates. They produce a protein isolate referred to as Puratein© by applying a saline solution to the RM which enhances protein solubilization as the protein known as cruciferin is a salt-soluble globulin protein. Vacuum filtration combined with centrifugation and a filtration step is used to remove fine meal particles and UF using a 10 kDa membrane is used to concentrate the protein. Protein is then diluted with cold water of 4 °C which causes a drop in the ionic strength of the protein solution and a protein micellar mass (PMM) forms. This mass is spray dried to form the protein isolate Puratein©. The rapeseed protein known as Supertein™ is then isolated from the remaining supernatant. A PMM is again formed by the addition of cold water to the supernatant and UF is used to remove impurities and concentrate the protein. Puratein© protein contains globulin and cruciferin and Supertein™ contains mainly albumin and napin proteins. Puratein© has good emulsifying and gelling properties while Supertein has good foaming properties and excellent solubility at different pH values (Hadnađev 2002).

A designed aqueous extraction process was developed at the University of Wageningen. This process involves soaking de-hulled rapeseed seeds with an alkaline solution at pH 9 and subsequent mixing of the seeds and then screw-pressing the mixture to remove seeds and fiber. Centrifugation is then used to produce three different layers – where the middle layer, known as the subnatant is an aqueous solution where soluble proteins are dissolved (cruciferin and napin). Dia-filtration is then applied to the subnatant and the resultant product is freeze-dried (Dekker 2018). The intended use of this protein is to replace emulsifiers of animal origin. A yield of 39% protein was achievable using this process.

A substance known as Kaempferol 3-O-(2″′-O-sinapoyl-β-sophoroside) was recently identified as the compound that makes rapeseed protein extracts bitter tasting and inedible for humans. A German research group used trained sensory panels in combination with activity guided fractionation to identify the rapeseed protein known as Cruciferin as the bitter causing protein group due to the fact it contains a large amount of this substance (~390 mg/Kg). They also identified that rapeseed napin protein contains a tenth less of this compound (Hald et al. 2019). Future protein extraction methods should focus on removal of this compound from protein extracts to enhance potential techno-functional applications in food products (Hald et al. 2019).

10.3 Hydrolysis of Rapeseed Proteins and Rapeseed Meal to Produce High Value Bioactive Compounds

Rapeseed proteins have limited value as food ingredients as at neutral pH values they can be insoluble which limits their application as techno-functional ingredients. However, they have potential for use as functional food ingredients with health benefits and enzymatic hydrolysis and fermentation processes are useful methods to realize this potential by improving solubility and creating bioactive peptides. Bioactive peptides are sequences of between 2 and 30 amino acids in length that are largely inactive within the parent protein source, but, once released using hydrolysis with proteolytic enzymes or lactic acid bacteria (LAB) can impart a health benefit to the consumer that goes above and beyond basic, human nutrition (Hayes 2018). They are considered to have hormone-like activities and can impart immune-modulatory, antimicrobial, heart health benefits, and mental health benefits to the consumer.

RM protein hydrolysates can be produced using different enzymes including pepsin, trypsin, chymotrypsin, papain alcalase, protamex, flavorzyme, and several other commercially available offerings. The enzymes selected can break down RM proteins into smaller oligopeptide and tetra, tri and di-peptide hydrolysates. Modified proteins (or high molecular weight proteins) can be produced at the early stages of a hydrolysis process and the size of the peptides produced depends on the temperature, pH, stirring, and degree of hydrolysis (DH) of the protein that is achieved during the process (Alashi et al. 2013). Likewise, LAB rich in proteolytic enzymes can be used to produce bioactive peptides. pH or temperature are normally used to terminate a hydrolysis or fermentation process and the resultant hydrolysate can be freeze-dried or generated into additional fractions which enrich bioactivity using filtration processes – normally MWCO filtration. As bioactive peptides range in size from 2 to 30 amino acids 1- or 3- kDa MWCO membranes are normally used to generate a bioactive peptide-rich permeate. Identification of bioactive peptides from a protein hydrolysate usually involves the following steps: filtration (for enrichment of peptides); RP-HPLC or chromatography to generate a further fraction; mass spectrometry analysis to identify the size and sequence of bioactive peptides contained in the hydrolysate permeate fraction; assessment of novelty of the sequenced peptides by consultation of databases such as BIOPEP-UWM (http://www.uwm.edu.pl/biochemia/index.php/en/biopep), and conformation of bioactivity using bioassays at each fractionation step. Alternatively, identified peptides can be chemically synthesized and assessed for bioactivities *in vitro*.

A number of factors can determine the potential bioactivity and indeed sensory and safety characteristics of a protein hydrolysate. In silico methods can be very useful before design of the protein hydrolysis process. Programs such as enzyme peptide cutter and databases including BIOPEP-UWM (Minkiewicz et al. 2019), PeptideRanker (http://bioware.ucd.ie/~compass/biowareweb/Server_pages/help/peptideranker/help.php) and ToxinPred (http://crdd.osdd.net/raghava/toxinpred) can determine the best enzymes to use to produce bioactive peptides from any protein source and the novelty, sensory, toxicity, and potential allergenicity of resultant peptides (Bleakley et al. 2017).

Fermentation with proteolytic bacteria has also been widely used for the removal of glucosinolates from RM. Fermentation is cheaper than hydrolysis and provides an effective approach to degrade glucosinolates without loss of proteins (Xie et al. 2015). Previously it was found that mixed, solid-state fermentation by *Bacillus* sp. of RM degraded glucosinolates by 53.4% and increased protein and peptide content of resultant rapeseed extracts by 5.58 and 8.1%, respectively (Wang et al. 2011).

10.4 Techno-Functional Attributes of Rapeseed Proteins

10.4.1 Emulsifying Properties

Alternatives to animal-derived protein ingredients are required currently based on consumer demand for ethically produced and sustainable proteins and also their moral preferences associated with consuming plant rather than animal by-products. In order to increase the value and market integration of rapeseed protein a greater understanding of their structure–function relationships is required (Karaca et al. 2011). Protein emulsifying activity is the ability of the protein to participate in emulsion formation and to stabilize the newly created emulsion. Emulsifying capacity is the ability of a protein or protein suspension to emulsify an oil. It is a useful functional characteristic and plays an important role in the development of new sources of plant protein products. Proteins dominate in most food emulsions. A significant number of foods are emulsions or foams and proteins, in combination with lipids and carbohydrates, are important stabilizers (Zayas 1997). Oil in water emulsions give a creamy texture while water in oil emulsions give an oily or greasy texture to a food formulation.

The method of production of rapeseed proteins and isolates was found previously to greatly influence both the physicochemical and emulsifying properties of canola protein isolates. Salt-extracted isolates were found to have higher solubility and interfacial activity compared to those produced by isoelectric precipitation (Karaca et al. 2011). More recently, Östbring and colleagues recovered rapeseed proteins from industrially cold-pressed rapeseed press cake (RPC) with a higher protein recovery yield as compared with hot-pressed RM and solvent-extracted RM. These proteins displayed better emulsifying properties when compared to either the hot-pressed and solvent-extracted rapeseed proteins and were in the same range as commercial emulsifying agents at concentrations up to 16 mg protein/ml oil. When heat was applied the ability to stabilize emulsions was significantly reduced (Östbring et al. 2020).

10.4.2 Digestibility of Rapeseed Proteins

Protein digestibility is primarily dependent on the accessibility of enzymes to it and the protein molecular structure. Rapeseed protein has poor digestibility values due largely to the glucosinolates and their breakdown products – isothiocyanates (ITC) such as 5-vinyloxazolidine-2-thione (VOT, goitrin), butenyl-ITC, and pentenyl-ITC, and phenylethyl-ITC in addition to phytates and phenolics. These phenolics bind proteins and limit enzyme access to rapeseed proteins. However, the production of rapeseed protein isolates can, to a great extent, eliminate the interference of non-protein components. Work by Savoie et al. (1988) stated that rapeseed protein concentrate had lower *in vitro* digestibility values than casein (87% *v* 97%) and attributed this to the rigidity of rapeseed proteins that can resist acid-induced denaturation. Digestibility studies carried out in rats, pigs, and humans suggest that rapeseed proteins have poor digestibility *in vivo* and produce hydrolysis resistant protein fragments. Indeed, the protein efficiency ratio (PER) for rapeseed flour, 2S and 12S rapeseed fractions were reported as 2.64, 2.49 and 2.12 *versus* casein which had a PER score of 3.23 (Delisle et al. 1983).

The protein digestibility–corrected amino acid score (PDCAAS) is a method previously recommended by the FAO/WHO as the method to measure protein value in human nutrition. It is based on comparison of the concentration of the first limiting essential amino acid in the test protein with the concentration of that amino acid in a reference (scoring) pattern. This scoring pattern is derived from the essential amino acid requirements of the preschool-age child. The chemical score obtained in this way is corrected for true fecal digestibility of the test protein. PDCAAS values higher than 100% are not accepted as such but are truncated to 100% (Schaafsma 2000). Rapeseed protein isolates have better values and the PDCASS values reported for napin-rich Supertein and cruciferin-rich Puratein (Burcon Nutrascience protein products) were 0.61 (61%) and 0.64 (64%), respectively. When updated in accordance with FAO/

WHO/UNU guidelines in 2007 (WHO/FAO/UNU 2007) values of 0.83 and 0.71 for Supertein and Puratein, were obtained (Wanasundara et al. 2016).

Based on a scoring pattern for a 0.5–3 year-old child, the Digestible Indispensable Amino Acid Score (DIAAS) of a rapeseed protein isolate produced by DSM was calculated to be 95%, which is comparable to estimated DIAAS for soy protein isolate. A protein with a DIAAS between 75 and 99% is considered a good source of protein according to FAO recommendations (FAO 2013).

10.4.3 Solubility

Solubility of a protein is related to the functionalities of the colloidal structure development and characteristics including gelation, foaming, emulsification, and liquid or oil holding capacity. The solubility of cruciferin and napin is different from each other and varies according to changes in pH, temperature, and salt levels (Wanasundara et al. 2012). RM proteins have solubility between pH 3.0 and 4.0 (Wanasundara et al. 2012). Rapeseed proteins cruciferin is insoluble at pH3–4 while napin is soluble at this pH (Wanasundara et al. 2012). Cruciferin and napin are soluble above pH 5.5 and only napin shows solubility in a wider pH range of 2–10. Solubility differences are due to the abundance of basic amino acids in napin proteins. Examples of foods where rapeseed protein isolates or concentrates, especially Puratein® or Supertein™ are used as casein or other protein substitutes are shown in Table 10.1.

Table 10.1 Application of rapeseed protein isolates or concentrates in foods.

Food carrier	References
Baked goods	
Rapeseed protein isolate included at 5% level in bread dough to improve emulsification	Kodagoda et al. (1973)
92.2% rapeseed protein isolate used to replace wheat flour in bread at an inclusion level of 18%	Mansour et al. (1999)
2% inclusion of Puratein and Supertein in various baked goods to enhance moisture content	GRAS (2010)
Beverages	
Puratein and Supertein can be added to fruit and vegetable drinks up to 10%	GRAS (2010)
Dairy and Egg substitute	
Rapeseed protein isolate can replace egg whites in meringue (3%) and can be added at a level for 5% to dairy and 60% to eggs	GRAS (2010) and Kodagoda et al. (1973)
Meat substitutes	
Rapeseed protein concentrate can be added at an inclusion level of 3.8% to sausages to improve peelability of sausage casings (wieners) and can reduce cooking loss and replace sodium caseinate. Has also been used in hot dog formulations (Puratein)	GRAS (2010), Thompson et al. (1982)

10.5 Bioactivities of Rapeseed Protein Hydrolysates and Identified Bioactive Peptides

10.5.1 Heart Health Benefits – Inhibition of Enzymes Associated with Cardiovascular Disease

Metabolic syndrome is a term given to a combination of risk factors occurring simultaneously including obesity, insulin resistance, high blood pressure, and dyslipidemia (Quinn 2019). Cumulatively these factors can lead to the development of high blood pressure, which can result in stroke and heart attack as well as type-two diabetes mellitus (T2DM). Central to the control of blood pressure and salt water balance in the body is the renin-angiotensin-aldosterone system (RAAS). Two enzymes are central to control of RAAS – the enzyme Renin (EC 3.4.23.15), which is a 37 kDa protein that catalyzes the rate-limiting step in the RAAS cascade through hydrolysis of angiotensinogen to angiotensin I (Ang I), which is an inactive decapeptide. Secondly, the enzyme ACE-I (EC 3.4.15.1) can remove a dipeptide from the vasodilatory peptide Ang I to yield a vaso-constricting peptide angiotensin II (Ang II), which binds to the vascular wall to cause blood vessel contraction. ACE also breaks down bradykinin, a vasodilatory peptide. ACE-I and renin can therefore reduce vasodilatory functions and increase sodium retention which can cause hypertension or high blood pressure. Hypertension is defined as systolic blood pressure (SBP) ≥140 mmHg with diastolic blood pressure (DBP) ≥90 mmHg (Aluko 2018). Peptides that can inhibit the action of renin and ACE-I can help to prevent hypertension and maintain normal blood pressure. ACE-I and renin inhibitory peptides have been identified from numerous protein sources previously, including rapeseed. There are over 958 ACE-I inhibitory peptides listed in BIOPEP-UWM and 19 renin inhibitory peptides, most of which are derived from either rapeseed or pea proteins and identified by the same group based in Canada.

10.5.1.1 Angiotensin-I-Converting Enzyme (ACE-I; EC 3.4.15.1)

Rapeseed proteins were identified as a source of several ACE-I inhibitory peptides. ACE-I is an enzyme found in the RAAS which is responsible for the control and maintenance of high blood pressure and salt water balance in humans and many animals. The di-, tetra- and tri- peptides with the amino acids sequences LY, RALP, and GHS from rapeseed proteins were identified to be ACE-I and renin inhibitors (He et al. 2013). More recently, these peptides were individually evaluated in spontaneously hypertensive rats (SHR), and the mRNA and/or protein levels of angiotensin-converting enzyme (ACE-I), renin, ACE2, angiotensin II and angiotensin-(1–7) in myocardial tissues were assessed (He et al. 2019). At a dose of 30 mg peptides/kg body weight every two days for five weeks a significant decrease in SBP of SHRs was found. Dipeptide LY also inhibited angiotensin II expression in heart tissue (He et al. 2019). In addition, Pedroche et al. (2004) generated an ACE-I inhibiting hydrolysate using the enzyme alcalase. The rapeseed protein isolate possessed the ability to inhibit ACE by 23%. Similarly, Mäkinen and colleagues (Mäkinen et al. 2012) also identified ACE-I inhibitory peptides from RM. Proteins were extracted from the meal using the method described by Berot and colleagues (Berot et al. 2005). RM was mixed with 50 mM Tris–HCl pH8.5, 750 mM NaCl, 5 mM EDTA (ethylenediaminetetraacetic acid), 0.3% $Na_2O_5S_2$ and stirred for 1 hour at room temperature and subsequently centrifuged. Supernatants were then dried and subjected to hydrolysis with alcalase. Two peptides with the amino acid sequences Ser-Leu-Val-Ser-Pro-Ser-Ala-Ala-Ala-Ala-Ala-Ala-Ala-Ala-Pro-Gly-Gly-Ser and Lys-Lys-Arg-Ser-Lys-Lys-Lys-Ser-Phe-Gly were identified using RP-HPLC and mass spectrometry and the IC_{50} value of the hydrolysate determined as 156 µg protein/ml +/−1.0. In addition, the peptides Val-Ser-Val and Phe-Leu were purified and located in the primary structures of the napin and cruciferin native proteins by Wu and colleagues (Wu et al. 2009). ACE-I inhibitory peptides along with other bioactive peptides derived from RM are shown in Table 10.2 along with the hydrolysis conditions used to generate and purify them. The anti-hypertensive properties of rapeseed proteins have been also reported (Yamada et al. 2010). The peptide Arg-Ile-Tyr which is known as rapakinin (Arg-Ile-Tyr), and has an IC_{50} value of 28 µM, was isolated from rapeseed protein digested with subtilisin. It was found to have an EC_{50} of 5.1 µM in the mesenteric artery of SHRs and the antihypertensive effect is thought to be due to the PGI(2)-IP receptor, followed by CCK–CCK(1) receptor-dependent vaso-relaxation (Yamada et al. 2010).

10.5.1.2 Dipeptidyl Peptidase III and IV Inhibitory Peptides from Rapeseed Meal

Diabetes mellitus is a complex metabolic disorder that affects approximately 387 million people worldwide and this number is expected to double globally within the next 20 years (Zaccardi et al. 2016). T2DM is also known as non-insulin dependent diabetes and is increasing globally due to a sedentary lifestyle, poor dietary choices, and lack of physical activity. It is characterized by insulin resistance that results in impaired glucose uptake into muscle and adipose tissues and patients may also suffer from cancer, liver disease, and heart health problems as well as infections (Quinn 2019; Zaccardi et al. 2016).

Dipeptidyl peptidase-IV (DPP-IV) (EC 3.4.14.5) is a homodimeric serine peptidase, consisting of 700 amino acids in each subunit. It cleaves substrate peptides with Pro or Ala at the penultimate position of peptides and is responsible for the degradation of GLP-1 and GIP. Therefore, inhibition of DPP-IV is a key target in the treatment of T2DM. In fact, DPP-IV inhibitors were one of the first classes of oral anti-diabetic drugs proposed as anti-hyperglycemic agents.

Sitagliptin was the first gliptin to be approved by the United States Food and Drug Administration (FDA). However, synthetic DPP-IV inhibitors may have adverse side effects including allergic reactions, gastrointestinal disorders, skin-related side effects, and musculoskeletal disorders. Alternatives from food-derived sources for prevention of T2DM are therefore favorable. Many DPP-IV inhibitory peptides have been identified from meat, milk and dairy sources, as well as wheat. Recently, the rapeseed protein napin was hydrolysed by alcalase and trypsin and the DPP-IV inhibitory peptides IPQVS and ELHQEEPL identified. The rapeseed protein hydrolysate generated had a DPP-IV IC_{50} value of 0.68 mg/ml (Xu et al. 2019).

Dipeptidyl peptidase-III (EC 3.4.14.4) is an important enkephalin degrading enzyme and its inhibitors are promising agents for pain management (Khaket et al. 2015). The rapeseed protein cruciferin CRU4 was observed previously to have several potential DPP-III inhibitory peptides within its protein sequence and warrants further investigation.

Table 10.2 Bioactive peptides derived from rapeseed proteins.

Bioactivity	Peptide sequence	Hydrolysate conditions	IC_{50} value	References
ACE-I-inhibition	Ser-Leu-Val-Ser-Pro-Ser-Ala-Ala-Ala-Ala-Ala-Ala-Ala-Pro-Gly-Gly-Ser	Addition of extraction buffer to rapeseed meal, centrifugation and hydrolysis of supernatant with alcalase	156+/−1 μg protein/ml	Berot et al. (2005); Mäkinen et al. (2016)
	Lys-Lys-Arg-Ser-Lys-Lys-Lys-Ser-Phe-Gly	Addition of extraction buffer to rapeseed meal, centrifugation and hydrolysis of supernatant with alcalase	156+/−1 μg protein/ml	Berot et al. (2005) and Mäkinen et al. (2016)
	Arg-Ile-Tyr	Subtsilin hydrolysis of rapeseed protein isolate	28 μM	Yamada et al. (2010)
	Val-Ser-Val	Alcalase 2.4 hydrolysis of defatted rapeseed meal and purification and identification from napin and cruciferin	Not determined	Wu et al. (2009)
	Phe-Leu	Alcalase 2.4 hydrolysis of defatted rapeseed meal and purificationand identification from napin and cruciferin	Not determined	Wu et al. (2008, 2009)
Antioxidant peptides	Pro-Ala-Gly-Pro-Phe	Pentapeptide derived from rapeseed flour hydrolysed with alcalase 2.4 L – peptide found in napin sequence	ED_{50} values were between 4.06 and 4.69 mg/mL	Zhang et al. (2009)
Di-peptidyl peptidase-IV (DPP-IV) inhibition	Ile-Pro-Phe-Val-Ser	Alcalase and trypsin hydrolysis of napin rapeseed protein	0.68 mg/ml	Xu et al. (2019); Mäkinen et al. (2016)
	ELHQEEPL	Alcalase and trypsin hydrolysis of napin rapeseed protein	0.68 mg/ml	Xu et al. (2019)
Anti-thrombotic activity	Not determined	Crude rapeseed peptide fractions prepared from an aqueous extraction of rapeseed proteins digested with Alcalase 2.4 L	ED50 value of 0.07 mg/ml	Zhang et al. (2008)
Cell growth – Effect on Chinese hamster ovary cells (CHO)	Not determined	Rapeseed protein hydrolysates generated by hydrolysis with Alcalase, Esperase, Neutrase, Orientase and Pronase – the 1 kDa fraction.	Shown to increase the maximal cell density of Chinese hamster ovary (CHO) C5 cells	Chabanon et al. (2008)

(*Continued*)

Table 10.2 (Continued)

Bioactivity	Peptide sequence	Hydrolysate conditions	IC_{50} value	References
Bile acid binding capacity	not determined	Rapeseed protein isolates generated by iso-precipitation or ultrafiltration, and their pepsin and pepsin/pancreatin hydrolysates	5.77–12.6% of sodium cholate and sodium deoxycholate were bound by the rapeseed precipitated protein isolate and hydrolysates	Yoshie-Stark and Wäsche (2004))
Renin inhibition	LY, RAPL, GHS	Alcalase hydrolysis of rapeseed meal	Oral administration of 30 mg peptides/kg body weight every 2 days for five weeks significantly decreased the systolic blood pressure and the myocardial mRNA and protein levels of ACE and renin in SHR. LY, RALP and GHS also increased the expression of ACE2	He et al. (2013)

10.5.2 Anti-Proliferative Activity of Rapeseed Meal Hydrolysates/Fermentates

RM fermentates prepared using solid-sate fermentation with a mixed bacterial culture of *Bacillus subtili* and *Acrogenotheca elegans* were shown previously to inhibit proliferation capability of human HepG2 liver cancer, and human MCF-7 breast cancer cells (Xie et al. 2015).

10.6 Safety of Rapeseed Proteins and Hydrolysates

To date, the use of rapeseed protein as a human food ingredient is limited to only a few products produced in Canada, Japan, and the United States where it is found in processed meats, cheeses, pizza, and bagels (Mejia et al. 2009). Rapeseed protein has generally recognized as safe (GRAS) status in the United States but applications to the EU for its use as a novel protein food were rejected previously due to the lack of limit values for relevant undesirable compounds (EFSA NDA Panel [EFSA Panel on Dietetic Products, Nutrition and Allergies] 2013). Issues of concern regarding the use of rapeseed proteins include the level of anti-nutritional factors found in protein isolates, heavy metal content, and the risk of protein allergy. Rapeseed contains erucic acid, glucosinolates, phytic acid, and phenolic compounds including sinapinic acid, tannins, and it is also rich in fiber (Kozlowska et al. 1990). Napins are considered allergens as they have cross-reactivity with related Brassica species (such as mustard) (Monsalve et al. 2001), and have sequence similarity with 2S albumins from different seeds. The 2S albumin seed storage proteins in mustard seeds share 94% sequence similarity with 2S albumins from rapeseed (Monsalve et al. 2001). Several allergic reactions in humans have been reported after consumption of mustard and rapeseed products previously (Meding 1985). It is likely that hydrolysis could lessen the allergenic potential of napin proteins although no data on how processing may affect the allergenic potential of rapeseed protein has been carried out to date (Campbell et al. 2016).

Table 10.3 highlights animal studies carried out to determine the safety of rapeseed proteins. The safety of rapeseed protein isolates was demonstrated with a 13 week toxicity study in rats fed either a cruciferin-rich protein isolate (Puratein from ADM/Burcon) or a napin-rich protein isolate (Supertein from ADM/Burcon) with low content of anti-nutritional factors (Mejia et al. 2009, 2010).

10.7 Conclusion

Rapeseed protein is of excellent quality and is well known to be comparable to soy protein in terms of amino acid content, PDCAAS and DIAAS scores, and is produced in large volumes for use in the animal, and in particular, ruminant nutrition sector. However, there are some concerns regarding its safe use due to the glucosinolate content of the protein extracts. Despite this, commercial rapeseed protein concentrates and isolates exist and are approved for use in the United States in human foods as outlined in Table 10.1. The price of rapeseed protein product is an important point concerning its positioning in the alternative protein market. It has several potential health benefits as outlined in this chapter and by others and it is likely it will need to move to higher value

Table 10.3 Safety of rapeseed protein products in animal trials.

Type of study	Protein preparation	Results	Remarks	References
Feeding, rats	Industrial rapeseed press cake	"Satisfactory protein source"	Toxic glucosides removed	Matet et al. (1949)
Feeding, dairy calves	Mustard oil meal	As good as soybean meal	No toxic effect	Huang (1950)
Feeding, rats	Mustard and rapeseed meals	As good as soybean meal	Toxic factors removed	Goering et al. (1960)
Toxicity test, laying hens	Extracted Algerian rapeseed meal and French rapeseed meal. 1.6, 3.2, 4.8, 6. and 8% protein of Algerian rapeseed meal diet with crude protein 39.6, 1.7, 3.4, 5.2, 6.9 and 8.6% protein of French rapeseed meal diet with crude protein 37.6%	Thyrotoxic	Weight increase of the thyroid gland	Jackson (1970)
Toxicity test, rats	Rapeseed lipid-protein concentrate	Similar to control	Higher incidence of renal calculi	Agren and Eklund (1972)

or volume markets in the future. Another challenge concerning rapeseed protein is the availability of suitable feedstock rapeseed for protein production (Campbell et al. 2016). This depends largely on the extraction process employed for rapeseed oil production. Conventionally processed RM which has gone through a high temperature process is not an efficient starting material for protein recovery. Currently, there is not a viable oil extraction technique that can replace this process without using solvent. Possible solutions are a low temperature vacuum method or use of cold methods with expeller pressing of oil. The economics of any alternative technology should be assessed and products need to be competitively produced. A higher value product with health benefits and a unique selling point compared to other plant proteins such as soy might enable economic production of rapeseed proteins going forward.

References

Agren, G. and Eklund, A. (1972). Rapeseed protein fractions. 3. The nutritive value of a detoxified protein concentrate prepared from rapeseed (*Brassica napus* L.) by hydraulic processing. *Journal of the Science of Food and Agriculture* 23 (12): 1457–1462.

Alashi, A.M., Blanchard, C.L., Mailer, R.J. et al. (2013). Technological and bioactive functionality of canola meal protein and hydrolysates. *Food Reviews International* 29 (3): 231–260.

Aluko, R.E. (2018). Food protein-derived renin-inhibitory peptides: in vitro and in vivo properties. *Journal of Food Biochemistry* 43 (1): e12648.

Berot, S., Compoint, J.P., Larre, C. et al. (2005). Large scale purification of rapeseed proteins (Brassica napus L.). *Journal of Chromatography B. Analytical Technologies in the BioMedical and Life Sciences* 818: 35–42.

Bleakley, S., Hayes, M., O' Shea, N. et al. (2017). Predicted release and analysis of novel ACE-I, renin, and DPP-IV inhibitory peptides from common oat (*Avena sativa*) protein hydrolysates using *in silico* analysis. *Food* 6: 108.

Campbell, L., Rempel, C.B., and Wanasundara, J.P.P. (2016). Canola/rapeseed protein: future opportunities and directions – workshop proceedings of IRC2015. *Plants* 5 (2): 17. https://doi.org/10.3390/plants5020017.

Chabanon, G., Alves da Costa, L., Farges, B. et al. (2008). Influence of the rapeseed protein hydrolysis process on CHO cell growth. *Bioresource Technology* 99: e7143–e7151.

Cosgrove, D.J. (1980). *Inositol Phosphates: Their Chemistry, Biochemistry and Physiology*. New York;: Elsevier Science 191 pp.

Dekker, K. (2018). Process design for sustainable extraction of rapeseed protein mixtures. BSc Thesis Biotechnology, Wageningen University, YBT-80324, 1–60.

Delisle, J., Amiot, J., Brisson, G.-J. et al. (1983). Improvement of rapeseed protein nutritional value. *Plant Food for Human Nutrition* 33: 173–177.

EFSA NDA Panel (EFSA Panel on Dietetic Products, Nutrition and Allergies) (2013). Scientific opinion on the safety of "rapeseed protein isolate" as a novel food ingredient. *EFSA Journal* https://doi.org/10.2903/j.efsa.2013.3420.

FAO (2013). Dietary protein quality evaluation in human nutrition. In: *Report of an FAO Expert Consultation in Rome*. Rome: FAO.

Goering, K., Thomas, O., Beardsley, D. et al. (1960). Nutritional value of mustard and rapeseed meals as protein source for rats. *Journal of Nutrition* 72: 210–216.

GRAS (2010). GRAS Notice 327. GRAS notification for cruciferin-rich and napin-rich protein isolates derived from canola/rapeseed (Puratein and SuperteinTM). https://www.cfsanappsexternal.fda.gov/scripts/fdcc/index.cfm?set=GRASNotices&id=327 (accessed March 27, 2016).

Hadnađev, M.S. (2002). Progress in vegetable proteins isolation techniques: a review. *Food Feed Resarch* 44 (1): 11–21.

Hald, C., Dawid, C., Tressel, R. et al. (2019). Kaempferol 3-O-(2‴-O-Sinapoyl-β-sophoroside) causes the undesired bitter taste of canola/rapeseed protein isolates. *Journal of Agricultural and Food Chemistry* 67: 372–378.

Hayes, M. (2018). Food proteins and bioactive peptides, new and novel sources, characterisation strategies and applications. *Food* 7 (3): 38. https://doi.org/10.3390/foods7030038.

He, R., Malomo, S.A., Girgih, A.T. et al. (2013). Glycinyl-histidinyl-serine (GHS), a novel rapeseed protein-derived peptide has blood pressure-lowering effect in spontaneously hypertensive rats. *Journal of Agricultural and Food Chemistry* 61: 8396–8402.

He, R., Yang, Y., Wang, Z. et al. (2019). Rapeseed protein-derived peptides, LY, RALP, and GHS, modulate key enzymes and intermediate products of renin–angiotensin system pathway in spontaneously hypertensive rat. *npj Science of Food* 3: 1. https://doi.org/10.1038/s41538-018-0033-5.

Höglund, A.S., Rödin, J., Larsson, E. et al. (1992). Distribution of napin and cruciferin in developing rapeseed embryos. *Plant Physiology* 98: 509–515.

Huang, W.Y. (1950). The feeding of mustard oil meal in the concentrate mixture to dairy cows. MSc Thesis, Montana State College.

Jackson, N. (1970). Algerian and French rapeseed meals as a protein source for caged laying hens, with observations on their toxic effects. *Journal of the Science of Food and Agriculture* 21 (10): 511–516.

Karaca, A.C., Low, N., and Nickesson, M. (2011). Emulsifying properties of canola and flaxseed protein isolates produced by isoelectric precipitation and salt extraction. *Food Research International* 44 (9): 2991–2998.

Khaket, T.P., Redhu, D., Dhanda, S. et al. (2015). *In silico* evaluation of potential DPP-III inhibitor precursors from dietary proteins. *International Journal of Food Properties* 18 (3): 499–507. https://doi.org/10.1080/10942912.2013.787626.

Kodagoda, L.P., Nakai, S., and Powrie, W.D. (1973). Some functional properties of rapeseed protein isolates and concentrates. *Canadian Institute of Food Science and Technology* 6: 266–269.

Kozlowska, H., Narzk, M., Shahidi, F. et al. (1990). Phenolic acids and tannins in rapeseed and canola. In: *Canola and Rapeseed: Production, Chemistry, Nutrition and Processing Technology* (ed. F. Shahidi), 193–210. New York: Springer.

Mäkinen, S., Johansson, T., Vegarud, G.E. et al. (2012). Angiotensin-I-converting enzyme inhibitory and antioxidant properties of rapeseed hydrolysates. *Journal of Functional Foods* 4: 575–583.

Mäkinen, S., Streng, T., Larsen, L.B. et al. (2016). Angiotensin-I-converting enzyme inhibitory and antihypertensive properties of potato and rapeseed protein derived peptides. *Journal of Functional Foods* 25: 160–173.

Mansour, E.H., Dworschak, E., Pollhamer, Z. et al. (1999). Pumpkin and canola seed proteins and bread quality. *Acta Alimentaria* 28: 59–70.

Matet, J., Montagne, R., and Buchy, A. (1949). Food value of the rapeseed press cake. *Oleagineux* 4: 145–154.

Meding, B. (1985). Immediate hypersensitivity to mustard and rape. *Contact Dermatitis* 13 (2): 121–122.

Mejia, L.A., Korgaonkar, C.K., Schweizer, M. et al. (2009). A 13-week dietary toxicity study in rats of a napin-rich canola protein isolate. *Regulatory Toxicology and Pharmacology* 55 (3): 394–402.

Mejia, L., Korgaonkar, C., Schweizer, M. et al. (2010). Safety profiles of cruciferin-rich (puratein) and napin-rich (supertein) canola protein isolates following subchronic dietary toxicity studies in Sprague Dawley rats. *International Journal of Toxicology* 29 (1): 92.

Minkiewicz, P., Iwaniak, A., and Darewicz, M. (2019). BIOPEP-UWM database of bioactive peptides: current opportunities. *International Journal of Molecular Sciences* 20: 5978. https://doi.org/10.3390/ijms20235978.

Monsalve, R.I., Villalba, M., and Rodrıguez, R. (2001). Allergy to mustard seeds: the importance of 2S albumins as food allergens. *Internet Symptoms Food Allergens* 3 (2): 57–69.

Östbring, K., Malmqvist, E., Nilsson, K. et al. (2020). The effects of oil extraction methods on recovery yield and emulsifying properties of proteins from rapeseed meal and press cake. *Food* 9: 19. https://doi.org/10.3390/foods9010019.

Pedroche, J., Yust, M.M., Lqari, H. et al. (2004). Brassica carinata protein isolates: chemical composition, protein characterization and improvement of functional properties by protein hydrolysis. *Food Chemistry* 88 (3): 337–346.

Quinn, L. (2019). Generation of sinapinic-acid containing extracts from Irish rapeseed meal and characterisation of their bioactive properties for potential human health applications. PhD thesis, Department of Clinical Medicine, Trinity College Dublin. 1–195.

Rodrigues, I.M., Coelho, J.F.J., and Carvalho, M.G.V.S. (2012). Isolation and valorisation of vegetable proteins from oilseed plants: methods, limitations and potential. *Journal of Food Engineering* 109 (3): 337–346.

Savoie, L., Galibois, I., Parent, G. et al. (1988). Sequential release of amino acids and peptides during in vitro digestion of casein and rapeseed proteins. *Nutrition Research* 8: 1319–1326.

Schaafsma, G. (2000). The protein digestibility-corrected amino acid score. *The Journal of Nutrition* 130 (7): 1865S–1867S.

Thompson, L.U., Liu, R.F.K., and Jones, J.D. (1982). Functional properties and food applications of rapeseed protein concentrate. *Journal of Food Science* 47: 1175–1180.

Tzeng, Y.-M., Diosady, L.L., and Rubin, L.J. (1990). Production of canola protein materials by alkaline extraction, precipitation, and membrane processing. *Journal of Food Science* 55 (4): 1147–1151.

Uppstrom, B. (1995). Seed chemistry. In: *Brassica Oilseeds: Production and Utilization* (eds. D.S. Kimber and D.I. McGregor), 217–242. Wallingford, UK: CAB Interntional.

Wanasundara, J.P.D., Abeysekara, S.J., McIntosh, T.C. et al. (2012). Solubility differences of major storage proteins of Brassicaceae oilseeds. *Journal of the American Oil Chemists' Society* 89: 869–881.

Wanasundara, J.P.D., McIntosh, T.C., Perera, S.P. et al. (2016). Canola/rapeseed protein functionality and nutrition. *Oilseeds and Fats, Crops and Lipids* 23 (4): D407.

Wang, G., Cai, G.Q., and Lu, J. (2011). Improving the quality of rapeseed meal by microbial fermentation. *China Oils and Fats* 36: 24–28.

WHO/FAO/UNU (2007). Protein and amino acid requirements in human nutrition. Report of a joint FAO/WHO/UNU expert consultation. WHO Technical Report Series 935. https://www.who.int/nutrition/publications/nutrientrequirements/WHO_TRS_935/en/ (accessed March 20, 2016).

Wu, J., Aluko, R.E., and Muir, A.D. (2009). Production of angiotensin I-converting enzyme inhibitory peptides from defatted canola meal. *Bioresource Technology* 100: 5283–5287.

Xie, H., Wang, Y., Zhang, J. et al. (2015). Study of the fermentation conditions and antiproliferative activity of rapeseed peptides by bacterial and enzymatic cooperation. *International Journal of Food Science and Technology* 50 (3): 619–625.

Xu, L. and Diosady, L. (2002). Removal of phenolic compounds in the production of high-quality canola protein isolates. *Food Research International* 35 (1): 23–30.

Xu, F., Yao, Y., Xu, X. et al. (2019). Identification and quantification of DPP-IV inhibitory peptides from hydrolysed rapeseed protein derived napin with analysis of the interactions between key residues and protein domains. *Journal of Agricultural and Food Chemistry* 67 (13): 2679–2690.

Yamada, Y., Iwasaki, M., Usui, H. et al. (2010). Rapakinin, an anti-hypertensive peptide derived from rapeseed protein, dilates mesenteric artery of spontaneously hypertensive rats via the prostaglandin IP receptor followed by CCK (1) receptor. *Peptides* 31 (5): 909–914.

Yoshie-Stark, Y. and Wäsche, A. (2004). in vitro binding of bile acids by lupin protein isolates and their hydrolysates. *Food Chemistry* 88: 179–184.

Zaccardi, F., Webb, D.R., Yates, T. et al. (2016). Pathophysiology of type 1 and type 2 diabetes mellitus: a 90-year perspective. *Postgraduate Medical Journal* 92 (1084): 63–69.

Zayas, J.F. (1997). Emulsifying properties of proteins. In: *Functionality of Proteins in Food*, 134–227. Berlin, Heidelberg: Springer.

Zhang, S.B., Wang, Z., and Xu, S.U. (2008). Antioxidant and antithrombotic activities of rapeseed peptides. *Journal of American Oil Chemists' Society* 85: 521–527.

Zhang, S.B., Wang, Z., Xu, Z.Y. et al. (2009). Purification and characterization of a radical scavenging peptide from rapeseed protein hydrolysates. *Journal of the American Oil Chemists' Society* 86: 959–966.

11
Oils and Oilseeds in the Nutraceutical and Functional Food Industries

Manuel Suárez, Andreu Gual-Grau, Javier Ávila-Román, Cristina Torres-Fuentes, Miquel Mulero, Gerard Aragonès, Francisca Isabel Bravo and Begoña Muguerza
Universitat Rovira i Virgili, Departament de Bioquímica i Biotecnologia, Nutrigenomics Research Group, Tarragona, Spain

11.1 Introduction

Non-communicable diseases (NCDs), also known as chronic diseases, are responsible for the death of 40 million people every year, representing 70% of the total deaths worldwide (World Health Organization 2018). In 2018, the World Health Organization (WHO) reaffirmed its commitment to reduce by one-third premature mortality from NCDs by 2030 (United Nations General Assembly 2018). This group of pathologies includes cardiovascular diseases (CVD), chronic respiratory diseases, cancer, and diabetes, among others. Despite the fact that these diseases distinctively impact on health status and, consequently, must be differently approached, they share four predisposing risk factors that are highly present in modern societies: physical inactivity, overconsumption of alcohol, tobacco use, and unhealthy diets (The World Health Organization 2018). Therefore, clinical and nutritional strategies able to effectively combat the onset and progression of NCDs are on the rise, thus constituting a major challenge for the scientific community.

Whereas only drugs from pharmacological treatments can control or palliate advanced stages of the NCDs, an adequate diet is known to prevent and counteract some of these pathologies, particularly those leading to undesired metabolic disturbances. For instance, a high adherence to the Mediterranean diet is widely accepted to prevent the onset of CVD (Dinu et al. 2018). In this sense, the PREDIMED study

Oil and Oilseed Processing: Opportunities and Challenges, First Edition.
Edited by Tomás Lafarga, Gloria Bobo, and Ingrid Aguiló-Aguayo.

(*PREvención con DIeta MEDiterránea*) confirmed the efficacy of the Mediterranean diet in reducing the incidence of major cardiovascular events in people at high risk (Estruch et al. 2018). Specifically, several scientific publications have evidenced a strong relationship between dietary fatty acids and CVD. Thus, a simple shift in the dietary fatty acids profiles might have a profound relevance to the prevention of these diseases. However, the mechanisms and pathways by which this diet exerts beneficial effects on health have been only partially elucidated.

In addition to dietary lipids, other bioactive compounds from food including (poly) phenols, vitamins, peptides, and fiber, together with prebiotics and probiotics, are broadly considered to be effective tools for the management of several metabolic disorders from a nutritional perspective (Ghazani and Marangoni 2016). However, some food companies prioritize taste and sales rather than supply nutrients that contribute to the maintenance of a healthy human life. From this perspective, demands for high-value foods regarding their nutritive and functional properties have increased in the recent years. Altogether, a substantial change in dietary habits complemented with the consumption of some bioactive compounds, which can be contained in functional foods or nutraceuticals, may constitute an effective way to reduce the incidence of these pathologies.

11.2 Functional Food and Nutraceuticals

11.2.1 Definition

The concept of functional foods was coined in the mid-1930s by the Japanese Dr. Shirota, who developed for the first time a bacterial product, called Yakult, able to improve the consumer's intestinal health (Heasman and Mellentin 2001). Since then, many products have been marketed due to increased consumers' interest for the relationship between diet and disease prevention/amelioration. Moreover, functional foods are recognized as healthier and more beneficial products. However, this concept differs among countries. In 2007, the Food and Agricultural Organization of the United Nations (FAO) defined functional foods as those similar in appearance to conventional foods and consumed as a part of a usual diet but that have demonstrated physiological benefits and can reduce the risk of chronic diseases beyond basic nutritional functions. In the same year, South Korea defined this concept as dietary supplements that have to be marketed in measured doses such as tablets or pills (Kim et al. 2007). On the other hand, Health Canada defined functional foods as similar to traditional foods but having demonstrated physiological benefits (Shahidi 2009). More recently, functional foods are referred to as those foods that are consumed as part of a normal diet and contain biologically active components (added or not) which have the potential to enhance health and/or reduce the risk of diseases (Gul et al. 2016).

Besides functional foods, in 1989, nutraceutical products were defined by DeFelice and the Foundation for Innovation in Medicine as pharmaceutical forms containing food phytochemicals as active principles (Espín et al. 2007). Specifically, this concept refers to a dietary supplement with a potential to deliver a concentrated form of a bioactive agent from a food. Moreover, these products provide medical or health benefits, encompassing prevention and treatment of disease. These products can be sold in a non-food matrix such as pills, extracts, or tablets (Zeisel 1999) and used with the purpose of enhancing health or preventing illness in dosages that exceed those that could be obtained from regular food. More recently, nutraceuticals have involved a great variety of products such as isolated nutrients, dietary supplements, genetically engineered designer foods, herbal components, and processed foods such as cereals, soups, or beverages (Dudeja and Gupta 2017).

11.2.2 Regulation

Currently, although the differences between the functional foods and nutraceuticals concepts are unclear, there is an increasing interest and consumption of these products by the population. Nevertheless, there is still not an international legislation for these products. Therefore, globalization entails the need for a unified definition and a common legislative control.

Japan was the first country to legislate functional foods in 1952. It proclaimed a regulation about the improvement of health throughout food consumption (248/1952). The European Food Safety Authority (EFSA) was created in 2002 aiming to evaluate all the aspects related to food in Europe and provide scientific support to the legislation process. Currently, Regulation 1924/2006 refers to functional foods health claims. In this sense, nutritional claims are defined as "any claim, which states, suggests or implies that a food has particular beneficial nutritional properties due to the presence, absence, increased or reduced levels of a particular nutrient or other substance." On the other hand, health claims are defined as "any claim that states, suggests or implies that a relationship exists between a food category, a food or one of its constituents and health" (Khedkar et al. 2017). In the framework of this regulation, food companies in Europe have had to adapt and are continuously developing and launching new types of product to the food market, introducing in their formulations natural compounds with demonstrated bioactivity.

11.3 Vegetable and Seed Oils as Source of Bioactive Compounds

Vegetables and seed oils are an excellent source of bioactive compounds able to be used in the nutraceuticals and functional foods field. The chemical composition of the oils can be divided in two complementary fractions: saponifiable and unsaponifiable based on their property to be hydrolysed in a soap-generating alkaline medium.

11.3.1 Saponifiable Fraction

From these two fractions, the saponifiable is the most important, representing approximately the 98–99% of the content of the oil. The main components of this fraction are triglycerides of long chain fatty acids. These fatty acids are classified depending on the occurrence of double bonds in the chain. Hence, saturated fatty acids do not contain double bounds whereas monounsaturated fatty acids (also known as MUFAs) contain a single double bound and polyunsaturated fatty acids (PUFAs) can have from two to up to six double bonds. Moreover, based on the configuration, unsaturated fatty acids can be distinguished in *cis* or *trans* isoforms. In addition to the regular International Union of Pure and Applied Chemistry (IUPAC) nomenclature, the unsaturated fatty acids can be also denominated following the omega nomenclature depending on the position of the first double bond starting from the fatty acid methyl-end (namely, omega-3, omega-6, and omega-9) (Orsavova et al. 2015).

The proportion of the different fatty acids (saturated and unsaturated) in the oils is very important as it has been demonstrated that MUFAs and PUFAs are usually associated with more healthy benefits than the saturated species. The content of the different fatty acids in the oils is highly variable and depends on multiple factors. For example, there is a great effect of agronomic features, such as the variety of the plant, the cultivar system, environmental conditions of the production area, and the harvesting time (which can be related with the ripening state). In addition, the extraction procedure used to obtain the oil has also a deep influence in its composition (Akkaya 2018). Table 11.1 shows the main fatty acids present in the triglycerides of some edible vegetable and seed oils and their percentage abundance range. In general, these oils stand out by their high content in unsaturated fatty acids. As it can be seen, oleic acid (a MUFA) is the main component of extra virgin olive oil (EVOO) (up to 83%), avocado (up to 65%), rapeseed and peanut oils (up to 65 and 71%, respectively). This fatty acid is also found in high concentration in palm (up to 50%) and sunflower oils (up to 39%). Interestingly, the content of oleic acid in some sunflower oils (the high-oleic acid containing type) can be increased up to 82%, being the main fatty acid in this type of oils. Especially remarkable is also the high content of PUFAs (omega-6 and omega-3 fatty acids), in some oils. These essential fatty acids play key roles in human metabolism but cannot be synthetized by the organism so they have to be incorporated from exogenous sources. In this regard, as it can be seen in Table 11.1, linoleic acid (omega-6) is the main fatty acid in edible oils such as sunflower (up to 71%), corn (up to 66%), cottonseed (up to 60%), and grapeseed (up to 75%). Sesame oil also contains an important amount of this fatty acid, reaching up to 43%. Regarding the omega-3 subfamiliy, specifically linolenic acid, linseed oil stands out by the high content of this fatty acid (up to 63%). However, due to the high degree of unsaturation (three unsaturations), this fatty acid is highly unstable, which deeply conditions the potential applications of this oil.

Although in general, vegetables and seed oils are eminently rich in unsaturated fatty acids, some of them also present high amounts of saturated fatty acids. In this sense, it is important to highlight the composition of palm oil, which contains up to 44% of palmitic acid. This oil is one of the most used in the elaboration of different processed foods due to its physicochemical properties, provided by its composition (50% saturated/50% oleic acid). Moreover, other oils with a relevant amount of palmitic acid are cottonseed, avocado, and corn (up to 31, 25, and 17%, respectively).

Finally, the saponifiable fraction also contains in very low proportion other lipid molecules such as phospholipids, glycolipids, sphingolipids, and waxes that can also play essential roles in the metabolism after ingestion (Ghazani and Marangoni 2016).

11.3.2 Unsaponifiable Fraction

The unsaponifiable fraction represents only 1–2% of the oil. Despite its low concentration, some of its constituents are determining factors when evaluating the nutritional quality of the oils, their oxidative stability, and their organoleptic characteristics. In addition, its quantitative analysis can significantly contribute to the authentication of various types of olive oils – even to the detection of adulterations. This fraction contain numerous compounds of a very varied chemical nature, such as hydrocarbons,

Table 11.1 Main fatty acid composition of some vegetable oils in percentage (g/100 g oil).

Fatty acid	Extra virgin olive oil	Sunflower	Corn	Palm	Peanut	Avocado (Hass variety)	Linseed/ Flaxseed	Sesame	Rapeseed/ Canola	Cottonseed	Grapeseed
Lauric acid (C12:0)	–	0.02	ND-0.3	0.2–0.4	–	–	–	–	–	1.0	–
Miristic acid (C14:0)	0.004–0.01	0.1	ND-0.3	0.9–1.7	0.04–0.7	–	–	–	<0.1	0.5–1.3	<0.1
Palmitic acid (C16:0)	7.5–18.3	5.0–9.2	8.6–16.5	27.1–44.0	7.2–26.5	13.4–25.6	3.7–11.7	9.2–9.7	4.0–5.6	17.0–31.0	6.4–7.2
Palmitoleic acid (C16:1)	0.3–3.5	0.1–1.2	ND-0.5	–	0.1–0.2	3.4–11.6	0.02–0.1	0.1–0.1	0.2–0.4	<1.0	<0.1
Stearic acid (C18:0)	0.5–5.4	2.6–6.0	ND-3.3	2.9–7.5	2.1–4.0	0.2–0.9	2.3–8.7	3.9–6.5	1.4–2.5	1.0–3.3	3.1–4.8
Oleic acid (C18:1 cis omega-9)	55.0–83.0	14.0–39.4	20.0–42.2	32.2–49.8	31.1–71.1	42.6–65.3	11.5–37.9	31.9–41.5	56.8–64.9	13.0–21.0	14.3–31.0
Linoleic acid (C18:2cis omega-6)	1.0–25.7	22.0–71.3	34.0–65.6	9.2–15.1	13.3–30.3	10.6–20.9	7.6–31.0	32.9–42.9	17.1–20.6	34.0–60.0	56.8–74.7
Linolenic acid (C18:3 omega-3)	0.2–1.3	0.2–1.6	ND-2.0	0.2–2.0	0.1–0.4	0.6–3.2	34.1–63.4	0.2–0.4	0.7–3.4	0.3–3.9	0.1–0.2
Arachidic acid (C20:0)	0.2–0.8	0.2–0.5	–	0.1–0.3	0.3–2.0	–	0.6	0.6–0.6	–	0.2–0.5	0.2
Eicosenoic acid (C20:1 omega-9)	0.2–0.4	0.2–0.3	–	–	0.1–2.0	–	1.2	0.2–0.3	1.3–10.0	–	–

(Continued)

Table 11.1 (Continued)

Fatty acid	Extra virgin olive oil	Sunflower	Corn	Palm	Peanut	Avocado (Hass variety)	Linseed/ Flaxseed	Sesame	Rapeseed/ Canola	Cottonseed	Grapeseed
Behenic acid (C22:0)	0.1–0.3	0.6–1.2	–	–	0.1–4.0	–	–	0.2	–	–	–
References	Crizel et al. (2020), El Qarnifa et al. (2019), Boskou (2009), Fregapane and Salvador (2019)	Ghazani and Marangoni (2016), Aung et al. (2018), Devi and Khatkar (2016), Orsavova et al. (2015), Konuskan et al. (2019), Akkaya (2018), Alves et al. (2019)	Barrera-Arellano et al. (2018), Codex Alimentarius (1999)	Aung et al. (2018), Devi and Khatkar (2016), Alves et al. (2019), Mancini et al. (2015)	Ghazani and Marangoni (2016), Aung et al. (2018), Devi and Khatkar (2016), Orsavova et al. (2015), Konuskan et al. (2019), Martini et al. (2018)	Flores et al. (2019), Yanty et al. (2011)	Ghazani and Marangoni (2016), Nykter et al. (2006)	Aung et al. (2018), Orsavova et al. (2015)	Orsavova et al. (2015), Konuskan et al. (2019), Matthaus et al. (2016)	Ghazani and Marangoni (2016), Devi and Khatkar (2016), Alves et al. (2019)	Orsavova et al. (2015), Alves et al. (2019), Martini et al. (2018)

tocopherols, triterpenic alcohols, sterols, pigments, phenolic compounds, and volatiles responsible for the aroma belonging to the group of aldehydes, alcohols, and esters, among others.

Phytosterols, which are molecules that derive from squalene, are the main components of this fraction. Regarding their structure, they have a tetracyclic cyclopenta[α]-phenanthrene form with a hydroxyl group at C-3 and a flexible side chain with 8–10 carbons at C-17. They can be found in free forms or esterified in the 3β-OH group (fatty-acid esters, steryl glycosides, acylated steryl glycosides, and hydroxycinnamic acid steryl esters). Phytostanols, the saturated forms of phytosterols, can also be found in the oils, although in lower concentrations (Moreau et al. 2002). Especially remarkable, it has been seen that these phytosterols exert hypocholesterolemic effect when they are consumed (García-Llatas and Rodríguez-Estrada 2011). In general, in vegetable oils their concentration range varies between 1000 and 5000 mg/kg. However, in some oils such as rapeseed and corn, their concentration can be clearly higher (5583–14069 and 5102–14336 mg/kg, respectively) (Yang et al. 2019).

More than 200 types of phytosterols have been identified in plants, although the most abundant are β-sitosterol, campesterol, stigmasterol, and Δ5-avenasterol (Moreau et al. 2002). Table 11.2 shows the main phytosterol concentration of some vegetable and seed oils in mg/kg. As can be seen in Table 11.2, β-sitosterol is the most widely distributed phytosterol. Sunflower oil has the highest concentration (up to 9537 mg/kg) followed by corn, sesame, rapeseed, and cottonseed oils (up to 4000–5000 mg/kg). On the other hand, palm and avocado oils stand out by their low concentration (lower than 400 mg/kg). Campesterol content is especially important in rapeseed (up to 3020 mg/kg), corn (up to 2150 mg/kg), linseed and sesame (up to 1000 mg/kg) oils. In addition, brassicasterol, which can be found in high concentration in oils obtained from Brassicaceae species seeds (García-Llatas and Rodríguez-Estrada 2011), is an important phytosterol in the composition of rapeseed oil, reaching values of up to 2000 mg/kg.

In addition to phytosterols, other components in the unsaturated fraction are tocopherols. They constitute a chromane ring with a hydroxyl group and a hydrophobic side chain. Tocopherols can be found in four forms denominated as α, β, γ, and δ. Together with tocotrienols, which are analogous structures that contain three unsaturations in the side chain, tocopherols exert vitamin E activity. Remarkably, not all the isoforms hold the same activity. Thus α, β, γ, and δ are described to have respectively 60, 30, 1.5, and 0.1% vitamin activity (Frankel 2005), while the antioxidant activity follows the decreasing order $\gamma > \delta > \beta > \alpha$ (Fan and Eskin 2015). These compounds are very used and highly demanded in the food industry to prevent the oxidation of the unsaturated fatty acids in oils, a fact that is related to their stability. Actually, vegetable and seed oils are the main source of tocopherols in the diet. As shown in Table 11.2, corn, linseed, peanut, and cottonseed oils are the oils with the highest quantities of tocopherols (over 1000–1700 mg/kg, considering the maximum values). In contrast, oils from avocado, palm, or grapeseed contain a low quantity of tocopherols, not exceeding a concentration of 250 mg/kg. Concerning the different tocopherol isoforms, in the oils the most abundant are the α and γ ones. Extra virgin olive and sunflower oils contain more α-tocopherol concentration than the other isoforms. However, in linseed, sesame, peanut, rapeseed, and cottonseed oils the γ isoform prevails. In this sense, it is import to highlight linseed oil, which can contain up to 1697 mg/kg of γ-tocopherol, much more than either peanut or sesame (up to 728 and 637 mg/kg, respectively), which are the second oils with the highest concentrations. Similarly to

Table 11.2 Main phytosterols and tocopherols identified in some vegetable oils. Data expressed in mg/kg oil.

Compounds	Extra virgin olive oil	Sunflower	Corn	Palm	Peanut	Avocado	Linseed/ Flaxseed	Sesame	Rapeseed/ Canola	Cottonseed	Grapeseed
PHYTOSTEROLS											
β-sitosterol	795.0–1200.0	1941.0–9537.3	3414.0–5150.0	240.0–395.0	426.6–2997.4	82.9–95.2	1517.5–2060.0	3100.0–4682.0	3760.0–4130.0	3902.4–4038.0	1367.8–2397.5
Apparent β-sitosterol	940.0–960.0	–	–	–	–	–	–	–	–	–	–
Campesterol	22.0–59.0	270.0–712.6	815.0–2150.0	90.0–174.0	102.6–860.1	0.3–198.0	885.6–1167.1	128.0–1000.0	2710.0–3020.0	442.0–480.0	210.1–316.5
Stigmasterol	4.0–15.0	176.0–753.7	383.0–660.0	60.0–87.0	43.2–577.8	0.2–12.2	334.0–450.5	360.0–484.0	31.0–49.0	5.2–5.4	375.0–491.5
5-avenasterol	47.0–230.0	95.9–156.5	210.0	–	45.0–825.4	0.2–115.3	314.8–590.0	470.0–709.0	290.0–320.0	152.0–158.0	–
7-stigmastenol	–	860.6–2028.0	–	–	–	–	–	–	–	–	–
7-stigmasterol	–	–	–	–	ND-225.9	–	15.4–19.8	63.0–65.0	–	33.0–35.0	–
7-avenasterol	2.4–9.8	123.9–383.7	–	–	ND-286.7	4.5–5.1	15.7–26.0	82.0–86.0	–	20.0–22.0	
Brassicasterol	ND	ND-18.3	ND-116.0	–	ND	–	ND-45.7	ND	703.0–2009.8	–	12.2–30.8
Sitostanol	–	29–51	131.0–290.0	ND-10	–	5.9–6.3	18.5–19.8	–	–	–	44.9–218.1
Squalene	2.0–10.5	–	–	–	270.0–1296.0	–	–	–	–	–	95.2–165.7
TOCOPHEROLS											
α-tocopherol	120.5–590.0	590.0–723.0	134.0–420.6	3.0–185.0	24.6–304.0	32.1–87.4	3.0–15.0	59.0–79.0	103.0–240.0	282.2–573.0	50.3–132.6
β-tocopherol	2.2–6.7	17.4–41.0	11.0–31.0	ND	ND-41.0	ND-2.1	<1.0	ND-4.1	<1.0–134.0	ND-40.0	0.2–2.6

γ-tocopherol	4.8–18.1	6.7–65.0	412.0–729.0	4.0–36.0	49.3–728.5	0.2–9.1	196.0–1697.0	360.0–637.2	174.0–420.0	317.0–572.0	7.2–41.2
δ-tocopherol	<1	ND-6.0	19.4–39.0	ND	ND-56.0	ND-0.1	3.0–9.5	7.0–12.0	9.8–41.0	1.3–50	0.1–1.6
References	Fregapane and Salvador (2019), Schwartz et al. (2008), Baccouri et al. (2008)	Yang et al. (2019), Schwartz et al. (2008), Maestri et al. (2006), Hassanien (2013), Aguirre et al. (2014), Mahmoud Hassanein et al. (2011)	Yang et al. (2019), Schwartz et al. (2008), Maestri et al. (2006), Hassanien (2013)	Maestri et al. (2006), Hassanien (2013)	Yang et al. (2019), Maestri et al. (2006), Akhtar et al. (2014)	Flores et al. (2019), Manaf et al. (2018), Tan (2019), Berasategi et al. (2012)	Matthaus et al. (2016), Yang et al. (2019), Schwartz et al. (2008)	Yang et al. (2019), Schwartz et al. (2008), Mariod et al. (2011)	Yang et al. (2019), Schwartz et al. (2008), Maestri et al. (2006)	Maestri et al. (2006), Mahmoud Hassanein et al. (2011), Mariod et al. (2011)	Yang et al. (2019), Wen et al. (2016)

what happens with the other components of the unsaponifiable fraction, the content of tocopherols of the oils is highly dependent of the extraction processes, and can be even completely lost in the process (De Greyt et al. 1999).

Finally, in addition to phytosterols and tocopherols, in the unsaponifiable fraction it is important to highlight the presence of phenolic compounds, which are molecules characterized by the presence of at least one hydroxylated aromatic ring that have several beneficial effects on the body. For example, its antioxidant, antihypertensive, antidiabetic, and anticarcinogenic properties, among others (Del Rio et al. 2013) have been described. Although the level of these metabolites in the oils is low as most of them remain in the raw material after the oil extraction process, their bioactivity is so important that some health claims related with oil consumption have been made due to the presence of phenolic compounds. For instance, following EFSA's advice, the European Commission stated that consumption of 20 mg of olive oil containing at least 5 mg of hydroxytirosol and its derivatives (which are characteristics of this type of oil) contributes to the protection of blood lipids from oxidative stress (European Commission 2013). Hence, it is clear that this family of compounds contains key elements when determining the bioactivity and potential application of either oils or its derived compounds in the field of functional foods and nutraceuticals.

11.4 Bioactivity of Vegetable Oils and Oilseeds

Vegetable and seed oils provide a broad range of components (including micro and macronutrients as well as phytochemicals) that can exert health benefits, related to beneficial bioactivity. The following sections provide a description of the main bioactivities associated with the consumption of some oils.

11.4.1 Olive Oil

Evidence shows that one of the best examples of a healthy oil is olive oil, which is one of the main components of the Mediterranean diet. Its beneficial properties are related to the fact that it is the oil with the highest MUFA content (75%). Additionally, olive oil presents a large variety of valuable antioxidants (hydroxytirosol is one of the main antioxidant compounds in olive oil) that play a significant role in the many health benefits associated with olive oil consumption (Caramia et al. 2012; García-González et al. 2008; Gunstone 2002; Harwood and Yaqoob 2002). MUFA consumption, particularly oleic acid, results in lowering of low-density lipoprotein cholesterol (LDL-c) and elevation of high-density lipoprotein cholesterol (HDL-c), which causes a significant decrease in the risk of suffering CVD (Bester et al. 2010; Choe 2008; St-Onge et al. 2007). Furthermore, olive oil phenolics exert important anti-inflammatory, antithrombotic, antihypertensive, and vasodilator effects (Turner et al. 2005). Interestingly, it has been observed that the least processed forms of olive oil, i.e. EVOO have more MUFA than other olive oils. Furthermore, EVOO contains the highest concentration of polyphenolic antioxidants (oleuropein or hydroxytirosol) that have been linked with increased arterial elasticity, and decreased reactive oxygen species (ROS) formation, which leads to decreased risk of stroke and heart attacks (Presti et al. 2017; Sánchez-Rodriguez et al. 2019). EVOO also shows a protective effect against certain malignant tumors in the breast, endometrium, prostate, and digestive tract, among others. This is mainly due to its ability to interfere with key points of cancer progression, such as: tumor initiation (EVOO acting as ROS scavengers); promotion (EVOO action decreasing estrogen receptors); proliferation (EVOO mediating cell cycle arrest and apoptosis induction); as well as on metastasis (EVOO induced angiogenesis inhibition) (Reboredo-Rodríguez et al. 2018). Interestingly, EVOO also contains metabolo-epigenetic inhibitors of cancer stem cells (Corominas-Faja et al. 2018). On the other hand, non-filtered EVOO can contain small amounts of peptides that possess angiotensin converting enzyme (ACE) inhibitory and antihypertensive activity (Alcaide-Hidalgo et al. 2020a, b). Regarding neurodegenerative disease, it has been shown that EVOO and its phenols (oleocanthal, oleuropein, tyrosol, and hydroxytirosol, among others) are able to prevent/counteract Alzheimer Disease, representing an emerging and very interesting research field (Angeloni et al. 2017).

11.4.2 Sunflower Oil

Due to its extended consumption worldwide, sunflower is another interesting oil. Sunflower oil is very rich in PUFAs and has a low content of saturated fatty acids. Specifically, this oil is high in the omega-6 linoleic acid and as a result, it is associated with an improved blood lipid profile and a reduced risk of CVD. Regarding its antioxidant activity, when compared with linseed oil and sesame seed oil, sunflower oil showed the highest antioxidant value when using the *in vitro* 2,2-diphenyl-1-picrylhydrazyl (DPPH) assay, but showed a poor value when Trolox method was used. This behavior is characteristic of an oil that contains low levels of phenolic compounds and high levels of vitamin-E-active compounds, respectively, such as sunflower oil (Kostadinović Veličkovska et al. 2018).

11.4.3 Corn Oil

Corn oil is frequently used in the food industry in different applications including the formulation of mayonnaise, mixtures, shortenings, technical fats, sauces, and margarines. Moreover, corn oil is also introduced in snacks, bakery products, and toppings (Barrera-Arellano et al. 2018). Corn oil is very rich in PUFAs and phytosterols such as sitosterol and campestanol. In fact, corn oil contains approximately five times more PUFAs (54 g/100 g oil) and plant sterols (989 mg/100 g oil) compared with EVOO (10 g/100 g oil and 198 mg/100 g oil, respectively) (Maki et al. 2015). Plants sterols are structurally similar to cholesterol and, therefore, this effect could be due to the competition between phytosterols and cholesterol for the micelle formation in the intestine, inhibiting the absorption of cholesterol (Lau et al. 2005; Nissinen et al. 2002). Indeed, corn oil intake reduced atherogenic lipoprotein cholesterol, apolipoprotein, and lipoprotein particle levels to a larger extent compared to EVOO in humans, demonstrating its potential benefits against CVD risk (Maki et al. 2017). Recently, the presence of nitrogenated compounds such as 2,5-diketopiperazines (DKPs) has been detected for the first time in corn oil (Alberdi-Cedeño et al. 2017). These are a group of compounds that have received very limited attention in oils but that have shown bioactive abilities such as antifungal (Nishanth Kumar et al. 2013), antioxidant (Ser et al. 2015), antimicrobial (Nishanth Kumar et al. 2014), and cytotoxic (Vázquez-Rivera et al. 2015) activities. Some of them have also shown a capacity to inhibit some virus such as the Influenza A (Kwak et al. 2013) or malaria (Pérez-Picaso et al. 2012). Moreover, corn oil contains phenolic compounds such as *p*-coumaric, ferulic, and sinapic acids, which have shown antioxidant properties (Siger et al. 2008).

11.4.4 Palm Oil

Palm oil is obtained from the palm tree fruit of *Elaeis guineensis* and is constituted by MUFAs (40% oleic acid), PUFAs (10% linoleic acid) and saturated fatty acids (45% palmitic acid and 5% stearic acid) (Vallyathan 2002). Despite the fact that palm tree fruit accounts for the smallest percentage of all the cultivated land for oils and fats globally, it produces the largest percentage of total output (32%), being considered, the most widely consumed vegetable oil worldwide (Mba and Ngadi 2015). This vegetable oil is characterized by its versatility, as it can be used in both food and non-food products (Aguiar et al. 2018). The most common use of palm oil is as the foundation of confectionery fillings. Interestingly, a wide range of filling textures can be obtained by making different blends of palm fractions because of their diverse physical properties (Hassim and Dian 2017). Nevertheless, palm oil is also highly suitable for frying, chocolate spread, shortenings, margarines, and for the production of non-dairy foods including ice-creams, cheese analogues, yogurt, and whipped toppings (Dian et al. 2017). Moreover, palm oil has demonstrated an important role in the cosmetic industry since some of its components, i.e. the triterpene squalene, have shown emollient, antioxidant, hydration, and antitumor activities on the skin (Huang et al. 2009). Furthermore, its long-chain and medium-chain triglycerides, and phytonutrients such as tocotrienol, tocopherol, and carotenes improve the bioavailability and solubility of the drug when incorporated in lipid-based formulations (Goon et al. 2019). In addition, the polyhydroxyalkanoate (PHA) present in this oil is considered a potential substitute for some petrochemical-based plastics since this biodegradable plastic is derived from microbial fermentation using various carbon substrates (Sudesh et al. 2011). However, while all these factors associated with palm oil offer clear advantages, the oil contains a much higher percentage of saturated fats in comparison with other vegetable oils (Sun et al. 2015). On this regard, a meta-analysis of increased palm oil consumption in 23 countries found a significant relationship with higher mortality from ischemic heart disease (Chen et al. 2011). In addition, the WHO and the FAO describe the relationship between saturated fat consumption with increased risk of CVD as convincing (Joint FAO/WHO 2003).

11.4.5 Peanut Oil

Most of the peanuts produced worldwide are utilized for the production of peanut oil, which is especially consumed in India. Across the countries, China, India, and Nigeria are estimated to produce almost 75% of the peanut oil in the world. Peanut oil is commonly used in cooking for frying, sautéing, and simply adding flavor. In this regard, peanut oil does not absorb flavors from foods during the cooking process, which prevents the cross-contamination of tastes between foods cooked in the same batch. Remarkably, the high content in oleic acid and the absence of linolenic acid give it a superior and long fry life. There are different types of peanut oils according to their extraction procedure: refined peanut oil (removing flavors and colors), cold-pressed peanut oil (retaining natural flavors and aromas), roasted peanut oil (having a deep nutty taste and dark golden color), and peanut oil blends (blending with soybean oil make it more affordable). Therefore, the properties and uses of distinct peanut oils highly depend on their production processes, frying being the most accepted use for peanut oil due to its high smoke point and flavor preservation properties. Clinical trials have shown cardiovascular protective effects of peanuts and peanut oil by decreasing total and LDL-c without reducing HDL-c (Sales et al. 2008). Moreover, peanut oil retarded the development of atherosclerosis in hamsters consuming an atherosclerosis inducing diet (Stephens et al. 2010).

11.4.6 Avocado Oil

Another interesting vegetable oil, due to its healthy properties, is avocado oil. In this sense, compared to olive oil, avocado oil possess a higher proportion of saturated fatty acids, with a predominance of palmitic acid, a lower proportion of MUFAs (with a predominance of oleic acid) and a higher proportion of PUFAs (with linoleic acid the most important). Interestingly, it has been shown that the phytosterol content can be higher in avocado oil (3.3–4.5 mg/g of oil) than in some varieties of olive oil (Berasategi et al. 2012). Because of this interesting composition (rich in MUFAs, PUFAs, tocopherols, tocotrienols, phytoesterols, carotenoids, and polyphenols) avocado oil has been used as nutraceutical in both human and animal models studies. In human subjects that were under an hypercaloric and hyperlipidic diet, the addition of avocado oil for a period of six days reflected an improvement in the postprandial profile of insulin, glycemia, total cholesterol, LDL-c, triglycerides, and inflammatory parameters, such as C-reactive protein (CRP) and interleukin-6 (IL-6) (Furlan et al. 2017). Using a diabetic rat model it has been shown that avocado oil is able to prevent mitochondrial dysfunction in the brain and liver by decreasing ROS, and enhancing the lipid damage (reducing the production of malondialdehyde (MDA) as well as the ratio between oxidized and reduced glutation (GSH/GSSG ratio) (Ortiz-Avila et al. 2015a). Beyond its anti-inflammatory and antioxidant activity, administration of avocado oil to hypertensive rats (1 ml/250 g rat weight) significantly decreased both systolic and diastolic pressure, but not in controls, acting similar to the antihypertensive drug losartan on blood pressure, vascular performance, and oxidative stress (Márquez-Ramírez et al. 2018). Moreover, reduction of the levels of triglycerides, very low-density lipoprotein cholesterol (VLDL-c) and LDL-c, without affecting the levels of HDL-c was shown in sucrose-fed rats (Carvajal-Zarrabal et al. 2014). In addition to these results, in preclinical studies with diabetic rats, avocado oil has shown antioxidant effects attenuating oxidative stress in the liver (Ortiz-Avila et al. 2015b). Interestingly, avocado pulp oil has shown a capability to inhibit the enzymes cyclooxygenase-1 (COX-1) and COX-2, similarly to ibuprofen and EVOO, demonstrating its anti-inflammatory capacity (Espinosa-Alonso et al. 2017). In addition, vitamin B12 cream containing avocado oil was well tolerated in the therapy of plaque psoriasis (Stücker et al. 2001). Finally, it has been shown that avocado oil enhances the carotenoid absorption from salads and salsa in humans (Unlu et al. 2005).

11.4.7 Linseed Oil

Linseed, also named flaxseed oil, has demonstrated beneficial effects on health due to its particular composition. The main components in this oil are linoleic and linolenic fatty acids, various dietary fibers and lignans suchs as secoisolariciresinol diglucoside. These compounds provide bioactivity of value to the health of animals and humans through their anti-inflammatory and antioxidant capacity, as well as their lipid modulating properties. Consumption of these components has shown to reduce the cancer incidence and the risk of CVD by decreasing cholesterol levels and relaxing the smooth muscle cells in arteries increasing the blood flow (Parikh and Pierce 2019). Moreover, linseed oil, which is rich in omega-3, as linolenic acid, has shown to suppress inflammation and modulate gut microbiota, which may potentially contribute to dietary control of diabetes mellitus in rats (Zhu et al. 2020). On the other hand, omega-6 linoleic acid has demonstrated a pivotal role in metabolic syndrome, risk of diabetes mellitus, and CVD. Linoleic acid promotes the linoleic acid-derived lipid mediator 9-hydroxy-octadecadienoic acid (9-HODE) that regulates FOXO family transcriptional factors, which acts as transducer of the insulin-signaling pathway. Thus, activating 9-HODE and FOXO provides a molecular conduit linking changes in dietary fatty acid balance, inflammation, and insulin resistance (Kwon et al. 2020). In addition, linseed is a rich source of fiber, whose consumption have been related with an increase of intestinal *Bifidobacterium* and a decrease of *Faecalibacterium*, which are linked to prevention of colon cancer and activation of insulin sensibility, respectively (Parikh et al. 2019). In addition, linseed lignans have been observed to reduce blood pressure in a deoxycorticosterone acetate (DOCA-salt)-induced renal hypertensive experimental model in rats (Sawant and Bodhankar 2016) and in high fat and high sugar-fed rats (Park and Velasquez 2012), besides showing capacities to inhibit colitis and colon cancer.

11.4.8 Sesame Oil

Sesame is a valuable oilseed crop that contains various nutritionally rich bioactive compounds including lignans, tocopherol homologs, phytosterols, etc. Sesame oil has a combination of glycosylated lignans and oil-dispersed lignans. The most important lignans are sesamin, sesamol, sesamolin, and sesaminol, which have CVD, antihypertensive, as well as anti-inflammatory properties. In addition to lignans and tocopherols, sesame is an important source of phytosterols, phytates, PUFAs, and bioactive peptides. Regarding this last group of molecules, another approach for obtaining bioactive compounds from oilseeds crops could be the elaboration of peptide hydrolysates from these seeds. These peptides harbor very interesting bioactivities that could strength the value of the raw material. For example, this approach has already been used for sesame seeds, allowing the production of peptides with antioxidant and fibrinolytic properties (Liu and Chiang 2008). Other bioactivities that have been obtained by using the same approach are antihypertensive (ACE inhibitors), as well as anti-inflammatory (COX-2 inhibitors).

11.4.9 Canola Oil/Rapeseed

Canola oil is one the most utilized cooking oils because of its versatility, which is mainly defined by its light flavor, soft texture, and high smoke point. Specifically, the use of canola oils predominates in northern states of India. Briefly, canola oil is widely used as cooking oil for sautéing, baking, grilling, and stir-frying. In addition, canola oil can be also included in salads, sauces, and marinades, as well as a healthy replacement of solid fats (such as butter and margarine) in distinct recipes. Therefore, canola oil, which is characterized by the balanced ratios of desirable fatty acids in human diets, shows positive effects on health that, in turn, can be enhanced by mixing it with other beneficial vegetable oils such as linseed oil (Sagan et al. 2019).

11.5 New Trends and Applications

The new understandings concerning the close relationship between nutrition and well-being have contributed to the consumers' growing interest in maintaining healthy food habits in order to preserve a good state of health and to prevent the onset of several diseases (Bimbo et al. 2017). In this context, the presence of health claims in the front of the package positively impacts on the consumer's preference for choosing and consuming specific foods (Dean et al. 2012). In fact, numerous studies have revealed that specific components of natural foods may exert health-promoting effects, thus preventing the pathogenesis caused by distinct human conditions such as environmental and genetic factors (Hu 2003; Shahidi and Ambigaipalan 2015). This scientific evidence highly contributes to improve the competitiveness of functional foods in the market. On the one hand, these trends have prompted the agricultural sector to characterize their products in the search of health claims, whereas, on the other hand, the food industries have become aware about the importance of enriching their products with functional compounds that may contribute to get the attention of the consumers (Klopčič et al. 2020). In this scenario, the increasing knowledge of oil's nutritional and functional aspects is favoring its introduction in distinct food matrices or their use as nutraceutics in order to take advantage of the health benefits. For example, linseed oils can be found in a multitude of nutraceutical formulations such as pills, capsules, or microencapsulated powders. Also, as linseed oil is susceptible to oxidation, recently a encapsulation consisting of a tertiary conjugate of oxidized tannic acid-gelatin has been developed, which is concluded to be a wall material in an encapsulation and delivery system of functional and sensitive oils (Mohseni and Goli 2019). Avocado oil is another example of oil that haa shown high potential for microencapsulation, as it has been observed that this improves its oxidative stability (Bae and Lee 2008).

Regarding the interests of the functional food industry, something that lately has gained attention is the relation of the impact of an excessive consumption of dietary fats, especially saturated fatty acids, as it leads to an increased circulating triglycerides levels (Mensink 2016). Therefore, limiting total fat intake and preferably choosing MUFAs and PUFAs instead of replacing saturated fatty acids with carbohydrates, which might also contribute to increase serum triglycerides concentrations, represent one of the most effective strategies for reducing the incidence of some diseases (Virtanen 2018). In this framework, it was found that the use of corn oil instead of milk fat in cheese production decreased its cholesterol content, thus indicating that corn oil can be potentially introduced in the cheese industry (Arslan et al. 2014). Moreover, the awareness among consumers of the deleterious health consequences of consuming *trans* fatty acids has increased the demand for using natural oils and fats free of *trans* fatty acids in food formulations (O'Brien and Wakelyn 2005). For example, the fractionation of palm oil into its two main components (olein and stearin), engages the use of palm stearin to produce margarine free of *trans*-FA (Adhikari et al. 2010; Ornla-ied et al. 2016). In this sense, palm oil is widely used commercially to replace butterfat in various food formulations (Prasanth Kumar et al. 2016).

Accordingly, new functional foods and nutraceuticals containing bioactive compounds from oils and oilseeds, such as saponifiable and non-saponifiable lipids, and phenolic compounds, are emerging in the food industry (Clark and Lands 2015). In this sense, vegetable oils may provide cardiovascular benefits and antioxidant activities due to their high content in linoleic and linolenic acids (Sokoła-Wysoczańska et al. 2018). Actually, olive and avocado oils are often regarded as functional ingredients and their addition to different types of food has been broadly tested in the food industry. Scientific research has proved that the supplementation of bread and milk products, which are highly consumed in our society, with highly-stable microencapsulated PUFAs successfully enhanced the nutritional properties of these products without altering their sensorial features (Rasti et al. 2017). In this regard, milk enriched with PUFA (mainly omega 3) and/or oleic acid was demonstrated to reduce the blood lipids in subjects with increased risk factors and CVD patients (Lopez-Huertas 2010). In addition, a fivefold increase on the PUFA levels was achieved after adding conjugated linolenic acid (CLA) and omega-3 obtained from extruded linseed to goat cheese, thus reducing its cholesterol relative content and maintaining its sensorial quality (Santurino et al. 2017). Naturally enriched milks with *cis*-9, *trans*-11-CLA are known to prevent cancer, heart strokes, atherosclerosis, diabetes, and obesity in animal models (Koba and Yanagita 2014).

Ice creams are other dairy products in which oil and oilseed can be used under a functional food perspective. For instance, new formulations to improve the nutritional properties of ice-creams by adding EVOO have recently emerged in Italy (Sacchi et al. 2019). In this regard, milk fats were successfully replaced by EVOO that, in turn, improved the nutritional profile of the product. Therefore, EVOO is postulated to favor the design of new foods with enhanced health benefits and innovative flavors in the ice-cream industry.

The inclusion of linseed oil in functional foods is emerging due to its healthy lipid profile, which is characterized by low saturated amounts (9%), moderate MUFA content (18%) and high PUFA (73%) levels (Cunnane et al. 1993). In this sense, linseed oil is used in the formulation of various foods as neat oils, stable emulsions, and micro- and nano-encapsulated powders (Bakry et al. 2016). Up to date, linseed oil has been incorporated into baked foods, juices, milk, and dairy products, muffins, dry pasta products, macaroni, and beef patties. For instance, the partial substitution of soy oil with flaxseed oil in bread formulations resulted in an increased α-linolenic acid (omega-3) content, thus reaching a reduced omega-6:omega-3 ratio without negative impacts on bread quality (de Aguiar et al. 2011). Furthermore, linseed oil is also introduced in lipid infusions for patients with certain disorders (Talbot 2015). Moreover, the production of omega-3-enriched eggs from laying hens fed a flax-based diet is on the rise, thus expanding the potential applications of linseed and its derivatives on the food industry (Coorey et al. 2015).

Other applications in functional foods include the use of phytosterols, which are naturally contained in vegetable oils and are able to reduce circulating cholesterol levels mainly by reducing its absorption in the gut due to their similar structure (Ogbe et al. 2015). In this sense, phytosterols can be incorporated in various fatty foods including margarine, creams, cheese, mayonnaise, yogurt, and milk (Yildiz 2010). The consumption of phytosterol-enriched milk induces a significant diminution of the circulating LDL levels (Mannarino et al. 2009). Other foods like chocolate (Polagruto et al. 2006), cereal bars (Sarkkinen et al. 2018), juices (Devaraj et al. 2006), meat (Grasso et al. 2018), and bakery products (Quílez et al. 2006), have also been supplemented with plant sterols, showing that the matrix to which phytosterols are added can alter their efficacy as a cholesterol-lowering agent. For instance, the Norwegian Mills DA company added cholesterol-lowering oil (rich in PUFA) obtained from sunflower and rapeseed oils to commercial breads (Hicks and Moreau 2001). Additionally, Finland's Huittestern Lihapojat created a series of cholesterol-lowering meat products including frankfurters, sandwich meats, sausages, and meatballs containing a patented rapeseed oil rich in omega-3, omega-6 and several PUFAs (Nollet and Toldrá 2006). Beyond the introduction of vegetable oils in dairy products, soybean and wheat germ oils, which contain elevated amounts of tocopherol, are highly appreciated in the elaboration of rejuvenating lotions and creams due to their proved antioxidant capacity (Kumar and Krishna 2013).

Aside from the bioactive compounds contained in oils and oilseeds, the oil industry generates large amounts of waste that could be potentially recovered and used as valuable dietary components. In this line, a recent study demonstrated that enriched burgers with 10% dry olive paste flour, which is a solid olive oil by-product, showed a considerable antioxidant activity (Cedola et al. 2017). Despite the fact that the detrimental impact of this supplementation to the sensory quality of food limited its inclusion to the market, new strategies focused on the revalorization of industrial by-products derived from the oil production should be further developed.

To sum up, due their high content in bioactive molecules of different nature, vegetable oils and oilseeds are excellent ingredients to be used in the field of functional foods and nutraceuticals, both directly or as source of bioactive compounds that could be used individually as pure ingredients. In this framework, food industries are continuously innovating and developing new products that include these compounds in their formulations.

References

Adhikari, P., Shin, J.A., Lee, J.H. et al. (2010). Production of trans-free margarine stock by enzymatic interesterification of rice bran oil, palm stearin and coconut oil. Journal of the Science of Food and Agriculture 90 (4): 703–711.

de Aguiar, A.C., Boroski, M., Monteiro, A.R.G. et al. (2011). Enrichment of whole wheat flaxseed bread with flaxseed oil. Journal of Food Processing and Presevation 35 (5): 605–609.

Aguiar, L.K., Martinez, D.C., and Caleman, S.M.Q. (2018). Consumer awareness of palm oil as an ingredient in food and non-food products. Journal of Food Products and Marketing 24 (3): 297–310.

Aguirre, M.R., Velasco, J., and Ruiz-Méndez, M.V. (2014). Characterization of sunflower oils obtained separately by pressing and subsequent solvent extraction from a new line of seeds rich in phytosterols and conventional seeds. Oilseeds and Fats, Crops and Lipids 21 (6): D605.

Akhtar, S., Khalid, N., Ahmed, I. et al. (2014). Physicochemical characteristics, functional properties, and nutritional benefits of peanut oil: a review. Critical Reviews in Food Science and Nutrition 54 (12): 1562–1575.

Akkaya, M.R. (2018). Fatty acid compositions of sunflowers (*Helianthus annuus* L.) grown in east Mediterranean region. Rivista Italiana delle Sostanze Grasse 95 (4): 239–247.

Alberdi-Cedeño, J., Ibargoitia, M.L., and Guillén, M.D. (2017). Bioactive compounds detected for the first time in corn oil: cyclic dipeptides and other nitrogenated compounds. Journal of Food Composition and Analysis 62: 197–204.

Alcaide-Hidalgo, J.M., Margalef, M., Bravo, F.I. et al. (2020a). Virgin olive oil (unfiltered) extract contains peptides and possesses ACE inhibitory and antihypertensive activity. Clinical Nutrition 39 (4): 1242–1249.

Alcaide-Hidalgo, J.M., Romero, M., Duarte, J. et al. (2020b). Antihypertensive effects of virgin olive oil (unfiltered) low molecular weight peptides with ACE inhibitory activity in spontaneously hypertensive rats. Nutrients 12 (1): e271.

Alves, A.Q., Amaro Da Silva, V., José, A. et al. (2019). The fatty acid composition of vegetable oils and their potential use in wound care. Advances in Skin and Wound Care 32 (8): 1–8.

Angeloni, C., Malaguti, M., Barbalace, M. et al. (2017). Bioactivity of olive oil phenols in neuroprotection. International Journal of Molecular Sciences 18 (11): e2230.

Arslan, S., Topcu, A., Saldamli, I. et al. (2014). Use of corn oil in the production of Turkish white cheese. Journal of Food Science and Technology 51 (10): 2382–2392.

Aung, W.P., Bjertness, E., Htet, A.S. et al. (2018). Fatty acid profiles of various vegetable oils and the association between the use of palm oil vs. peanut oil and risk factors for non-communicable diseases in Yangon Region, Myanmar. Nutrients 10 (9): 1193–1207.

Baccouri, O., Guerfel, M., Baccouri, B. et al. (2008). Chemical composition and oxidative stability of Tunisian monovarietal virgin olive oils with regard to fruit ripening. Food Chemistry 109 (4): 743–754.

Bae, E.K. and Lee, S.J. (2008). Microencapsulation of avocado oil by spray drying using whey protein and maltodextrin. Journal of Microencapsulation 25 (8): 549–560.

Bakry, A.M., Abbas, S., Ali, B. et al. (2016). Microencapsulation of oils: a comprehensive review of benefits, techniques, and applications. Comprehensive Reviews in Food Science and Food Safety 15 (1): 143–182.

Barrera-Arellano, D., Badan-Ribeiro, A.P., and Serna-Saldivar, S.O. (2018). Corn oil: composition, processing, and utilization. In: Corn: Chemistry and Technology, 3e (ed. S. Serna-Saldivar), 593–613. Elsevier.

Berasategi, I., Barriuso, B., Ansorena, D. et al. (2012). Stability of avocado oil during heating: comparative study to olive oil. Food Chemistry 132 (1): 439–446.

Bester, D., Esterhuyse, A.J., Truter, E.J. et al. (2010). Cardiovascular effects of edible oils: a comparison between four popular edible oils. Nutrition Research Reviews 23 (2): 334–348.

Bimbo, F., Bonanno, A., Nocella, G. et al. (2017). Consumers' acceptance and preferences for nutrition-modified and functional dairy products: a systematic review. Appetite 113: 141–154.

Boskou, D. (2009). Phenolic compounds in olives and olive oil. In: Olive Oil: Minor Constituents and Health (ed. D. Boskou), 12–36. CRC Press.

Caramia, G., Gori, A., Valli, E. et al. (2012). Virgin olive oil in preventive medicine: from legend to epigenetics. European Journal of Lipid Science and Technology 114 (4): 375–388.

Carvajal-Zarrabal, O., Nolasco-Hipolito, C., Aguilar-Uscanga, M.G. et al. (2014). Avocado oil supplementation modifies cardiovascular risk profile markers in a rat model of sucrose-induced metabolic changes. Disease Markers 2014: 1–8.

Cedola, A., Cardinali, A., Del Nobile, M.A. et al. (2017). Fish burger enriched by olive oil industrial by-product. Food Science and Nutrition 5 (4): 837–844.

Chen, B.K., Seligman, B., Farquhar, J.W. et al. (2011). Multi-country analysis of palm oil consumption and cardiovascular disease mortality for countries at different stages of economic development: 1980–1997. Globalization and Health 7 (45): 1–10.

Choe, E. (2008). Effects and mechanisms of minor compounds in oil on lipid oxidation. In: Food Lipids, Chemistry, Nutrition, and Biotechnology (eds. C.C. Akoh and D. Min), 449–474. CRC Press.

Clark, C. and Lands, B. (2015). Creating benefits from Omega-3 functional foods and nutraceuticals. Food and Nutrition Sciences 6 (17): 1613–1623.

Codex Alimentarius Commission (1999). Codex Alimentarius: Standard for Named Vegetable Oils. Rome: FAO/WHO.

Coorey, R., Novinda, A., Williams, H. et al. (2015). Omega-3 fatty acid profile of eggs from laying hens fed diets supplemented with chia, fish oil, and flaxseed. Journal of Food Science 80 (1): S180–S187.

Corominas-Faja, B., Cuyàs, E., Lozano-Sánchez, J. et al. (2018). Extra-virgin olive oil contains a metabolo-epigenetic inhibitor of cancer stem cells. Carcinogenesis 39 (4): 601–613.

Crizel, R.L., Hoffmann, J.F., Zandoná, G.P. et al. (2020). Characterization of extra virgin olive oil from southern Brazil. European Journal of Lipid Science and Technology 22 (4): 1900347.

Cunnane, S.C., Ganguli, S., Menard, C. et al. (1993). High α-linolenic acid flaxseed (*Linum usitatissimum*): some nutritional properties in humans. British Journal of Nutrition 69 (2): 443–453.

De Greyt, W.F., Kellens, M.J., and Huyghebaert, A.D. (1999). Effect of physical refining on selected minor components in vegetable oils. European Journal of Lipid Science and Technology 101 (11): 428–432.

Dean, M., Lampila, P., Shepherd, R. et al. (2012). Perceived relevance and foods with health-related claims. Food Quality and Preference 24 (1): 129–135.

Del Rio, D., Rodriguez-Mateos, A., Spencer, J.P.E. et al. (2013). Dietary (poly)phenolics in human health: structures, bioavailability, and evidence of protective effects against chronic diseases. Antioxidants and Redox Signaling 18 (14): 1818–1892.

Devaraj, S., Autret, B.C., and Jialal, I. (2006). Reduced-calorie orange juice beverage with plant sterols lowers C-reactive protein concentrations and improves the lipid profile in human volunteers. The American Journal of Clinical Nutrition 84 (4): 756–761.

Devi, A. and Khatkar, B.S. (2016). Physicochemical, rheological and functional properties of fats and oils in relation to cookie quality: a review. Journal of Food Science and Technology 53 (10): 3633–3641.

Dian, N.L.H.M., Hamid, R.A., Kanagaratnam, S. et al. (2017). Palm oil and palm kernel oil: versatile ingredients for food applications. Journal of Oil Palm Research 29 (4): 487–511.

Dinu, M., Pagliai, G., Casini, A. et al. (2018). Mediterranean diet and multiple health outcomes: an umbrella review of meta-analyses of observational studies and randomised trials. European Journal of Clinical Nutrition 72 (1): 30–43.

Dudeja, P. and Gupta, R.K. (2017). Nutraceuticals. In: Food Safety in the 21st Century (eds. R.K. Gupta, P. Dudeja and A.S. Minhas), 491–496. Elsevier.

El Qarnifa, S., El Antari, A., and Hafidi, A. (2019). Effect of maturity and environmental conditions on chemical composition of olive oils of introduced cultivars in Morocco. Journal of Food Quality 2019: 1–14.

Espín, J.C., García-Conesa, M.T., and Tomás-Barberán, F.A. (2007). Nutraceuticals: facts and fiction. Phytochemistry 68 (22–24): 2986–3008.

Espinosa-Alonso, L.G., Paredes-López, O., Valdez-Morales, M. et al. (2017). Avocado oil characteristics of Mexican creole genotypes. European Journal of Lipid Science and Technology 119 (10): 1600406.

Estruch, R., Ros, E., Salas-Salvadó, J. et al. (2018). Primary prevention of cardiovascular disease with a mediterranean diet supplemented with extra-virgin olive oil or nuts. New England Journal of Medicine 378 (25): e34.

European Commission (2013). Commission Regulation (EU) No 1018/2013 of 23 October 2013 amending Regulation (EU) No 432/2012 establishing a list of permitted health claims made on foods other than those referring to the reduction of disease risk and to children's development and health. Official Journal of the European Union L282 56 (1924): 43–45.

Fan, L. and Eskin, N.A.M. (2015). The use of antioxidants in the preservation of edible oils. In: Handbook of Antioxidants for Food Preservation (ed. F. Shahidi), 373–388. Woodhead Publishing.

Flores, M., Saravia, C., Vergara, C.E. et al. (2019). Avocado oil: characteristics, properties, and applications. Molecules 24 (11): e2172.

Frankel, E.N. (2005). Lipid Oxidation. The Oily Press.

Fregapane, G. and Salvador, M.D. (2019). Chemical and sensory characteristics of extra virgin olive oils produced in Central Iberian Peninsula under the protected designation of origin quality scheme. European Journal of Lipid Science and Technology 121 (3): 1800134.

Furlan, C.P.B., Valle, S.C., Östman, E. et al. (2017). Inclusion of Hass avocado-oil improves postprandial metabolic responses to a hypercaloric-hyperlipidic meal in overweight subjects. Journal of Functional Foods 38: 349–354.

García-González, D.L., Aparicio-Ruiz, R., and Aparicio, R. (2008). Virgin olive oil – chemical implications on quality and health. European Journal of Lipid Science and Technology 110 (7): 602–607.

García-Llatas, G. and Rodríguez-Estrada, M.T. (2011). Current and new insights on phytosterol oxides in plant sterol-enriched food. Chemistry and Physics of Lipids 164 (6): 607–624.

Ghazani, S.M. and Marangoni, A.G. (2016). Healthy Fats and Oils. In: Encyclopedia of Food Grains (eds. C. Wrigley, H. Corke, K. Seetharaman, et al.), 257–267. Academic Press.

Goon, D.E., Hamimah, S., Abdul, S. et al. (2019). Palm oil in lipid-based formulations and drug delivery systems. Biomolecules 9 (64): 1–19.

Grasso, S., Harrison, S.M., Monahan, F.J. et al. (2018). The effect of plant sterol-enriched turkey meat on cholesterol bio-accessibility during in vitro digestion and Caco-2 cell uptake. International Journal of Food Sciences and Nutrition 69 (2): 176–182.

Gul, K., Singh, A.K., and Jabeen, R. (2016). Nutraceuticals and functional foods: the foods for the future world. Critical Reviews in Food Science and Nutrition 56 (16): 2617–2627.

Gunstone, F.D. (ed.) (2002). Vegetable Oils In Food Technology: Composition, Properties and Uses. Blackwelll Publishing.

Harwood, J.L. and Yaqoob, P. (2002). Nutritional and health aspects of olive oil. European Journal of Lipid Science and Technology 104: 685–697.

Hassanien, M.F.R. (2013). Plant sterols and tocols profile of vegetable oils consumed in Egypt. International Journal of Food Properties 16 (3): 574–585.

Hassim, N.A.M. and Dian, N.L.H.M. (2017). Usage of palm oil, palm kernel oil and their fractions as confectionery fats. Journal of Oil Palm Research 29 (3): 301–310.

Heasman, M. and Mellentin, J. (2001). *The Functional Foods Revolution: Healthy People, Healthy Profits?* London: Earthscan.

Hicks, K.B. and Moreau, R.A. (2001). Phytosterols and phytostanols: functional food cholesterol busters. Food Technology 55 (1): 63–67.

Hu, F.B. (2003). Plant-based foods and prevention of cardiovascular disease: an overview. The American Journal of Clinical Nutrition 78 (3SUP): 544S–551S.

Huang, Z., Lin, Y., and Fang, J. (2009). Biological and pharmacological activities of squalene and related compounds: potential uses in cosmetic dermatology. Molecules 10: 540–554.

FAO/WHO (2003). Diet, nutrition and the prevention of chronic diseases. FAO/WHO.

Khedkar, S., Bröring, S., and Ciliberti, S. (2017). Exploring the nutrition and health claims regulation (EC) no. 1924/2006: what is the impact on innovation in the EU food sector? International Journal of Food Sciences and Nutrition 68 (1): 10–17.

Kim, Y.A., Xiao, D., Xiao, H. et al. (2007). Mitochondria-mediated apoptosis by diallyl trisulfide in human prostate cancer cells is associated with generation of reactive oxygen species and regulated by Bax/Bak. Molecular Cancer Therapeutics 6 (5): 1599–1609.

Klopčič, M., Slokan, P., and Erjavec, K. (2020). Consumer preference for nutrition and health claims: a multi-methodological approach. Food Quality and Preference 82 (103863): 1–10.

Koba, K. and Yanagita, T. (2014). Health benefits of conjugated linoleic acid (CLA). Obesity Research and Clinical Practice 8 (6): e525–e532.

Konuskan, D.B., Arslan, M., and Oksuz, A. (2019). Physicochemical properties of cold pressed sunflower, peanut, rapeseed, mustard and olive oils grown in the Eastern Mediterranean region. Saudi Journal of Biological Sciences 26 (2): 340–344.

Kostadinović Veličkovska, S., Catalin Moţ, A., Mitrev, S. et al. (2018). Bioactive compounds and "in vitro;" antioxidant activity of some traditional and non-traditional cold-pressed edible oils from Macedonia. Journal of Food Science and Technology 55 (5): 1614–1623.

Kumar, G.S. and Krishna, A.G.G. (2013). Studies on the nutraceuticals composition of wheat derived oils wheat bran oil and wheat germ oil. Journal of Food Science and Technology 52 (2): 1145–1151.

Kwak, M.K., Liu, R., Kwon, J.O. et al. (2013). Cyclic dipeptides from lactic acid bacteria inhibit proliferation of the influenza a virus. Journal of Microbiology 51 (6): 836–843.

Kwon, S.Y., Massey, K., Watson, M.A. et al. (2020). Oxidised metabolites of the omega-6 fatty acid linoleic acid activate dFOXO. Life Science Alliance 3 (2): e201900356.

Lau, V., Journoud, M., and Jones, P. (2005). Plant sterols are efficacious in lowering plasma LDL and non-HDL cholesterol in hypercholesterolemic type 2 diabetic and nondiabetic persons. The American Journal of Clinical Nutrition 81 (6): 1351–1358.

Liu, B.-L. and Chiang, P.-S. (2008). Production of hydrolysate with antioxidative activity and functional properties by enzymatic hydrolysis of defatted sesame (*Sesamum indicum* L.). International Journal of Applied Science and Engineering 6 (2): 73–83.

Lopez-Huertas, E. (2010). Health effects of oleic acid and long chain omega-3 fatty acids (EPA and DHA) enriched milks. A review of intervention studies. Pharmacological Research 61 (3): 200–207.

Maestri, D.M., Nepote, V., Lamarque, A.L. et al. (2006). Natural products as antioxidants. In: Phytochemistry: Advances in Research (ed. F. Imperato), 105–135. Research Signpost.

Mahmoud Hassanein, M.M., El-Shami, S.M., and Hassan El-Mallah, M. (2011). Investigation of lipids profiles of nigella, lupin and artichoke seed oils to be used as healthy oils. Journal of Oleo Science 60 (3): 99–107.

Maki, K.C., Lawless, A.L., Kelley, K.M. et al. (2015). Corn oil improves the plasma lipoprotein lipid profile compared with extra-virgin olive oil consumption in men and women with elevated cholesterol: results from a randomized controlled feeding trial. Journal of Clinical Lipidology 9 (1): 49–57.

Maki, K.C., Lawless, A.L., Kelley, K.M. et al. (2017). Corn oil intake favorably impacts lipoprotein cholesterol, apolipoprotein and lipoprotein particle levels compared with extra-virgin olive oil. European Journal of Clinical Nutrition 71 (1): 33–38.

Manaf, Y.N., Rahardjo, A.P., Yusof, Y.A. et al. (2018). Lipid characteristics and tocopherol content of the oils of native avocado cultivars grown in Indonesia. International Journal of Food Properties 21 (1): 2758–2771.

Mancini, A., Imperlini, E., Nigro, E. et al. (2015). Biological and nutritional properties of palm oil and palmitic acid: effects on health. Molecules 20 (9): 17339–17361.

Mannarino, E., Pirro, M., Cortese, C. et al. (2009). Effects of a phytosterol-enriched dairy product on lipids, sterols and 8-isoprostane in hypercholesterolemic patients: a multicenter Italian study. Nutrition, Metabolism and Cardiovascular Disases 19 (2): 84–90.

Mariod, A., Matthäus, B., and Hussein, I.H. (2011). Fatty acids, tocopherols and sterols of Cephalocroton cordofanus in comparison with sesame, cotton, and groundnut oils. Journal of the American Oil Chemists' Society 88 (9): 1297–1303.

Márquez-Ramírez, C.A., Hernández de la Paz, J.L., Ortiz-Avila, O. et al. (2018). Comparative effects of avocado oil and losartan on blood pressure, renal vascular function, and mitochondrial oxidative stress in hypertensive rats. Nutrition 54: 60–67.

Martini, W.S., Porto, B.L.S., De Oliveira, M.A.L. et al. (2018). Comparative study of the lipid profiles of oils from kernels of peanut, babassu, coconut, castor and grape by GC-FID and raman spectroscopy. Journal of the Brazilian Chemical Society 29 (2): 390–397.

Matthaus, B., Özcan, M.M., and Al Juhaimi, F. (2016). Some rape/canola seed oils: fatty acid composition and tocopherols. Zeitschrift fur Naturforschung, C, Journal of Biosciences 71 (3–4): 73–77.

Mba, O. and Ngadi, M.O. (2015). Palm oil: processing, characterization and utilization in the food industry – a review. Food Bioscience 10: 26–41.

Mensink, R.P. (2016). Effects of saturated fatty acids on serum lipids and lipoproteins: a systematic review and regression analysis. World Health Organization.

Mohseni, F. and Goli, S.A.H. (2019). Encapsulation of flaxseed oil in the tertiary conjugate of oxidized tannic acid-gelatin and flaxseed (*Linum usitatissimum*) mucilage. International Journal of Biological Macromolecules 140: 959–964.

Moreau, R.A., Whitaker, B.D., and Hicks, K.B. (2002). Phytosterols, phytostanols, and their conjugates in foods: structural diversity, quantitative analysis, and health-promoting uses. Progress in Lipid Research 41 (6): 457–500.

Nishanth Kumar, S., Mohandas, C., and Nambisan, B. (2013). Purification of an antifungal compound, cyclo(l-Pro-d-Leu) for cereals produced by *Bacillus cereus* subsp. *thuringiensis* associated with entomopathogenic nematode. Microbiological Research 168 (5): 278–288.

Nishanth Kumar, S., Nath, V.S., Pratap Chandran, R. et al. (2014). Cyclic dipeptides from rhabditid entomopathogenic nematode-associated *Bacillus cereus* have antimicrobial activities. World Journal of Microbiology and Biotechnology 30 (2): 439–449.

Nissinen, M., Gylling, H., Vuoristo, M. et al. (2002). Micellar distribution of cholesterol and phytosterols after duodenal plant stanol ester infusion. American Journal of Physiology: Liver Physiology 282 (6): G1009–G1015.

Nollet, L.M.L. and Toldrá, F. (2006). Advanced Technologies for Meat Processing. CRC Press.

Nykter, M., Kymäläinen, H.R., Gates, F. et al. (2006). Quality characteristics of edible linseed oil. Agricultural and Food Science 15 (4): 402–413.

O'Brien, R.D. and Wakelyn, P.J. (2005). Cottonseed oil: an oil for trans-free options. Inform: International News Fats, Oils Related Materials 16 (11): 677–679.

Ogbe, R.J., Ochalefu, D.O., Mafulul, S.G. et al. (2015). A review on dietary phytosterols: their occurrence, metabolism and health benefits. Pelagia Research Library: Asian Journal of Plant Science and Research 5 (4): 10–21.

Ornla-ied, P., Sonwai, S., and Lertthirasuntorn, S. (2016). Trans-free margarine fat produced using enzymatic interesterification of rice bran oil and hard palm stearin. Food Science and Biotechnology 25 (3): 673–680.

Orsavova, J., Misurcova, L., Vavra Ambrozova, J. et al. (2015). Fatty acids composition of vegetable oils and its contribution to dietary energy intake and dependence of cardiovascular mortality on dietary intake of fatty acids. International Journal of Molecular Sciences 16 (6): 12871–12890.

Ortiz-Avila, O., Esquivel-Martínez, M., Olmos-Orizaba, B.E. et al. (2015a). Avocado oil improves mitochondrial function and decreases oxidative stress in brain of diabetic rats. Journal of Diabetes Research 2015: 1–9.

Ortiz-Avila, O., Gallegos-Corona, M.A., Sánchez-Briones, L.A. et al. (2015b). Protective effects of dietary avocado oil on impaired electron transport chain function and exacerbated oxidative stress in liver mitochondria from diabetic rats. Journal of Bioenergetics and Biomembranes 47 (4): 337–353.

Parikh, M. and Pierce, G.N. (2019). Dietary flaxseed: what we know and don't know about its effects on cardiovascular disease. Canadian Journal of Physiology and Pharmacology 97 (2): 75–81.

Parikh, M., Maddaford, T.G., Austria, J.A. et al. (2019). Dietary flaxseed as a strategy for improving human health. Nutrients 11 (5): e1171.

Park, J.B. and Velasquez, M.T. (2012). Potential effects of lignan-enriched flaxseed powder on bodyweight, visceral fat, lipid profile, and blood pressure in rats. Fitoterapia 83 (5): 941–946.

Pérez-Picaso, L., Olivo, H.F., Argotte-Ramos, R. et al. (2012). Linear and cyclic dipeptides with antimalarial activity. Bioorganic & Medicinal Chemistry Letters 22 (23): 7048–7051.

Polagruto, J.A., Wang-Polagruto, J.F., Braun, M.M. et al. (2006). Cocoa flavanol-enriched snack bars containing phytosterols effectively lower total and low-density lipoprotein cholesterol levels. Journal of the American Dietetic Association 106 (11): 1804–1813.

Prasanth Kumar, P.K., Jeyarani, T., and Gopala Krishna, A.G. (2016). Physicochemical characteristics of phytonutrient retained red palm olein and butter-fat blends and its utilization for formulating chocolate spread. Journal of Food Science and Technology 53 (7): 3060–3072.

Presti, G., Guarrasi, V., Gulotta, E. et al. (2017). Bioactive compounds from extra virgin olive oils: correlation between phenolic content and oxidative stress cell protection. Biophysical Chemistry 230: 109–116.

Quílez, J., Ruiz, J.A., Brufau, G. et al. (2006). Bakery products enriched with phytosterols, α-tocopherol and β-carotene. Sensory evaluation and chemical comparison with market products. Food Chemistry 94 (3): 399–405.

Rasti, B., Erfanian, A., and Selamat, J. (2017). Novel nanoliposomal encapsulated omega-3 fatty acids and their applications in food. Food Chemistry 230: 690–696.

Reboredo-Rodríguez, P., Varela-López, A., Forbes-Hernández, T.Y. et al. (2018). Phenolic compounds isolated from olive oil as nutraceutical tools for the prevention and management of cancer and cardiovascular diseases. International Journal of Molecular Sciences 19 (8): 2305.

Sacchi, R., Caporaso, N., Squadrilli, G.A. et al. (2019). Sensory profile, biophenolic and volatile compounds of an artisanal ice cream ("gelato") functionalised using extra virgin olive oil. International Journal of Gastronomy and Food Science 18 (2): 100173.

Sagan, A., Blicharz-Kania, A., Szmigielski, M. et al. (2019). Assessment of the properties of rapeseed oil enriched with oils characterized by high content of α-linolenic acid. Sustainability 11 (5638): 1–11.

Sales, R.L., Coelho, S.B., Costa, N.M.B. et al. (2008). The effects of peanut oil on lipid profile of normolipidemic adults: a three-country collaborative study. Journal of Applied Research 8 (2): 216–225.

Sánchez-Rodriguez, E., Biel-Glesson, S., Fernandez-Navarro, J. et al. (2019). Effects of virgin olive oils differing in their bioactive compound contents on biomarkers of oxidative stress and inflammation in healthy adults: a randomized double-blind controlled trial. Nutrients 11 (3): 561.

Santurino, C., Calvo, M.V., Gómez-Candela, C. et al. (2017). Characterization of naturally goat cheese enriched in conjugated linoleic acid and omega-3 fatty acids for human clinical trial in overweight and obese subjects. PharmaNutrition 5 (1): 8–17.

Sarkkinen, E., Lyyra, M., Nieminen, S. et al. (2018). Cereal-based snack bar with added plant stanol ester (Benecol®) consumed between meals lowers serum total and LDL cholesterol effectively in mildly to moderately hypercholesterolemic subjects. Cholesterol 2018 (1463628): 1–9.

Sawant, S.H. and Bodhankar, S.L. (2016). Flax lignan concentrate reverses alterations in blood pressure, left ventricular functions, lipid profile and antioxidant status in DOCA-salt induced renal hypertension in rats. Renal Failure 38 (3): 411–423.

Schwartz, H., Ollilainen, V., Piironen, V. et al. (2008). Tocopherol, tocotrienol and plant sterol contents of vegetable oils and industrial fats. Journal of Food Composition and Analysis 21 (2): 152–161.

Ser, H.L., Palanisamy, U.D., Yin, W.F. et al. (2015). Presence of antioxidative agent, Pyrrolo[1,2-a]pyrazine-1,4-dione, hexahydro- in newly isolated *Streptomyces mangrovisoli* sp. nov. Frontiers in Microbiology 6 (854): 1–11.

Shahidi, F. (2009). Nutraceuticals and functional foods: whole versus processed foods. Trends in Food Science and Technology 20 (9): 376–387.

Shahidi, F. and Ambigaipalan, P. (2015). Phenolics and polyphenolics in foods, beverages and spices: antioxidant activity and health effects – a review. Journal of Functional Foods 18: 820–897.

Siger, A., Nogala-Kalucka, M., and Lampart-Szczapa, E. (2008). The content and antioxidant activity of phenolic compounds in cold-pressed plant oils. Journal of Food Lipids 15 (2): 137–149.

Sokoła-Wysoczańska, E., Wysoczański, T., Wagner, J. et al. (2018). Polyunsaturated fatty acids and their potential therapeutic role in cardiovascular system disorders – a review. Nutrients 10 (10): e1561.

Stephens, A.M., Dean, L.L., Davis, J.P. et al. (2010). Peanuts, peanut oil, and fat free peanut flour reduced cardiovascular disease risk factors and the development of atherosclerosis in Syrian golden hamsters. Journal of Food Science 75 (4): H116–H122.

St-Onge, M.-P., Aban, I., Bosarge, A. et al. (2007). Snack chips fried in corn oil alleviate cardiovascular disease risk factors when substituted for low-fat or high-fat snacks. The American Journal of Clinical Nutrition 85 (6): 1503–1510.

Stücker, M., Memmel, U., Hoffmann, M. et al. (2001). Vitamin B12 cream containing avocado oil in the therapy of plaque psoriasis. Dermatology 203 (2): 141–147.

Sudesh, K., Bhubalan, K., Chuah, J. et al. (2011). Synthesis of polyhydroxyalkanoate from palm oil and some new applications. Applied Microbiology and Biotechnology 89 (5): 1373–1386.

Sun, Y., Neelakantan, N., Wu, Y. et al. (2015). Palm oil consumption increases LDL cholesterol compared with vegetable oils low in saturated fat in a meta-analysis of clinical trials. Journal of Nutrition 145 (7): 1549–1558.

Talbot, G. (2015). Specialty Oils and Fats in Food and Nutrition. Woodhead Publishing.

Tan, C.X. (2019). Virgin avocado oil: an emerging source of functional fruit oil. Journal of Functional Foods 54: 381–392.

The World Health Organization (2018). NCDs Noncommunicable Diseases and Their Risk Factors. World Health Organization.

Turner, R., Etienne, N., Garcia Alonso, M. et al. (2005). Antioxidant and anti-atherogenic activities of olive oil phenolics. International Journal for Vitamin and Nutrition Research 75 (1): 61–70.

United Nations General Assembly (2018). Political declaration of the third high-level meeting of the General Assembly on the prevention and control of noncommunicable diseases. A/ RES/73/2.

Unlu, N.Z., Bohn, T., Clinton, S.K. et al. (2005). Carotenoid absorption from salad and salsa by humans is enhanced by the addition of avocado or avocado oil. The Journal of Nutrition 135 (3): 431–436.

Vallyathan, V. (2002). Third international conference on "oxygen/nitrogen radicals: cell injury and disease" Morgantown, West Virginia, June 1–5, 2002. Antioxidants and Redox Signaling 4 (6): 957–960.

Vázquez-Rivera, D., González, O., Guzmán-Rodríguez, J. et al. (2015). Cytotoxicity of cyclodipeptides from pseudomonas aeruginosa PAO1 leads to apoptosis in human cancer cell lines. BioMed Research International 2015: 197608.

Virtanen, J.K. (2018). Randomized trials of replacing saturated fatty acids with n-6 polyunsaturated fatty acids in coronary heart disease prevention: not the gold standard? Prostaglandins, Leukotrienes and Essential Fatty Acids 133: 8–15.

Wen, X., Zhu, M., Hu, R. et al. (2016). Characterisation of seed oils from different grape cultivars grown in China. Journal of Food Science and Technology 53 (7): 3129–3136.

World Health Organization (2018). Noncommunicable diseases key facts. World Health Organization.

Yang, R., Xue, L., Zhang, L. et al. (2019). Phytosterol contents of edible oils and their contributions to estimated phytosterol intake in the Chinese diet. Foods 8: 334–346.

Yanty, N.A.M., Marikkar, J.M.N., and Long, K. (2011). Effect of varietal differences on composition and thermal characteristics of avocado oil. Journal of the American Oil Chemists' Society 88 (12): 1997–2003.

Yildiz, F. (2010). Development and Manufacture of Yogurt and Other Functional Dairy Products. CRC Press.

Zeisel, S.H. (1999). Regulation of "nutraceuticals". Science 285 (5435): 1853–1855.

Zhu, L., Sha, L., Li, K. et al. (2020). Dietary flaxseed oil rich in omega-3 suppresses severity of type 2 diabetes mellitus via anti-inflammation and modulating gut microbiota in rats. Lipids in Health and Disease 19 (20): 1–16.

12
Sensorial Evaluation and Aroma of Vegetable Oils

Gemma Echeverria[1], Chloe Leclerc[1], Jordi Giné-Bordonaba[1] and Agustí Romero[2]
[1] *IRTA, Postharvest Program, Edifici Fruitcentre, Parc Científic i Tecnològic Agroalimentari de Lleida, Lleida, Catalonia, Spain*
[2] *IRTA, Oliviculture, Oil Science, and Nuts, Mas de Bover, Constantí, Tarragona, Spain*

12.1 Introduction

The annual production of edible vegetable oils in the world is over 207 million metric tons (Statista 2020). Global vegetable oil production is expected to grow modestly, with palm, sunflower seed, palm kernel, and soybean oils (SOs) offsetting declines in olive, peanut, cottonseed, rapeseed, and coconut oils (COs). The forecast for global food oil consumption points to expansion after a year of slower growth due in part to the impact caused by the coronavirus pandemic. Palm, olive, and SOs are forecast to show strong food consumption growth of 3% or more in 2020–2021. Global vegetable oil trade will recover during 2020–2021 following a year marked by a decline of global palm oil trade. (https://apps.fas.usda.gov/psdonline/circulars/oilseeds.pdf n.d.).

Edible vegetable oils are important to our daily life since they provide energy, nutritional components, and pleasant flavors (Cert et al. 2000; Yeboah et al. 2012; Zhao et al. 2013). Indeed, edible vegetable oils such as sunflower oil, palm oil, rapeseed oil, and olive oil (OO) are important in human nutrition as essential sources of both omega-3 and omega-6 fatty acids. Some of them like soybean, rapeseed, sunflower seed, and peanut oils are the main cooking oils worldwide. Although a major part of them are used for food products such as shortenings, salads, large quantities are used for feed and multiple industrial applications. Some of the vegetable oils widely used within the industry include palm, palm kernel, coconut, and castor.

Vegetable oils are a group of fats that derive from some seeds, nuts, cereal grains, and fruits. Most of the vegetable oils have to be processed to remove their undesirable

Oil and Oilseed Processing: Opportunities and Challenges, First Edition.
Edited by Tomás Lafarga, Gloria Bobo, and Ingrid Aguiló-Aguayo.

constituents and hence to make them suitable for human consumption. During this refining oil process, as well as during the process of raw-material collection, oil production, or even during transport and storage, many undesired changes may occur deteriorating the oil sensory properties. These sensory properties include flavor, which comprises taste and aroma, and texture. In this regard, taste and flavor are terms often confused. Taste is perceived by taste receptor cells located mainly on the front, sides, and rear of the tongue, the soft palate, cheeks and esophagus while aroma perception occur in olfactory sensory neurons in the nose. These neurons are embedded in the olfactory epithelium, a small patch of specialized epithelium that in humans covers a region in the back of the nasal cavity about 5 cm^2 in size. Aroma comprises all the volatile compounds with active odor. Aroma of edible vegetable oils is an important quality criteria and is generated by a number of volatile constituents that are present at extremely low concentrations (Temime et al. 2006; Zhao et al. 2013). These compounds include hydrocarbons, alcohols, aldehydes, ketones, esters, and acids. The hydrocarbon fraction, which has been largely studied in crude oils, comprises saturated, unsaturated, aromatic, and sesquiterpenic compounds (Olías-Jiménez et al. 1980). The volatile composition of edible oils clearly depends on the type of oil but also within the same type on its geographical origin, environmental factors, and technological conditions of its production (Baccouri et al. 2008). Moreover, the volatile compounds are mostly formed from fatty acids and aminoacids (Morales and Tsimidou 2000) and small changes in oil composition can lead to significant modifications in the vapor phase (Guichard 2002). The volatile profile is in part responsible for oil flavor and sensory attributes and consequently has an influence on the consumer's acceptability (Luna et al. 2006). The headspace (HS) of edible oils can be used as a marker of authenticity, as well as geographical or botanical origin of the product (Aparicio and Aparicio-Ruíz 2000), and can be also used to monitor the quality of products from the manufacturer's point of view (age monitoring, oxidation state, etc.) However, as mentioned earlier, since most edible vegetable oils undergo refining, the majority of the volatile components are lost during the process (Grob et al. 1994).

Thus, this chapter will focus on the sensory properties of the 16 most significant edible vegetable oils, as well as their specific aroma or composition of volatiles with active odor. Moreover, the main factors affecting the sensory quality and aroma volatile compounds of those oils will be analyzed.

12.2 Olive Oil

OO has been one of the pillars of the Mediterranean diet since Ancient Greece and Rome, over 3000 years ago. The OO obtained by olive fruit milling is the only foodstuff that must be subjected to sensory analysis in order to verify if the labeling correctly addresses the different marketing categories: extra virgin olive oil (EVOO) and virgin olive oil (VOO) (Mariotti 2014). The attributes color, taste, aroma, mouth feeling such as oiliness and fluidity, and chemesthesic sensations (e.g. pungent and astringent) are of paramount importance in the evaluation of the sensory quality of OOs. The close relationship between volatiles and desirable and undesirable sensory perceptions is considered as a useful tool to describe and quantify oil flavor by several researchers (Aparicio and Morales 1994; Angerosa et al. 1996).

Qualitative and quantitative determinations of volatiles alone do not generally provide information about flavor because they cannot capture the interactions within the volatile compounds; of odor with taste, of color with flavor and taste, and of taste with viscosity; nor they are capable of depicting the interactions between oral trigeminal irritations from some compounds such as secoiridoid compounds and taste and odor.

The color of VOOs depends on the content of liposoluble photosynthetic pigments including carotenoids (i.e. ß–carotene, lutein, violaxanthin, neoxanthin and ß–cryptoxanthin, and more) and chlorophyll derivatives (i.e. chlorophyll A and B, phaeophytin a and b, and other minor derivatives) naturally occurring in the fruit. Chlorophylls provide oils their yellow/green color, while carotenoids determine shades between yellow and red. Chlorophylls can lose the central Mg atom becoming pheophytins with green-to-brown color. Then, chlorophylls and pheophytins can break down into pyropheophytins, through a process that involves decarbomethoxylation and that is promoted by temperature and the storage length (Hornero-Mendez et al. 2005).

The taste of VOOs is affected by the phenolic composition. Several studies have shown a correlation between the bitterness, astringency, and pungency of OOs and the concentration of total phenols (Angerosa et al. 2000; Servili et al. 2004). Some authors found a significant and positive correlation between secoiridoid derivatives and OO bitterness and pungency (Andrewes et al. 2003; Gutierrez-Rosales et al. 2003). Mateos et al. (2004) reported a linear relationship between intensity of bitterness, measured by a panel test, and oleuropein aglycon concentration. The level of the chlorophylls and carotenoids, as well as the phenolic composition of the OO, are responsible for the oil color and taste, respectively and their content is largely influenced by genetic and environmental factors, but also by the olives' maturity stage at the time of picking as well as the conditions used for olives' oil extraction.

Concerning OO aroma, the volatile compounds are primarily produced by oxidation of fatty acids. It is generally agreed that endogenous plant enzymes, through the lipoxygenase pathway, are responsible for the positive aroma perceptions in OO whereas chemical oxidation and exogenous enzymes, usually from microbial activity, are associated with sensory defects. As for the taste and color attributes, both the processing and the storage length of the fruit or oil contribute importantly to the flavor and overall quality of OO (Kiritsakis 1998; Angerosa 2002). The aroma of OO is attributed to aldehydes, alcohols, esters, hydrocarbons, ketones, furans, and, probably, other still unidentified volatile compounds. The major volatile compounds reported in VOOs are the C6 volatile compounds. Hexanal, trans-2-hexenal, and hexan-1-ol are found in most VOOs in Europe and Morocco (Kiritsakis 1998; Aparicio et al. 1999; Angerosa 2002; Mansouri et al. 2017). Figure 12.1 shows the schematic lypoxygenase pathway for the formation of major volatile compounds in VOOs. It should be however noted that even if these compounds are present at high concentrations they are not necessarily the major contributors of the OO odor. The contribution of a volatile compound to the aroma depends on its odor activity value. However, some volatile compounds that occur in OO below their olfactory threshold, and make no direct contribution to the aroma, may be important in understanding the formation and degradation of the volatiles with significant contribution to aroma, and they may provide useful quality markers (Buttery and Takeoka 2004).

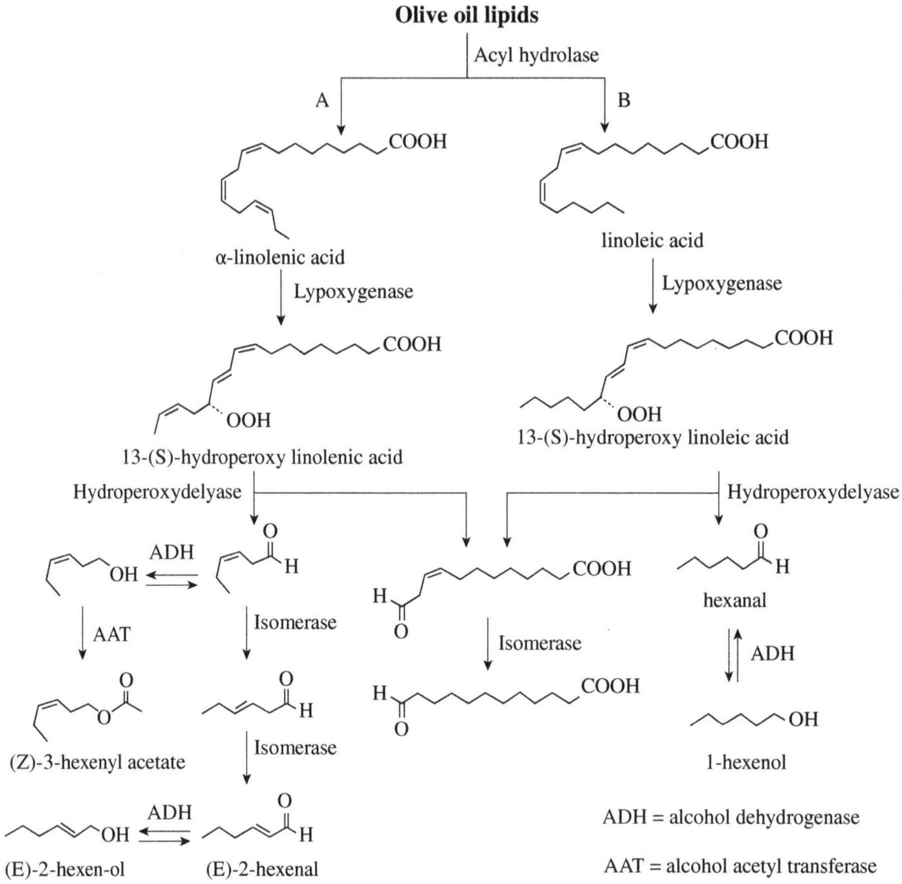

Figure 12.1 Lypoxygenase pathway for the formation of major volatile compounds.
Source: Benincasa et al. (2003).

This fraction includes, for instance, C5 carbonyl compounds, pentenols, hydrocarbons, and minor compounds not derived from fatty acid transformations (Buttery and Takeoka 2004).

Cultivar, geographic area, fruit maturity, and processing methods influence the volatile composition of OO. Fruit from different cultivars grown under the same environmental conditions produce oils with different volatile compounds, as does fruit of the same cultivar grown in different geographic regions (Sacchi et al. 1998; Angerosa et al. 1999; Prenzler et al. 2002; Ridolfi et al. 2002; Benincasa et al. 2003). The aroma compounds in oil are known to increase up to a certain point with the degree of fruit maturity (Aparicio and Morales 1998; Kiritsakis 1998; Ranalli et al. 1998). Apart from the condition of the fruit at harvest, differences in post-harvest handling of the fruit and oil can lead to different volatile profiles. Extraction methods and conditions, in particular the malaxation time and temperature, produce OOs with different flavors (Angerosa et al. 1998; Ranalli et al. 2001; Di Giovacchino et al. 2002; Ranalli et al. 2003). Storage of the fruit after harvest and of the oil before reaching the consumer changes the volatile composition of OO. Storage of the fruit decreases the aldehyde and ester content that is responsible for the positive OO aroma. Storage of either the fruit or the oil produces volatile compounds that are responsible for off-flavors (Kiritsakis 1998; Koprivnjak et al. 2000; Vichi et al. 2009; Vichi et al. 2014). The absence of the C6 aldehydes, alcohols, and esters from the lipoxygenase pathway and the presence of many aldehydes from chemical oxidation, including hexanal from both chemical and enzymatic reactions, are characteristic off-flavors of OO. Some off-flavor compounds are potentially toxic and have low odor thresholds (Angerosa 2000; Ha et al. 2004).

12.3 Palm Oil

Palm oil is rich in antioxidants such as tocopherols and tocotrienols (Kamisah et al. 2005). It is the most widely consumed edible vegetable oil in the world, with a consumption of 74.62 million metric tons in 2019–2020 (Statista 2020). It is used in household but its main use is for industrial deep frying purposes. In this latter use, the palm oil is heated, which causes changes in physical appearance such as darkening in color and increase in viscosity due to the alteration of the fatty acid composition (Rani et al. 2010). Heating also causes some chemical reactions in the oil such as oxidation, hydrolysis, and polymerization (Choe and Min 2007). During these processes, many oxidative products such as hydroperoxides and aldehydes are produced, which can be absorbed into the fried food (Choe and Min 2006). While some studies are available about the volatile profile of raw palm oil, very limited studies have been done on the volatile profile of palm oil after heating (Peng et al. 2017).

In fresh palm oil, seven volatile compounds have been reported, namely E–2-hexenal, heptanal, octanal, nonanal, decanal, Z-2-decenal, and 2-undecenal (Giuffrè et al. 2020). The most and less abundant volatiles were nonanal (32.75%) and Z-2-decenal (6.82%), respectively. In heated palm oil samples, the content of heptanal, octanal, and decanal tend to decrease significantly and nine new compounds were formed during heating. The E,E–2,4-decadienal was the most represented in all the heated samples, followed by Z-2-heptenal and E,E–2,4-decadienal. The formation of these compounds is due to the oxidation of the unsaturated fatty acids after heating: the higher the unsaturated fatty acid content, the higher the aldehydes formation (Peng et al. 2017). This is because polyunsaturated fatty acids such as linoleic and linolenic acids are more susceptible to oxidation because of the presence of multiple double-bonds.

In this same study, Giuffrè et al. (2020) studied the relationship between the heating temperatures, the heating time, and the production of different aldehyde groups including alkanals, alkenals, and alkadienals. They reported that the total aldehyde profile showed a similar trend regardless of the heating temperature used, with a significant increment in total alkadienal compounds, and a decrease in total alkanal compounds. Total alkanals and total alkenals of palm oil were dependent on heating temperature and duration (Figure 12.2).

The color intensity of palm-based oils is relatively higher than that of other vegetable oils. In general, the palm-based oils darkened gradually when heated at 180 °C, except the red palm olein (Tarmizi et al. 2016).

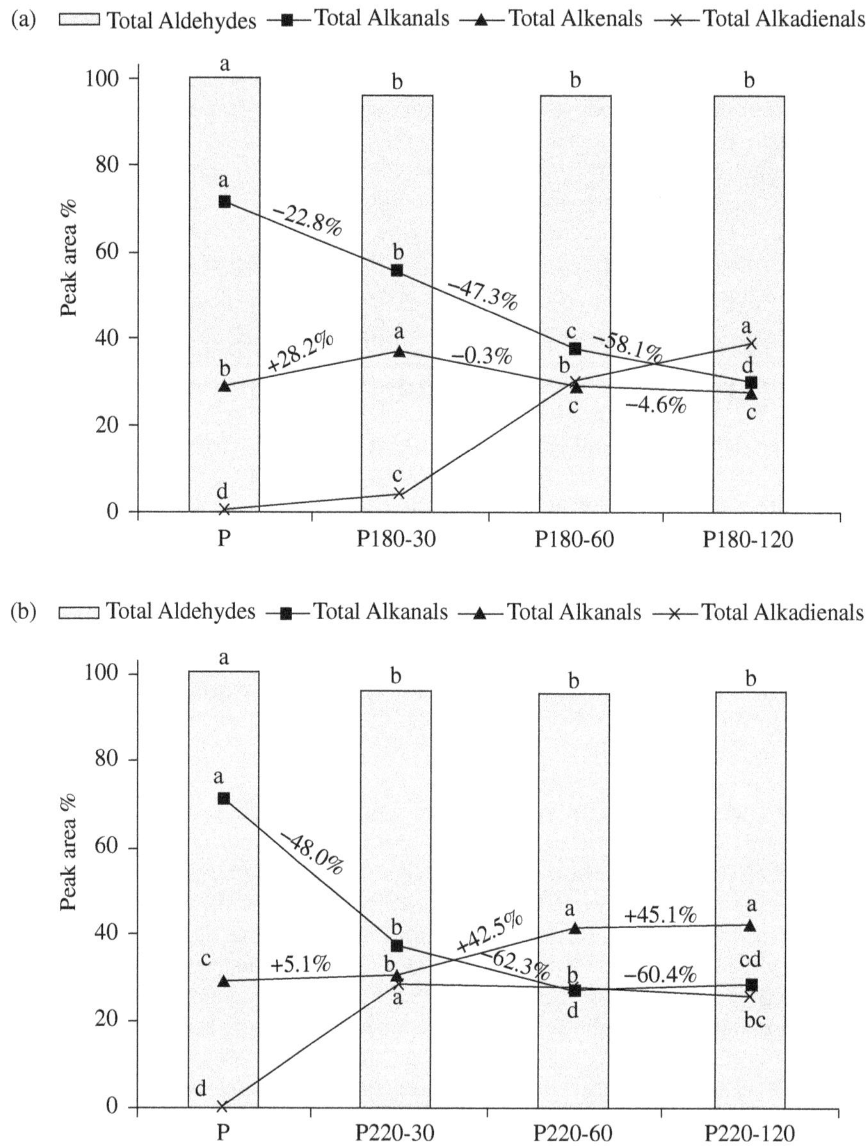

Figure 12.2 Variation in volatile profile compounds of palm oil with heating and time of heating. *Source:* Giuffrè et al. 2020).

12.4 Soybean Oil

SO is one of the most widely used vegetable oils due to its high nutritional and industrial attributes (Alexandratos and Bruinsma 2012). SO is commonly used in cooking, e.g. for pan frying and deep frying, and in the food industry, e.g. in the preparation of dressings, mayonnaise, and shortening. However, this oil suffers from a major disadvantage, namely, the production of an unpleasant beany, green, and hay-like odor under photooxidative conditions; this phenomenon, which is characteristic of SO, is known as flavor reversion (Smouse 1985). Because commercially available SO is stored in clear bottles, it is often blended with other oils to reduce the adverse effects of photooxidation. It would thus be highly advantageous to develop a strategy to prevent the off-odor formation in SO. While the source of off-odor production in SO has been widely discussed, the mechanisms underlying its formation remain yet to be clarified.

As mentioned, SO is vulnerable to oxidation, because it contains about 65–90% of unsaturated fatty acid (Jung and Min 1990; Hou et al. 2012). Most of the research relevant to frying SO focused on oil quality and on the effect of frying SO on animals and cells. It has been reported that frying SO could impair glucose metabolism, alter mRNA expressions of vitamin and causes oxidative damage to pancreatic islet cells (Naz et al. 2005; Liao et al. 2008; Huang et al. 2014).

It has been reported that oil deterioration is accompanied by changes in the volatile profiles (Chang et al. 2019). The volatile characteristics could provide the effective information for assessing the quality of oil or fat during frying, because concentration of volatile markers could determine the oxidation degree of different oils (Zhang et al. 2015; Barden and Decker 2016). Volatile compounds in frying oils are mainly generated from thermal oxidation of the unsaturated fatty acids linked to triacylglycerols. Volatile organic compounds in oils usually have low molecular weights (less than 300Da) and some of them could produce an odor sensation due to their relatively low odor threshold values (Temime et al. 2008). Thirty-seven volatile compounds were detected in fried SO, revealing a complex aroma profile, composed of 15 aldehydes, 8 alcohols, 4 ketones, 4 acids, 5 alkanes, and 1 furan. The volatiles hexanal, heptanal, (E)-2-hexenal, octanal, (E)-2-heptenal, nonanal, (E)-2-octenal, undecanal, (E,E)-2,4-heptadienal, (E)-2-decenal, 2-undecenal, (E,E)-2,4-decadienal, 1-pentanol, 2,2-dimethyl-3-hexanol, (Z)-2-dodecenol, 1-octen-3-ol, pentanoic acid, octanoic acid, nonanoic acid and 2-pentyl-furan have been referred to as potential markers for evaluating lipid oxidation of frying SO (Liu et al. 2020). The results obtained by Sano et al. (2017) strongly suggest that the light-induced off-odor in SO results from the photooxidation of furan fatty acids. 3-methyl-2,4-nonanedione formation, which was linearly associated with the concentration of furan fatty acids, appears to be crucial for the development of light-induced off-odor. In addition, the contribution of 3-hydroxy 3, methyl 2,4nonanedieone to this off-odor was demonstrated for the first time. However, the possibility that other components in SO may also contribute to light-induced off-odor cannot be ruled out; further research is necessary to examine the contribution of such potential factors.

12.5 Sun Flower Seed Oil

Applications in the industrial world of sunflower seeds depend on the nutrients contained in them and their usefulness (Bernard 2016). Industrial products from sunflower seed oil (SSO) that have been commercialized in the world include raw materials for vegetable oils, margarine, cosmetics, pharmaceuticals, and for various food needs (Byrareddy et al. 2008). Three typologies of sunflower oil, characterized by different percentage of oleic acid are available on the market: low, mid and high oleic sunflower oil (Bendini et al. 2011). One of the extraction methods to get virgin oils is cold-pressed. As prescribed by Codex Alimentarius (Codex standards for named vegetable oils 2005) cold pressed oils are obtained by mechanical procedures only, e.g. expelling or pressing, without altering the oil and without the application of heat. They may have been purified by washing with water, settling, filtering, and centrifuging only. Virgin oils are produced as the cold pressed oils apart for the possibility to apply heat. The color, odor, and taste of each product shall be characteristic of the designated product, which has to be free from foreign and rancid odor and taste. Cold pressed or virgin oils are generally obtained by small and medium-sized plants by means of a screw press or an expeller only in a single- or double-step process. While refined sunflower oil, especially high-oleic, is very versatile and due to its neutral flavor and heat stability it can be used in many ways in the kitchen, such as frying, cooking, or as ingredient (mayonnaise), cold pressed or virgin sunflower oil. In contrast, is usually preferred as condiment for salads or cold dishes. Cold pressed sunflower oils, obtained without application of heat retain some aromatic components that are characteristic of the original seeds (Bocci and Frega 1996). Lipid oxidation is the main deteriorative process lowering fats and oils quality, particularly in relation to the off-flavors that develop as an outcome of autoxidation during storage (Bendini et al. 2009). Unsaturated fatty acids are particularly susceptible to oxidation and it is well known that some of the secondary oxidation compounds are volatile aldehydes, as E-2-heptenal, characterized by very low odor thresholds and responsible for the unpleasant rancid or fried odor (Kalua et al. 2007). On the other hand, many volatile compounds that remain in the sunflower oil after cold pressing the seeds play a key role providing the oil's peculiar aroma (Bocci and Frega 1996; De Leonardis et al. 2001). As suggested by Grob et al. (1994) many edible oils labeled as cold pressed or not-refined are indeed deodorized, through a mild deodorization process which eliminates almost completely the monoterpene (α-pinene) and the sesquiterpene (calarene), both compounds considered volatile markers of sunflower oil. Regular raffination removes almost all terpenes. Bocci and Frega (1996) analyzed the volatile fraction from sunflower oil extracted under pressure and reported that the principal compounds were terpenes (about 225 mg/l) consisting of α-pinene (80%), limonene (5%), sabinene (3.3%), β-pinene (1.8%), 1,2,6,6-tetramethyl-1,3-cyclohexadiene (1.5%), and camphene (1.2%). Small amounts of hexanal (1.3%) and traces (<1%) of unidentified compounds, which probably corresponded to other terpenic hydrocarbons and to terpenic alcohols and/or aldehydes, were also detected in the volatile fraction of SSO by the same authors (Bocci and Frega 1996). Bendini et al. (2011) also confirmed that the main terpenic compound found in cold-pressed sunflower seed oil was the monoterpene α-pinene. Accordingly, Grob et al. (1994) identified this specific compound as a useful marker to distinguish between cold pressed sunflower oils obtained by mechanical processes from those subjected to a physical treatment (mild deodorization).

Regarding sensory attributes, Bendini et al. (2011) reported that the odor/taste attributes described as sunflower seed-like and nutty, together with the volatile compounds α-pinene and 3-methyl-1-butanol were the most important vectors to qualify the good sensory quality of virgin or cold pressed sunflower oils. In general, the yellow color is considered positive except for samples characterized also by roasted/burnt seeds sensations. In fact, an excessive heating during the oil extraction process can lead to the formation of several Maillard reaction products like the formation of some volatile compounds responsible for these olfactive notes and some polymeric molecules responsible for the darkening of the color (Whitfield 1992).

12.6 Corn Oil

Crude corn oil is usually fully refined by alkaline refining, bleaching, de-waxing, and deodorization processes. However, the cold press method is currently gaining attention for the quality of the resulting oil. Aydeniz Güneşer et al. (2017) compared the volatile profile of corn oil obtained by refining or by cold press method and detected that the cold pressed corn oil emitted 22 volatiles versus the 7 volatiles plus hexane released by the refined oil. This is a clear example that most volatiles are lost during the refining process. The most abundant volatiles present in the cold pressed corn oil were hexanal, acetic acid, 3-methyl-butanal, hexanoic acid and butanoic acid. In the refined corn oil only minor amounts of hexanal, heptanal, hexanoic acid, limonene, 2-ethyl hexanol, 3-carene, and nonanal remained. It is well known that in any food product, the characteristic aroma is provided by the volatile blend, but not in an order of higher concentrations, rather in the order of aroma threshold value. In the study carried out by Tuan (2011), 29 volatile compounds were identified in corn oils. Some of them showed some similarities with those found by Aydeniz Güneşer et al. (2017) in cold pressed corn oil as octane, 2,6-dimethyl-3-heptanone, 2-heptenal, 2-pentylfuran, nonanal, hexedecanoic acid, 9-octadecenoic acid ethyl ester. However, other volatiles detected were usually different isomers of the same base compounds, but acetic acid, a-pinene, benzaldehyde, limonene, and 3-carene were identified only in the study of Aydeniz Güneşer et al. (2017). These volatiles distinctly provide almond, citrus, pine, buttery, and terpenic aroma notes. It is quite possible that cold pressing of the corn germ yielded the advantage of the presence of these various volatiles for the oil. In another study (Goicoechea and Guillén 2014), the volatile compounds generated in corn oils during storage at room temperature were measured. Usually oxidized samples included many secondary oxidation compounds including alkanals, 2-alkenals, 2,4-alkadienals, acids, alcohols, ketones, alkylfurans, lactones, and a,b-unsaturated aldehydes. As storage time extended, the kinds and amounts of volatiles were generally enhanced.

12.7 Peanut Oil

Peanut (*Arachis hypogaea*) oil (PO), also known as groundnut oil or arachis oil, displays unique organoleptic properties which made it one of the most important edible oil in the international market. It is a common ingredient in Asian gastronomy (Akhtar et al. 2014). As for other nut-based oils, the proportion of oil increases along the maturation of peanuts while the carbohydrate and moisture content decreases, hence taste and physical properties of PO largely depend on the peanut maturity stage at the time of harvest.

The peanut oil color is usually light yellow due to the presence of β-carotene and lutein (Davis et al. 2016). It can be slightly darker depending on cultivar, endogenous sugar content, nut maturity, and processing conditions (Davis et al. 2008). Roast color is a quality control parameter in PO industry and is associated with flavor development. Cultivars with lower sugar content are less brown, leading to lighter yellow oil. Darker peanut oil is also associated to longer roasted time and more intense nutty/roasted flavor. Some flavor defects (fruity, moldy) in PO are associated to aldehydes formation, mainly from secondary oxidation products of linoleic acid (O'Keefe et al. 1993). As linoleic acid is 40–50 times more oxidatively labile than oleic acid

(Choe and Min 2006), the oil from cultivars with low oleic/linoleic ratio also tend to get slightly browner over time. Differences in the fatty acid composition, as well as in the sugar content, have important implications on the physical properties of the raw and roasted peanut oil. Cultivars with a high oleic/linoleic fatty acid ratio (>9.0) produce oils of slightly higher density (Davis et al. 2008).

Concerning aroma, about 20 pyrazines with different odor thresholds were identified by headspace solid-phase microextraction/Gas chromatography–mass spectrometry (HS-SPME/GC–MS) as the main contributors of the characteristic positive nutty roasty peanut oil aroma. Among the most important are the 2-ethyl-5-methylpyrazine, 2,6-diethylpyrazine, 2,5-dimethyl-3-ethylpyrazine and 2,3-dimethyl-5-ethylpyrazine (Liu et al. 2011). Acetylpyrazines evokes popcorn-like aromas, while benzenemethanol and benzeneethanol offer fresh bread-like and sweet rose-like aroma notes (http://www.thegoodscentscompany.com n.d.). Cyclohex-1-ol, 3-nonen-2-one, 2-pentanone, 2-hexenal, and (E,E)-2,4-nondienal are four specific volatile compounds that allow peanut oil discrimination (Hu et al. 2014). Based on processing quality, nutritional, and physicochemical properties data, Shi et al. (2016) identified some suitable cultivars for peanut oil production. Those having elevated oleic acid levels tended to exhibit more intense roasted and nutty odor notes (Isleib et al. 2006). On the other hand, the choice of these cultivars led to questions especially about emanation of off-flavors such as fruity/fermented/rancid (stale/cardboard and painty), mainly due to aldehyde formation. Higher hexanal concentration gives a greener note to oil from less mature peanut seed. The nut's growing environment and post-harvest time required for drying are also perceptible factors in the PO aroma (freshness or nutty notes). Ethanol, the most important alcohol in raw peanut, and methanol, derived from alcohol dehydrogenase and lipoxygenase activity, increase their concentration during peanut maturation.

PO is available in refined, unrefined, cold pressed, and roasted forms. Slight differences have been noticed depending on the type of extraction including hydraulic, expeller pressing, and/or solvent extraction, but also more recently, when using supercritical CO_2 and aqueous enzymatic extraction (Akhtar et al. 2014). Pure 100% PO has an intense roasted peanut flavor and aroma; usually it is used as flavoring, but also in confection, sauces, and baked goods. Refined oils are lower in saturated fats and some aromatic compounds like D-limonene (Liu et al. 2011) and (E,E)-2,4-decadienal and (E,E)-2,4-nonadienal, ending up in a neutral odorless oil (Liu et al. 2011).

Microwave blanching of peanuts sometimes leads to stale/floral, bitter, ashy off-note aromas generally associated with an enhancement of the guaiacol, phenylacetaldehyde, and 2,6-dimethylpyrazine content in the oil (Schirack et al. 2006). During peanut processing either in a pan, oven, or microwave prior to oil making, the specific temperature of drying and roasting, air flow, and moisture parameters strongly influence the sensorial quality of the final product (Smith et al. 2014). Significant differences were observed with roasted, sweet, dark roast, bitter, and raw beany green sensory attributes detected. High oven temperature gives higher roasted peanut, sweet, and raw beany green (hexanal) odors and lower bitterness. A long period of pan roasting (30–50 minutes) is needed for the formation of some volatile's compounds, accelerating Maillard reactions and Strecker degradations between amino acids and sugars but also by provoking caramelization of sugars and degradation of lipids (pyrolysis and oxidation). As roasting time lengthens for a given temperature, pyrazine concentrations increase (2,3-diethyl-5-methylpyrazine, dimethylpyrazine, 2-methyl-3-ethylpyrazine, trimethylpyrazine, and especially 2-2-acetyl-1-pyrroline).

This makes nutty, roasted, popcorn/potato-like, and earthy notes more intense in the PO (Liu et al. 2011). Although pyrazines are associated with roasted flavor, bitter flavors may also develop. Furan derivatives are also formed providing roasted (dihydro-2-methyl-3(2H)-furanone), caramel-like and sweet (furaneol, maltol), fruity and nutty (2-furaldehyde), and some coffee (2-furanethanol) aroma nuances. Aromatic aldehydes, benzaldehyde, and phenylacetaldehyde from the accelerated Strecker reaction add some almond- and honey-like notes. Some alcohols, alkanes, ketones, and aldehydes formed during lipid oxidation are responsible for off-flavors in peanuts (Schirack et al. 2006). Oxidative rancid/cardboard flavors are associated with pentanal, hexanal, octanal, and nonanal. Unsaturated aldehydes gain importance during the final stage of roasting: hexanal and trans, trans-2,4-decadienal from oxidation and peroxidation of linoleate and (E)-2-decenal, 2-undecenal and decanal from oleic acid peroxidation accentuated green, grassy, slightly fruity, oily, or fatty aromas (Ho and Shahidi 2005). Butyrolactone and pantolactone were also reported in roasted peanut oil (Neta and Ramos 2010).

12.8 Coconut Oil

CO aroma is recognized for its pleasant flavor, cooling effect in the mouth, and unctuous texture when used in food preparations (salads, confectionary, ice cream, biscuits, etc.). It is more commonly use in Indonesia, and the Indian Kerala and Karnataka southern coastal regions. It displays unique physical characteristics as it has a solid soft waxy-creamy texture and milky-white color under 25 °C.

Sensory descriptors of CO include bacony, burnt, buttery, nutty, rancid, soapy, and waxy flavors. Virgin and refined, bleached, and deodorized COs were analyzed by trained panelists who used the following attributes to characterize coconut aroma: acid (associated with acetic acid solution), cocojam (associated with sweetish burnt/roasted coconut), latik (associated with cooked sweet coagulated coconut milk), nutty (associated with the 2nd layer of fresh coconut kernel with testa), and rancid (associated with old stored oil) (Villarino et al. 2007; Dayrit et al. 2011). Faiza and Debnath (2016) also used the sensorial adjectives freshness, pleasant, oily, heated oil, fruity, sweetish, copra-like, coconut like, to characterize classical and enriched-CO.

Main volatiles responsible for the positive cocojam, sweet, lactic aroma are unsaturated lactones (mainly δ-octolactone and y/S-decalactone). The major component identified apart from the two mentioned lactones was 6-pentyl-α-pyron. The nutty note comes from 2-heptanone and ethyl acetate. Rancid off-flavors in coconut aroma were associated to octanoic and acetic acid content. Dodecanoic acid represents almost half of all the fatty acids present in CO. This said, given the high odor threshold of this compound, it was not found to contribute significantly to typical coconut aroma. Methyl ketone (2-pentanone and 2-heptanone) and acetic and lactic acids resulting from free fatty acid conversion by endogenous coconut microflora were associated to ketonic rancidity/ and microbial fermentative defaults in smell (Kinderlerer 1994). These compounds are perceived at relatively low sensory thresholds (3–4 ppm) and mask the pleasant flavor of CO, thereby reducing consumer acceptability. Santos et al. (2011) evaluated the effect of coconut maturation on oil quality and found that a better quality oil was obtained from young coconuts having lower peroxide value and free fatty acid content associated with off-flavors. Higher contents of unsaturated fatty acids and antioxidant content could suggest lower browning risk.

Villarino et al. (2007), with the help of 11 trained panelists, described and differentiated, virgin CO (VCO) and refined, bleached, and deodorized (RBD) CO samples and reported that the RBD samples were distinctly yellow, slightly salty, and had no perceptible aromas and flavors. In contrast, the VCO samples were nearly colorless, had slightly detectable acid aroma, sweet and salty tastes, and perceptible nutty aroma and flavor (Villarino et al. 2007). The RBD process alters oil quality, reduces desirable natural odor, and generates high content of free fatty acids shortening the oil shelf-life by promoting peroxidation (Nevin and Rajamohan 2004; Villarino et al. 2007; Raghavendra and Raghavarao 2010). The method of obtaining the oil also influences fatty acid composition and consequently coconut aroma. Trained panelists found that VCO obtained from fermentation process had stronger fermented off-odors than the expeller or centrifuge methods (Santos et al. 2011). Dayrit et al. (2011) found that wet processes (centrifuge and fermentation) had higher percentage of free fatty acids due to microbial hydrolysis of the vegetable oil and peroxide value than vacuum-dry expeller. The heating process to obtain CO causes an increase in its transparency, as well as makes it liquid and allows sweet subtle odors emanation. The typical odor notes tend to decrease upon heating (Faiza and Debnath 2016). Roasted and smoky notes can arise from strong or overpowered heat.

12.9 Linseed/Flaxseed Oil

Flaxseed and linseed names both refer to flax. Flaxseed is used for the flax consumed as food by humans while linseed describes flax processed by industry and for feed purpose (Gunstone 2011). Linseed oil (LO) is utilized as flavoring oil in salads. Because of the polymerizing properties of LO, it is employed as a texturing agent or shortening substitution in dairy (Veena et al. 2017) and baked-products (Rangrej et al. 2015).

Edible LO is usually produced by cold-pressing the oil from the seeds (Choo et al. 2007). The aromatic and color characteristics are less intense than hot-pressed oil. When extracted under heat and airless conditions, the oil is also more viscous (Wei et al. 2015). Higher oil viscosity has been related to lower linolenic acid concentration (Zhang et al. 2011). The extraction process of the oil influences the amount of aroma volatile compounds detected: 51 volatiles were identified from the hot-pressed LO, 47 from cold-pressed and 40 by solvent extraction. Some of these compounds showed good potential as biomarkers to differentiate processing methods, including hexanal, (E, E)-2,4-pentadienal, (E, E)-2,4-heptadienal, 6-hydroxy-2-hexanone, 1-hexanol, methylpyrazine, nonanal, 2,3-pentanedione, 1-butanol, acetic acid, hexanoic acid, and ethyl acetate. The percentage of compounds from the different chemical families also varies according to the extraction process, being the total amount of volatiles increasing from expelled, cold-pressed, to hot pressed-oil. Some volatiles seemed specific to an extraction process condition. As an example, pentanal, benzaldehyde, (E)-1, 3-butadienol, 1, 5-hexadien-3-ol, pentanoic acid, and tetradecane, were only found in hot-pressed oil. Numerous aldehydes in this oil were significantly correlated with its oxidative status. Cold-pressed oil was characterized by 1-heptanol, 2-methyl-1-butanol, acetic acid butyl ester and naphthalene, whereas 2, 3-dimethyl-pentanal, heptadecane were identified specifically in expeller-extracted oil. Higher percentages of the ketone family were found in cold and hot-pressed oil and they were in turn associated with desirable aroma. Thermal degradation of carotenoids due to heating leads to an increased amount of pyrazines in hot pressed oil and the lowest concentration of these compounds was found in expeller-extracted oil. Roasted, oily, herbaceous, green, and almond flavor attributes were more intense in hot than in cold pressed oil. The lowest perception of these attributes was found in expeller-extracted oil. Comparing hot-pressed oil to cold-pressed and expeller-extracted oils, the overall odor profile in the latter was weaker due to lower intensity of hexanal, (E, E)-2, 4-pentadienal, (E, E)-2, 4-heptadienal, 6-hydroxy-2-hexanone, 1-hexanol, methyl-pyrazine, nonanal, 2,3-pentanedione, 1-butanol, acetic acid, hexanoic acid, and ethyl acetate (Wei et al. 2015).

Freshly cold-pressed LO has low viscosity, exhibits a golden yellow color and is characterized by its subtle blend of nutty roasted potato, caramel-like, slightly spicy flavor with overtones of hay (Brühl et al. 2008). Some green odors can be also perceived. Panelists also described meat-like, umami, salty taste, mouthfulness, and continuity sensorial sensations. Bitterness and caramel traits were associated to bitter amino acids Tyr, Ile, Leu, and Phe, as well as peptides of low molecular weight (Wei et al. 2015). In a precedent study, Wei et al. (2015) identified methionine sulfoxide-containing, cyclic octapeptide cyclo (cyclolinopeptide E) as responsible for the lingering bitter off-taste. A fishy, rusty off-odor can diffuse from LO because of its high concentration in omega-3.

Ivanova-Petropulos et al. (2015) found that flaxseed oil was also characterized by the presence of sulfur-compounds, alkanes, and alcohol families. Indeed, Krist et al. (2006) identified 1-hexanol (herbaceous, green, woody, sweet), trans-2-butenal and acetic acid (strong, sour, pungent) as the main volatiles. Others compounds that also contribute to the overall flavor of flax oil were trans-2-pentenal (pungent, green, fruity), 2-heptanone (fruity-cinnamon), 1-octen-3-ol and trans, trans-2,4-hexadienal (moldy). Yalcin et al. (2011) detected limonene (citrus), 1-hexanol, styrene, benzaldehyde, p-cymene, and nonanol.

In addition to the oil extraction process, chemical composition and bitterness of LO is cultivar-dependent (Brühl et al. 2008; Nykter et al. 2008). As for most oils, the time of harvesting and weather conditions affect also the oil quality (Nykter et al. 2008). Hence, the moisture content of the seed at the time of harvest is determinant in the risk of mushroom-like off odors during processing and storage.

Linseed should generally not be cooked because of its fatty acid composition properties. Indeed, cooking accelerates the degradation of omega fatty acid and tends to cause off-flavors, such as bitter, and even rancid or putty-like.

Because of its biochemical composition, LO is highly perishable and has low storage capacity (Juita et al. 2012). Cold-pressed oil is usually sold on a small scale because of its propensity for oxidation of LO and color evolution. When exposed to sunlight, LO oxidizes and quickly becomes rancid. A model mouth system set up by Van Ruth et al. (2000) detected and identified propanal as a potential indicator for oil deterioration.

Bitterness develops after only one day of storage at room temperature and is detectable at a low recognition threshold concentration of 12.3 μmol/l water. To preserve the fresh taste of the oil, it can be stored in the freezer because of its low melting point (−16 to −20 °C). Faster deterioration of products containing LO was reported such as greater loss of texture, viscosity, firmness, consistency, color, and peroxide value (Veena et al. 2017).

12.10 Canola or Rapeseed

The oil is produced from the seeds of the bright yellow-flowering plant belonging to the Brassicaceae family. Although derived from the same plant but from two distinct varieties, canola is characterized by lower levels of erucic acid (<2%) and glucosinolates

(<30 umol/g) in comparison to rapeseed. Such characteristics provide canola oil a more pleasant neutral flavor than the rapeseed oil, which makes it acceptable to most consumers. It is usually fully refined, bleached, and deodorized to meet the edible grade but also by conferring excellent frying properties (Kraljić et al. 2018). Non-refined oil cold-pressed or virgin canola oil also exists and are preferred for salad dressing, seasoning, mayonnaise, and baking (Brühl and Matthäus 2008). Rapeseed oil is used for biofuel and in the chemical industries. Still, both names are often interchanged, and the information provided here reuses the same terms as in the original study (Brühl et al. 2008). Asparagus, cabbage, fresh-green vegetable, and some sulfuric, nutty, buttery notes characterize the aroma of canola/rapeseed oil (Brühl and Matthäus 2008). The volatiles responsible for these sensory attributes mainly result from the enzymatic degradation of glycosynolates, which are almost eliminated by thermal degradation during the heating process. The high content in monounsaturated fats makes canola oil one of the unique unblended oils resistant to high frying temperatures, conserving color or flavor quality. It is also appreciated for its resistance to cold-clouding effect when stored at low temperature, remaining thin, light in texture and transparent.

Ivanova-Petropulos et al. (2015) were able to separate rapeseed oil from other oils as flaxseed, pumpkin seeds, sunflower and sesame based on their volatile profile. Rapeseed oil was strongly correlated with alkenes, aldehydes, cycloalkanes, and furans families. Hu et al. (2014) compared the aroma of canola oil with other vegetable oils and detected 16 volatile compounds that were specifics to canola oil, including methyl thiocyanate, N-methylene-ethenamine, dimethyldisulfide (cabbage-like), 3-pentenenitrile, 2-hydroxy-2-methyl-propanoic acid, 2-furanmethanol, 5-cyano-1-pentene, 2(5H)-furanone, 5-methyl-hexanenitrile,1,5-hexadien-3-ol, hexanenitrile, 4-isothiocyanato-1-butene, dihydro-3-methyl-2,5-furandione, 2,3-dihydro-3,5-dihydroxy-6-methyl-4H-pyran-4-one, benzenepropanenitrile, and 2-methoxy-4-vinylphenol (smoky). The predominant compounds from canola oil were 2-methyl propanal, 3/2-methyl butanal (malty), 2,5-dimethyl pyrazine, 3-ethyl-2,5-dimethyl pyrazine, 2-furanmethanol and N-methylene-ethenamine. More recently, Matheis and Granvogl (2016) characterized and quantified the key aroma compounds in commercial native cold-pressed rapeseed oil by sensomics approach. A trained sensory panel rated the odors predominant notes of rapeseed oil such as fresh-green cabbage-like, nutty/fatty and earthy/pea-like with some malty hint. By aroma recombination, Matheis and Granvogl (2016) were able to recreate the canola oil aroma with 11 of the 54 active odorants identified, these were 3-hexanal (green), octanal (citrus like), dimethyl trisulphide (asparagus/cabbage-like), 2-isopropyl-3-methoxypyrazine (earthy, pea-like), 2-isobutyl-3-methoxypyrazine (bell pepper-like), 2-sec-butyl-3-methoxypyrazine (pea-like, roasty), butanoic acid (sweet), methylbutanal/propanal (malty), 3-hydroxy-4,5-dimethylfuran-2(5H)-one (seasoning-like) and (E,E)-2,4-nonadienal (cheesy, deep fried, fatty). Zhou et al. (2018) identified 1-isothiocyanato butane, allyl isothiocyanate, and 4-isothiocyanato-1-butene as the volatiles that cause the sulfur, pungent, and green odors in rapeseed oils.

From a sensory point of view, the most appreciated oils showed seed-like notes, while nutty, woody, astringent, and straw were only tolerated at low intensity (Brühl and Matthäus 2008).

Variations due to canola/rapeseed cultivar in the flavor profile and rheological characteristics of the oils are usually discrete. Zhou et al. (2018) found that *Brassica* sp. had different endogenous levels of glucosinolates precursors responsible for the intensity of bitter, pungent smell. Increasing concentration was found in *Brassica napus*,

Brassica campestris, and *Brassica juncea*. High-oleic acid canola oils are preferred for their sensory profiles, especially for frying. Syed (2015) reported the sensory profile of a classical and another extracted from a genetically modified cultivar for ω-9 enrichment. The 10 trained panelists did not find significant differences in both sensory profiles.

The oil quality is determined by climate during the growing period and at harvest time. Seeds harvested at about 7% moisture can develop musty odor. Straw smelling occurs when straw and other foreign matter are mixed during the seed cleaning and remain before pressing (5 or 10%). Fusty odor can diffuse when crushed seeds are stored at ambient temperature for about several weeks. Cold-pressed rapeseed oil is manufactured from either unpeeled or peeled rapeseeds (Pollner and Schieberle 2016). Thirty-five odor-active constituents were detected in the oil from unpeeled seeds. Sensory attributes were correlated to the chemical compounds: 2,5- isopropyl-3-methoxypyrazine (earthy, pea-like), and fatty predominated followed by 1-octene-6 3-one (mushroom-like) and (E,Z)-2,6-nonadienal (cucumber-like), all of them having high odor activity value. Other sensory attributes perceived were grassy (hexanal), citrus-like (octanal), bell-pepper (2-isobutyl-3-methoxypyrazine), sweaty and cheese-like (2- and 3-methylbutanoic acid), although the volatiles linked to them had lower odor activity values. Dimethyl sulfide (cabbage/asparagus-like) was present in higher concentration in the oil extracted from peeled seeds leading to a significantly different sensory profile. Seeds can even germinate if the relative humidity of the storage container is high, giving a sprouted seeds off-odor to the oil that can be detected even if mix with good quality seeds before pressing (Brühl and Matthäus 2008).

Kraljić et al. (2018) evaluated the effect of the application of increasing seed conditioning temperatures (60, 80, 100 °C), during canola oil processing, on sensory attributes and volatile compounds and showed that cold-pressed and virgin canola oil displayed similar sensory profile till 60 °C. As mentioned earlier, Kraljić et al. (2018) reported that the dominant volatile was the 4-isothiocyanato-1-butene, as also found by Wei et al. (2015). Isocyanates and epithionitriles give a pronounced seed-like flavor. Rising temperature then deactivated enzymes and caused thermal decomposition of seed components decreasing seed-like intensity while increasing nitriles, aldehydes, pyrazines, and furfurals (furfural and 5-methy-2-furfural) formation, responsible for the emanation of nutty and roasty flavors. Burnt off-flavor (2-furanmethanol) was identified in virgin oil conditioned at 100 °C.

Microwave processing before oil extraction intensified cabbage-like and decreased pungent-like odors due to glucosinolates (Zhou et al. 2018). Dimethyl sulfide, dimethyl trisulfide, and dimethyl sulfoxide suspected to derive from methionine or isothiocyanates during heating evoked onion, cabbage, garlic, and sulfurous-pungent odors.

Majchrzak et al. (2017) carried out a thermal degradation assessment of canola using ultra-fast GC coupled and showed that as temperature of incubation rose from 20 to 180 °C, palatability dropped from 3.5 to 1.2 on a 5-scale basis (5-very good, 4-good, 3- satisfactory, 2-unsatisfactory, 1-bad). In the same study, descriptors changed from "no flavor, nutty, grassy" (20, 60 °C), "grassy, rancid" (100 °C), to "grassy rancid, acidic, earthy (temperature above)" for flavors and smells. At this temperature, bread-like smell was also mentioned. High linoleic cultivar was more affected by roasting and could be distinguished by the presence of 2-methyl-propanal, 3-hydroxy-2-butanone and a larger variety of furan derivatives.

Higher content of sulfuric compounds was present in cold-pressed oil, likely due to enzymatic process. The compounds were suspected to evaporate under heat due to an acceleration of Strecker degradation reactions. Pyrazines and pyrols were identified only in virgin oil heated at 100 °C, in accordance with the roasty, nutty, woody or burnt odor mentioned by the sensory panel (2-methylpyrazine, 2,5-dimethylpyrazine, 2-ethylpyrazine, 2,6-dimethylpyrazine, 2-ethyl-6-methylpyrazine, 2-ethyl-3-methylpyrazine, 2-ethyl-5-methylpyrazine, and 2-ethyl-2,5-dimethylpyrazine, -(1H-pyrrol-2-yl)ethanone). These compounds formed during Brassica seed heating can provide popcorn-like and coffee-like pleasant notes compared to the pea-like aroma of raw seeds (Ortner et al. 2016).

Frandsen et al. (2017) demonstrated that the sensory profile could discriminate superfluid extracted and cold-pressed rapeseed oils by principal component analysis. The oil obtained by superfluid extraction exhibited a slighty lighter color than the other. Taste, mouthfeel, odor, and aftertaste were also significantly different when comparing both extraction methods. Superfluid extracted oil showed more intense chickpeas, roasted nuts, cabbage, sharp adjectives, oatmeal, fresh-nut, green, butter, and some flowers odor notes.

12.11 Hazelnut Oil

Hazelnut oil (HO) is usually extracted by pressing unroasted or roasted, shelled kernel. HO is commonly refined. Although HO has not a major production, it is incorporated into numerous confectionary and baking products such as spreads or chocolate. Extra-virgin cold-pressed HO can be found on the market and is usually used to enhance salads as a gourmet condiment.

HO is clear, yellow to yellow-brown. Fresh virgin HO from gently roasted nuts is used as reference in panel taste experiment for nutty descriptor calibration. Because of its high unsaturation characteristics, HO is prone to fast rancidity, despite of its high content in Vitamin E.

The aroma profile of HO can largely vary according to the cultivar. Slight differences exist in the endogenous array and concentration of aroma-active compounds. Alasalvar et al. (2004) evaluated five Turkish hazelnut varieties and found that even though all flavor qualities were not significantly different except for woody flavor, "Tombul" cultivar had more intense nutty, sweet, oily, aftertaste, woody, fruity and caramel-like flavor. "Tombul" and "Palaz" varieties were discernable by their oiliness and fruitiness. Discriminate Function Analysis on electronic nose reduced to seven volatiles (trichloroethylene, toluene, m-xylene, 5-methyl-(E)-2-hepten-4-one (filbertone), 2-nonanone, and (E,E)-2,4-hexadienal) the number of compounds needed for cultivar distinction. Castillo et al. (2002) found that the enantiomeric ratio of filbertone could be used to authenticate cultivars, since it seems to be unaffected by cold pressing during oil production and remains constant in raw hazelnuts oil of different geographical origins. Kiefl and Schieberle (2013) drew the aroma profiles based on the descriptors coffee-like, sulfury, nutty, fruity smoky, phenolic, malty, sweet, caramel-like, roasty, popcorn-like fatty, earthy, green. They saw that "Gentle" cultivar displayed naturally a sweeter-caramel like aroma than "Giffoni," "Romane" and "Akçakoca." Ojeda-Amador et al. (2019) analyzed the aroma profile of the cultivars "Tonda," "Pauetet" and "Negret." They reported that "Negret" cultivar showed higher hexanal and heptanal concentrations than the rest. "Pauetet" could be distinguished by the presence of furfural and benzaldehydes and for higher alcohol concentration, such as pentanol (fruity aroma). Pedoclimatic conditions also influence the sensory quality of the produced nut. Thus, "Avellana de Reus" (Spain) received the label of Protected Designation of Origin for its superiority and so have "Nocciola Romana," "Noisette du Piémont" (Italy) and "Noisette de Cervione-Nuciola di Cervoni" (France, Corsica).

The roasting of hazelnuts develops their odor intensity by enlarging the diversity of volatiles produced. Thirty-seven odor-active compounds were detected in raw nuts, whereas 46 were found in roasted nut material (Burdack-Freitag and Schieberle 2010). 2-methoxy-3-isopropylpyrazine and 2-methoxy-3,5-dimethylpyrazine as well as 5-methyl-4-heptanone were present in higher concentration and intensity in raw hazelnuts, while 2-acetyl-1-pyrroline, 2-propionyl-1-pyrroline, 2-furfurylthiol, and 2-thenylthiol (Z)-2-octenal, (Z)-2-nonenal, and (Z)-2-decenal were dominant in the roasted nuts. Kiefl and Schieberle (2013) studied the impact of roasting time, temperature, and applied techniques on the aroma generation of three hazelnut cultivars by correlating quantitative key odorant profiling with sensory evaluation. They reported that nutty and fruity smells were related to 5-methyl-(E)-2-hepten-4-one (filbertone), while 2-acetyl-1-pyrroline and 2-propionyl-1-pyrroline give roasty, popcorn-like-smelling odors. 3-(methylthio) propionaldehyde was responsible for earthy-smelling. 3,5-dimethyl-2-ethylpyrazine, 2,3-diethyl-5-methylpyrazine, 2-phenylacetaldehyde, 2-acetyl-1,4,5,6-tetrahydropyridine, 2-acetyl-3,4,5,6-tetrahydropridine, 4-hydroxy-2,5-dimethyl-3(2H)-furanone were responsible for the slight odorous variations of nut aromas. The roasty, nutty aroma of optimally roasted hazelnuts was found to be a combination 5-methyl-(E)-2-hepten-4-one (filbertone) and 3-methyl-4-heptanone. Donno et al. (2013) studied the effect of roasting temperature on some sensory characteristics of hazelnut and reported that low temperature (122 °C) of roasting displayed nuts with lighter color, having lower chroma and higher hue value. Samples roasted at high temperature roasting (170 °C) had more intense hazelnut odor and also exhibited oily sweet taste. Raw hazelnut had more caramel odor/flavor and woody notes. The over-roasted smell was the result of the sum of two 2-acyl-1-pyrrolines and two pyrazines. The analysis of off-odors in hazelnuts revealed that the prenyl ethyl ether was the responsible for a metallic, solvent-like off-note.

12.12 Avocado Oil

Avocado oil (AO) still has a limited market as an edible oil. It is mostly used by the cosmetics industry. The high lipid content makes volatile analysis somewhat more complex and so far, more information is currently available on oil content and texture than volatiles for AO. The oil is obtained from the flesh of the fruit of *Persea americana* Mill. and then usually refined. Extra virgin oil production started in New Zealand in 2000 based on cold-press extraction techniques inspired from the olive oil industry. The oil is sold for culinary use on salads and for cooking.

Extra virgin and cold-pressed oils are unrefined and better retain the flavor and color characteristics of the flesh fruit (Costagli and Betti 2015). The crude AO displays a brilliant emerald green to dark green color due to high level of chlorophyll. The oil pigmentation is even greater if avocado skin is included in the pulp during malaxing. The same author detailed the requirements for avocado to meet the different grades in oil: extra virgin avocado oil (EVAO) must be extracted from high-quality fruit having minimal levels of defects; whereas inferior categories tolerated a proportion of damaged fruits including color defects. Market standards about quality indices, composition, and sensory properties have been proposed for AO, including extra virgin, virgin (VAO), and pure grades. They follow the same guidelines as mentioned by the International Olive Council (IOC). Indeed, similarities between avocado and OO have already been established (Berasategi et al. 2012).

Moreno et al. (2003) analyzed AO samples from different quality fruit and highlighted nine sensory attributes. High quality new-season freshly extracted oil had greater percentages of E-2-hexenal and n-/cis/E-2-hexanol related to desirable freshness and greenness aroma. The trace terpenes, α/β-pinene, β-myrcene (musty, wet-soil), cis-trans-β ocimene, also contributed to this green-plant sensation. Smoky, grassy, mushroom/butter-like notes are usually moderate in intensity (above 40 on a sensory panel 100-point scale for EVAO and above 20 for VAO and others) (Costagli and Betti 2015). Hoppy, aniseed-like, greasy/oily traits are also used for the description of the oil. Default includes painty, fishy, and glue-like which should be below 20 and 35 out of 100-point scale intensity in case of EVAO, and inferior to 50 for VAO, pure and blends. The last three adjectives are considered good markers of AO sensory quality.

The region of production is also a factor responsible for sensory profile modification in AO (Moreno et al. 2003). "Hass" cv. oil from New Zealand was described as having higher grassy, hoppy, painty, and mushroom/butter-like qualities. The oil from the same variety produced in Mexico displayed higher unappreciated glue-like smell. Consideration of growing, handling, extraction, storage and retail conditions should be taken in consideration for more reliable conclusions. "Fuerte" and "Hass" cultivars fruit volatile composition was analyzed respectively by El-Mageed (2007) and Obenland et al. (2012), and the cultivar "Fuerte" was found mushroomier than "Hass" cultivar.

The maturation degree also influences AO's visual (Ashton et al. 2006) and aroma qualities (El-Mageed 2007; Obenland et al. 2012). During ripening and softening, the initial green color of skin turns to purple/black due to cyanidin 3-O-glucoside and chlorophyllide degradation (increasing hue angle, H°). Simultaneously, chroma (C*) and lightness values in flesh tissue and the hue angle remains almost identical. The total chlorophyll level decreased in the oil extracted from the skin and stays almost constant in flesh-extracted-oil (Ashton et al. 2006).

Concerning AO aroma, 43 compounds were identified including 15 terpenes. The proportion of these terpenes with respect to total volatile concentration increased from 9% to more than 25% during fruit maturation. The major terpenes were D-limonene (citrus), Z-nerolidol (woddy-floral, waxy), alpha-farnesene, and caryophyllene. Esters accounted for 32–34% of the total volatile composition. Ethyl ethanoate, ethyl acetate, and ethyl hexanoate were characteristics of green avocado and ethyl methanoate of ripe and overripe samples giving a fruity note. The aroma profile also comprised 8 aldehydes, including benzaldehyde (cherry) and nonanal (fatty, citrus), alcohols, ketones (1-penten-3-one [earthy, green]), thiazoles, pyridines, toluene, and acetic acid (El-Mageed 2007). Obenland et al. (2012) also found that 12 of the 25 volatiles detected by GC–MS/SPME significantly changed their concentration during maturation. Their results confirmed that E-2-hexenal, E-3-hexanol, hexanol, hexanal, and 2,4-hexadienal were associated with the grassy note of avocado aroma. All of them declined over maturation, except E-3-hexanol displaying a peak at mature stage. Acetaldehyde (pungent, solvent-like), nonanal (fatty, citrus), methyl acetate (ether, sweet), pentanal (fruity, nutty), beta-myrcene (musty, wet-soil) and 2,4-heptadienal (fatty, green) increased during maturation and were associated with creaminess and richness and loss of grassiness. Over ripe avocado fruits had the highest content of esters (ethyl ethanoate, ethyl acetate, and methyl propanoate) and terpenoids.

Moreno et al. (2003) reported that the volatile profile of AO was highly dependent on the extraction procedure. In their study, 21 descriptors were employed to qualify AO with semi-trained panelists and re-used in further studies. Mechanical pressing of 60 °C-dried-pulp in a vacuum oven was found to give best quality AO, preserving bioactive properties and meeting quality parameters recommended by current legislation (Krumreich et al. 2018). The refined, bleached, and deodorized oil was odorless, yellow to pale yellow-green, and had low viscosity. Concerning the volatile components determined in virgin AO, the cold-pressed method resulted in less volatile diversity. (E)-2-hexenal (fruity, almond-like), 1-hexanol (herbaceous green), α-pinene (pine, turpentine-like), b-pinene (sweet-lemon), and 3-carene (sweet-turpentine like) were specifically found in cold pressed oil (Haiyan et al. 2007). In comparison, 2-butenal, 1-penten-3-ol, pentanal isomer, toluene, and furfural were detected in refined oil. Hexanal, heptanal, acetic acid, pentanal, and nonanal were found in both oils regardless of the extraction method.

The edible AO could be also extracted by supercritical CO_2 (SCO_2) and CO_2/ethanol methods. Tan et al. (2018) reported that 1-hexanol was not found in both SCO_2 and ultrasound assisted avocado oil extraction (UAAE). The oils extracted by the two latter methods had numerous aldehydes (heptanal (fatty, green, nut lije), hexanal (fatty, green, grassy), nonanal (fatty, floral, grassy), octanal (strong fruity), and (E)-2-decenal (fatty). Farnesol (floral), acetoin (buttery, creamy), 1-methoxy-2-propyl acetate (Sweet ether-like), and several odorless compounds were found specifically in SCO_2 while Z-nerolidol (wooddy-floral, waxy) and pentanal (slightly fruity and nut-like) were only identified in UAAE oil. As temperature and pressure increase in the SCO_2 experiment, the oil color starting from straw-yellow (37 °C/350 atm) gradually acquired a greenish tint. Hexane extracted oil is dark green. Octanal (strong fruity), nonanal (fatty, floral, grassy), and beta caryophyllene (cloves-turpentine-like) were also found in microwave extracted AO from avocado pulp (Moreno et al. 2003). Greater deterioration of the oils was produced with solvents rather than microwaves. A larger number of volatiles was listed for solvent extraction (hexane, acetone) but could be attributed to thermal and decomposition effect on the fatty acids. Microwave + hexane and acetone extracted-oils contained numerous terpenoids, among them α-bergamotene, α-humulene, α-copaene, α-cubebene, α-farnesane, β-caryophyllene, and β-bisabolene. Additionally, dodecane and 2,4-decadienal were also noticed. Microwave–squeezing techniques totalized less volatiles (4 compared to 12–15 of the other oils). Guzmán-Gerónimo et al. (2008) found that avocado aroma qualities could be improved by addition of leaves to fresh and microwaved avocado (the ratio of terpenoids like α-pinene, estragole, and (E)-2-hexenal/hexanal increased). Although the results for organoleptic qualities were positive, the microwave technique still needs to be improved as it affects lipid acid modification and thus healthfulness, flavor, and shelf-life.

Volatile profile changes during storage also depend on the storage time, temperature, light, and air exposure conditions. Acetic acid was found responsible for glue-like smell providing rancid off-flavors. Although the high content in tocopherol and carotenoids inhibits to some extent oxidation, the high chlorophyll content makes AO photosensitive and hence the breakdown of chlorophyll is a common visual marker of quality.

12.13 Almond Oil

Almond oil is obtained from the *Prunus dulcis* (Miller) D.A. Webb var. dulcis or var. amarga (D.C.) or from a mixture of the two. It generally has discrete-medium flavor intensity (≤5 out of 15-point scale), a clear appearance, and low viscosity. Almond oil is rarely used as edible oil, but is appreciated for confectionery. Cultivar and maturity of the kernel, harvest, shelling, processing, and storage conditions are well-known factors that affect the almond oil edible quality.

Non-volatile compounds strongly influence the taste and textural sensation of sweet almonds. Bitterness in almond kernels depends on the amounts of amygdalin, a diglucoside that is broken down into hydrogen cyanide and benzaldehyde when the kernel is crushed and exposed to water or saliva (King et al. 2019). Almond phenotypes are classified based on their amygdalin content as sweet, semi-bitter, or bitter (>3% amigdalin). Wild almonds are usually richer in amigdalin than cultivated varieties, that making them more appreciated for flavoring in the industry. The phenolic compounds contained in almond nut skin that are responsible for astringency are the flavanol monomers like (−)-epicatechin and (+)-catechin) and the oligomers such as proanthocyanidins or condensed tannins (Bolling 2017).

The typical almond-meat flavor attribute is found in all the cultivars, but some of them like "Butte" and "Nonpareil" are characterized by a unique squash-like flavor. The latest cultivar together with "Carmel" cultivar could be distinguished from other varieties thanks to a light tea-like flavor (Bolling 2017). "Butte" and "Carmel" varieties display a distinctive walnut-like flavor. "Carmel" aroma had more intense sweet aromatics and benzaldehydes flavors. Marzipan/benzaldehyde attributes related to amygdalin, as well as the volatile compounds benzaldehyde, phenylethyl alcohol, and benzyl alcohol were also found in Aldrich and Fritz varieties more than in Nonpareil (Bolling 2017). Other European cultivars as "Ferragnes" and "Marcona" differ in the aldehydes amount, especially benzaldehyde (bitter almond) and hexanal and alcohols, such as hexanol (Ojeda-Amador et al. 2019).

The odor thresholds of pyrazines depend on the position of the alkyl group. Almond varieties including "Butte/Padre", "Comuna", and "Marcona" almonds show threshold of 0.06 mg/l in water in the case of 2-methylpyrazine. Toasted "Marcona" and "Comuna" almonds included in their aroma profile the volatiles 2-ethyl-5-methylpyrazine (threshold 0.10 mg/l in water) and 2-ethyl-6-methylpyrazine (0.13 mg/l in water). Furans, such as furfural and furfuryl alcohol, were also frequently reported in heat-treated almonds and have a sweet almond/bready and cooked sugar odor, respectively. Though these compounds have a relatively high odor threshold (3 and 1.9 mg/l, respectively), furfural and furfuryl alcohol were reported in superthreshold levels in toasted "Comuna" almonds (Vázquez-Araújo et al. 2008). Roasting also provokes the apparition of aldehydes as 2- and 3-methylbutanal and methional (fatty) from Strecker degradation. The 2 butanals (fatty) were quantified at high concentration in "Butte/Padre" and "Nonpareil" and fried Spanish almonds. Methional compound was also part of "Nonpareil", "Marcona", and "Comuna" cultivars' volatile profile. The chocolate, malty, and cooked potato aroma can be detected at very low thresholds in almond oil (0.01, 0.0054, and 0.0002 mg/kg, respectively).

Almond oil is commonly extracted with or without the use of heat by enzyme-assisted procedures, mechanical pressing, and solvent extraction. The last method is known to affect the degradation of active-oxidative compounds and the safety and quality of products as residual solvent can remain in the oil. Pasteurization or heating process did not diminish raw almond kernels sensory attributes. Besides, roasting is often used to develop flavor and modify the oil's texture. Increasing browning indices are noticed when the roasting temperature and duration of nuts are high. Almond oil is fairly pale if refined and dark yellow in case of cold-pressed.

Franklin and Mitchell (2019) recently reviewed almond flavors and sensory quality. They found that alcohols were most frequently reported in raw almonds; however, their effect on almond aroma may be limited, as alcohols have comparatively high aroma thresholds and may not be present in superthreshold levels in raw almonds. Aldehydes, apart from benzaldehyde, were also widely reported, and these tend to have lower sensory thresholds in comparison to alcohols and in some cases may contribute significantly to raw almond aroma. Lipid degradation products were found to be the major contributors of raw almond aroma while Maillard products were the key odorants of roasted almond. Furfural, 2,5-dimethylpyrazine, 2-ethylpyrazine, 2-methylpyrazine, 2-acetyl-pyrrole, and 2- and 3-methylbutanal and alkylpyrazines giving nutty, cocoa, toasted, hazelnut, bread crust, peanut, and caramel, nuts, coffee, and chocolate hints appear in heat-roasted almonds. Sometimes, earthy and potato odors can arise from over-heating.

Almond oil is characterized by a high concentration of unsaturated fatty acids. It is photosensitive and prone to hydrolytic and oxidative rancidity and thus must be stored in a dark place. Compounds detected in fresh samples tend to intensify during storage (hexanal, 2-propanone, 3-methyl-3-butenol, 3-methyl-2-butenol, benzene-ethanol, 1-pentanol, and benzene-methanol). Newly-formed compounds mostly arise from secondary lipid oxidation (heptanal, heptane, octanal, 1-octanal, nonanal, and pentane) or tertiary products (2-propanone, 2-butanone, 1-hexanol, and 2-propenoic acid acetic, hexanoic, and pentanoic acids and other organic acids). Products of degradation of oleic acid products include decanal, 2-undecenal, nonanal, 2-decenal, octane, 1-octanol, ornonanal, heptane, 1-heptanol. Linoleic acid alteration give 2,4-decadiene, pentane, 1-pentanol, hexanal, 2-octene, -heptenal, 1-octen-3-one, 2-nonenal, and 2,4-nonadienal. Seventeen volatiles belonging to ketones, aldehydes, alcohols, oxiranes, and short-chain acid were markers of storage, some of them indicating rancidity: 2-octanone, 3-octen-2-one, and acetic acid as in the study by Franklin et al. (2018). In this study, the authors combined the identification of the oil volatiles compounds and consumer hedonic and acceptance evaluations to assess the flavor of "Nonpareil" almonds. Light or dark roasted samples were submitted to accelerated storage at 39 °C for 0–12 months. The experiment evidenced the role of oleic and linoleic acid decomposition during storage and the formation of volatiles. Heptanal, octanal, nonanal, 2-octenal, 2-heptanone, 2-pentylfuran, hexanal, and pentanal were markers of rancidity negatively correlated with consumer hedonic preferences. Maillard reaction products 2- and 3-methyl-butanal and 2,5-dimethylpyrazine could predict the clean roasted and nutty odor attributes. Heptanal and octanal were better predictors of consumer acceptance than the traditionally biomarker for rancidity, hexanal.

12.14 Pistachio Oil

Pistachios are mostly consumed as snacks but their oil is appreciated for its unique flavor and green color in the manufacture of cakes, biscuits, breads, chocolates, and ice creams.

Ojeda-Amador et al. (2018) retained 18 descriptors for the sensorial assessment of pistachio oil. Typical desired qualities including seed-like, nutty flavor, roasted nut, hay/grain, vegetal/herbaceous, and woody. Fresh pistachio nuts were also depicted with the following odor attributes, fresh hazelnut/walnut/apricot kernel, pine, wet or chocolate/coffee smells if roasted, green, sour, sweet, bitter taste, astringent, milky, and oily feeling. Musty/humid odors and sour/acid taste can develop during pistachio storage. Defects such as rancid fried oil, burnt seeds, were also investigated.

Ling et al. (2016) compared the oil obtain from cold press raw kernel, conventional, and microwave roasting. Refractive index was virtually identical, but color showed important variations. Raw kernel oil displayed a light green color. Oils from conventional and microwave roasted pistachios changed from yellow–green and yellow–brown color respectively due to the development of Maillard reaction products and chlorophylls degradation. Variations were also found by Sena-Moreno et al. (2015) who analyzed the effect of different drying temperatures (30, 50, 70 °C) on the pistachio oil color. The 30 °C-dried-hydraulic-pressed oil was the yellowest while 70 °C dried-screw pressed oil was the darkest one. Heating allowed a major transfer of the pigment to the oil. Consumers expressed a preference for the greenest oils. As far as aroma is concerned, oils from 70 °C dried pistachio were found tastier and more aromatic. Indeed, roasting pre-conditioning intensified pistachio flavor. During processing from whole kernel to pistachio flour, the defatting step of pistachios eliminated some lipid-soluble compounds such as carotenoids pigment, decreasing the intensity of green-yellow color.

Noguera-Artiaga et al. (2019) examined in detail the sensory properties of eight European pistachio cultivars ("Aegina", "Avdat", "Kastel", "Kerman", "Larnaka", "Mateur", "Napolitana", and "Sirora") combining the aroma and sensory profiles obtained by gas chromatography and trained panel, respectively. Bitterness and astringency were not significantly different but the overall chemical composition influenced the perception of the flavor. The volatile profile of the pistachio oil extracted from "Kastel" and "Larnaka" had the highest total proportion of terpenes (Ojeda-Amador et al. 2020). Several volatiles were associated to pleasant smells, such as pine (α- and β-pinene), citrus (limonene and 3-carene), floral (myrcene), vanilla (camphene), berry (α-terpinene), green (thujene), and camphor (camphene). Monoterpene α-pinene and limonene were found to be key odorants in pistachio aroma and potential markers for the volatile fingerprinting of certain cultivars (Ling et al. 2016). "Kastel" and "Kerman" showed the most attractive sensory properties especially for their flavor intensity. "Kerman" had stronger nut flavor, pistachio, aftertaste which was associated to high concentrations in α-pinene, and limonene (lemon, orange citrus), 1-ethylpyrrole (woody, smoky, herbaceous), and hexanal (herbaceous). "Avdat" was the sweetest cultivar, which could be explained by the highest volatile compounds concentration of α/β-pinene (woody), hexanal (fatty, green), and benzyl acetate (fruity, floral). Alcohols were responsible for flower (1-hexanol), fruity (1-pentanol), and rose/green (2-ethyl-1-hexanol) flavors. In addition, aldehydes such as furfural and nonanal were related to almond/green notes, while butyrolactone and α-p-dimethylstyrene give caramel/sweet and citrus/pine odor, respectively (Ojeda-Amador et al. 2020).

Ojeda-Amador et al. (2018) assessed the changes of 18 different sensory descriptors in the virgin pistachio oils belonging to the "Avdat", "Kerman", "Larnaka", and "Mateur" varieties. They saw that eight of them showed significant differences among the studied oils. They were the visual green color, seed-like, nutty, and roasted nut smells, oily sensation in the mouth, sweetness flavor intensity, and persistent odor. Globally, few differences were reported for oiliness, sweetness, and seed-like parameters. The most important criteria of variation were the odor persistence and the nutty note.

Among the four oils, the one from the "Mateur" variety displayed the greenest color and "Larnaka" oil the oiliest sensation. "Mateur" and "Larnaka" oils had the highest persistence and flavor intensities, obtaining the higher scores in roasted nut and nutty notes, while these attributes were more discrete in the oils obtained from the cultivars "Kerman" and "Avdat".

Ojeda-Amador et al. (2020) characterized the volatile profile of virgin pistachio oil from the same four cultivars in comparison to commercial oil under different processing conditions (temperature, roasting, use of whole nuts, screw speed, and nozzle diameter). Twice as much terpenes were found in the oil extracted from whole cultivar (unshelled). High roasted conditions reduce their concentration by almost twofold while promoting the appearance of pyrazines as in other nuts. Samples processed at high temperatures (>130 °C) showed a unique profile mainly determined by pyrazine derivates and terpenes. In Kendirci and Onoğur (2011), 2-methyl pyrazine, 2,6-dimethyl pyrazine, 2,5-dimethyl pyrazine, 2-ethyl-5-acetyl pyrazine, pyrrol, and 2-pentyl pyrrol were found to belong to the fresh pistachio flavor, giving strong nutty aroma. The sweet roasty aroma of roasted pistachio were associated to ethanol, 2-hexenal, 2-methyl-2-hexenal, acetic acid, 2-furfural, nonenal, 2-acetyl furan, 5-methyl furfural, furfuryl alcohol, 2-nonenal, trans-2,trans-4-decadienal. Ling et al. (2016) found less volatiles in pistachio submitted to short microwave roasting 6 minutes at 130 °C than longer conventional roasting (160 °C/20 minutes), highlighting the importance of not only the temperature but also the duration of the roasting process in the pistachio oil quality.

12.15 Sesame Oil

Sesame oil is frequently used in Asian dishes to aromatize salads, soups, and as an ingredient in sweets. In Europe, it is mainly used for baking products and seasoning.

The clear and golden oil is extracted from peeled seeds, because of the presence of bitter antinutrient oxalates. Non-drying cold-pressed oil has a nutty smell. Oils from unroastead seed or refined oil were almost odorless, more liquid, clearer tending toward light yellow.

The roasting conditions are determinant in sesame aroma and visual aspect. The color of the oil becomes red-brown when heated/fried or extracted from roasted seeds (Namiki et al. 2002). Shimoda et al. (1997) compared black and light-roasted sesame oil and, as it was observed for other nuts, heat produced a concentration of the browning compounds from Maillard reaction (Namiki et al. 2002). The relative amount of simple alkylpyrazines decreased while di- and trialkylpyrazines increased with roasting. 1H-Pyrrole-2-carboxyaldehyde volatile content was reduced in deep-roasted oil. On the contrary, guaiacol and 2-furanmethanethiol, 4,5-dimethylisothiazole, 4,5-dimethylthiazole, 2-propyl-4-methylthiazole, and 2-butyl-5-methylthiazole were found in higher proportions. Hexanal, (E)-2-heptenal, and (E,E)-2,4-decadienal were found almost unaffected. Xu-Yan et al. (2012) processed three varieties of sesame seeds. Their volatile profile was close when processed in the same conditions. Main distinctions was the higher intensity of (E,E)-,4-decadienal (fatty, waxy), 2-penylpyridine (fatty, tallowy), and 2-ethyl-3,5-dimethylpyrazine (fried potatoe like) in black seeds compared to white seeds. However, 2-furfurylthiol (coffee-like) and 2-phenylethylthiol (rubbery, meat-like) were more discrete. The differences only develop with temperature and depend on the method of extraction especially for pyrines and sesamol.

"Gentle", mild roasting condition preserved the typical roasted, sweet-sulfur notes of sesame while deep-roasting intensifies burnt, rubbery, fatty odors.

In general, sesame seed oil is processed through manual squeezing, hot water flotation, mechanical pressing, and solvent extraction (Xu-Yan et al. 2012). The traditional water flotation method leads to an oil having a higher number of volatiles and higher olfaction scores producing an oil appreciated for its freshness. It was characterized by high levels of methyl-pyrazine, 2, 5-dimethyl-pyrazine, trimethyl-pyrazine, 3-ethyl-2,5-dimethyl-pyrazine, furfural, and 2-methoxy-phenol. The aromas of solvent and mechanically extracted oil were found weaker or not typical from sesame oil. 4-hydroxy-3-methylacetophenone, 2-methoxy-phenol, and (2-methyloctyl)-benzene were found in higher concentration in solvent extracted oil while methyl-pyrazine, 2,5-dimethyl-pyrazine, trimethyl-pyrazine 3-ethyl-2,5-dimethyl-pyrazine, and 2-methoxy-phenol were dominant in the mechanical pressing method. The roasting temperature had a relevant effect on volatile composition and sensory profile. Minor variation occurred in volatile composition between 160 and 180 °C but burning and bitter tastes arise in the oils prepared at roasting temperatures over 220 °C. Park et al. (2011) also collected increasing number of volatiles at 213, 230, and 247 °C. Main volatiles associated with sesame oil flavor belong to pyrazines, pyrroles, thiazoles, thiophenes, furans, alcohols, and alkanes families followed by terpenes (Ivanova-Petropulos et al. 2015). Lower diversity of terpenoids is due to thermal degradation. In absence of preconditioning or under low-mild temperatures more terpenes can be found as hypothesized by (Ivanova-Petropulos et al. 2015). Recently, Jia et al. (2019) found that microwave processing (0–8 minutes with 1-minutes intervals) enhance the odor activity value of volatiles and could be implemented as a prior step to oil extraction to obtain intense fragrances. Among the 82 compounds identified, 15 had an odor activity value >100 : 2-methyl-propanal (pungent, malt, green), 2-methyl-butanal (cocoa, almond), furaneol (caramel), 1-octen-3-one (mushroom), 4-methyl-3-penten-2-one (sweet), 1-nonanol (fat, citrus, green), 2-methyl-phenol (phenol), 2-methoxy-phenol (smoke, sweet), 2-methoxy-4-vinylphenol (clove, curry), 2,5-dimethyl-pyrazine (cocoa, roasted nut, roast beef), 2-furfurylthiol (coffee, roast), 2-thiophenemethanethiol (sulfur), methanethiol (gasoline, garlic), methional (cooked potato), and dimethyl trisulfide (fish, cabbage).

Sesame seed oil contains lignans (sesamin, seamolin, sesamol) with powerful antioxidant properties (Namiki et al. 2002). Pure unroasted sesame oil and refined oils are appreciated for their stability against on oxidation of methyl linoleate during deep-frying, preserving the organoleptic and nutritional quality of the oil. Pentane, hexane, 1-pentanol, acetic acid, and furfuryl alcohol, hexanal, heptanal, or 1-pentanol were target compounds in the accelerated storage assessment of sesame seed oil (4 weeks, 70 °C in the dark). Although the conditions are artificial, the last three volatiles appeared to be good markers of degradation.

12.16 Walnut Oil

The major market of walnut is in Germany where it is mainly used for baking, salad dressings, or sautéing but is usually preferred uncooked because of the development of a slightly bitter taste when heated.

Commercial oils show important variations in organoleptic qualities that can be explained by the cultivar, choice, and process. A degustation of 10 commercial walnuts oils demonstrated that weak aroma intensity and rancid note were responsible for low hedonic scores. The intensity of the aroma had greater importance than the color. Subtle notes of spices, pronounced aromatic herbs smells, and roasted hint were cited to describe the oil. The preferred product had in addition a green and persistent aroma and it was produced by cold-pressing. As far as texture is concerned, walnut oil is one of the less viscous vegetable oils, less than pistachio or almond (Diamante and Lan 2014). These physical qualities are correlated with its fatty acid profile rich in double bonds and concentration of specific triglycerides.

Concerning differences between cultivars, Warmund et al. (2009) defined a lexicon for walnut kernel sensory description based on 22 terms to compare wild Black (*Juglans nigra* L.) and Persian walnuts (*Juglans regia* L). Black walnuts' overall aroma obtained higher score than Persian walnuts. This was also reported by Lee et al. (2011) who were able to identify 34 volatiles in Black walnuts. This species is gaining industrial interest because of its original flavor – opening perspectives for innovative products. Indeed, they displayed more fruity, woody, sweet and musty flavors whereas Persian walnuts were characterized by their nutty, woody, and astringent qualities. Hexanal was associated with the rancid and acrid aromas of Persian walnuts and could be used as a marker to distinguish dark cultivars from the medium and light colored kernels (Lee et al. 2011).

The interaction between the environment and the cultivars gives a particular chemical composition. Persian walnut oils were found darker than Argentina cultivars. Chinese and Ukrainian walnuts contained more lipid-derived volatiles (hexanal, pentanal, 1-hexanol, and 1-pentanol from linoleic acid degradation, and 1-penten-3-ol from α-linolenic acid breakdown) (Elmore et al. 2005). Alkylbenzenes were present in higher concentration in cultivars from Chile. Martínez et al. (2011) assessed the sensory characteristics of walnut oil. Panelists were asked to evaluate odor attributes (nutty, oxidized, painty notes), basic tastes such as sweetness, bitterness, pungency, and astringency as mouthfeel and oily film texture. The cultivar "Chandler" produced an oil slightly more intense in color, having higher blue-green and lower lightness values than "Lara" cultivar. The variety "Hartley" was intermediate. The profile of the studied virgin walnut oils was characterized by aldehydes (31–37%), acids (20–38%), and alcohols (16–36%). "Chandler" oil has a higher proportion of aldehydes, 2-heptenal, octanal, nonanal, 2-methyl-2-butenal, and especially hexanal (green). For "Lara" and "Hartley" the major aldehyde was nonanal (citrus). 2-decenal and 2-undecenal was also important in "Hartley," which also had the lowest amount of alcohols. 1-dodecanol, 1-nonen-3-ol, and 1-pentanol (fruity) were identified. Terpenes and ketones were detected in the samples analyzed, as well as pyrazines and furans in commercial oils. "Chandler", "Tulare", "Criolla", and "Franquette" walnuts had a nutty aroma in coherence with its content in furans. However, "Franquette" was also used as a reference for its astringency (Martínez et al. 2011).

The oil viscosity and color are physical parameters that varied according to walnut cultivars, environment, and processing of the final product (Gharibzahedi et al. 2014). Walnut oil can be obtained by solvent extraction, supercritical fluid extraction (CO_2), and pressure systems (hydraulic and screw press). Cold pressed oil must range between straw-like yellow with light green reflects to golden yellow color and tend to be fluid in texture. It is characterized by persistent and intense aroma of dry walnut kernels, bread crumble with plant and fresh walnuts hints. Its delicate aromas have a good persistence in the mouth. Cold pressed oil is considered the most affordable and convenient methods to produce high quality oil. Martínez et al. (2008) found that the color of the oil changed as temperature increased from a whitish clear product to a yellow one. Darker oils were obtained by supercritical CO_2 extraction, probably due to higher pressure able to extract a greater proportion of carotenoids (Martínez et al. 2008). Through hydraulic pressing, oil presents less bright yellow colors than when extracted by screw pressing or supercritical CO_2.

Storage of the raw material under light and room temperature conditions, affect its sensory quality and shelf life. High oxygen level can also favor rancidity and development of bitterness. Specific-material, vacuum-packaging or oxygen absorber in combination with dark and low temperature for storage guarantee the stability of sensory attributes of walnuts (Mexis et al. 2009). Nutty and sweet tastes were reduced during long term storage (13 months at 21 °C). Roasting pre-processing contributes to oil stability and help to maintain organoleptic qualities (Vaidya and Eun 2013). Martínez et al. (2011) followed the evolution of sensory qualities of walnut oil stored in glass-bottles during two months at room temperature and under fluorescent light. They detected that oxidized, painty, oily film, and astringent flavors attributes rose significantly in correlation with peroxide value, conjugated dienes and trienes. The intensity of nutty, pungency, bitterness, and sweetness flavor remained stable. Pungency and astringency were attributed to phenolic compounds, trans-2-heptenal, and nonanal.

References

Akhtar, S., Khalid, N., Ahmed, I. et al. (2014). Physicochemical characteristics, functional properties, and nutritional benefits of peanut oil: a review. *Critical Reviews in Food Science and Nutrition* 54: 1562–1575.

Alasalvar, C., Odabasi, A.Z., Demir, N. et al. (2004). Volatiles and flavor of five Turkish hazelnut varieties as evaluated by descriptive sensory analysis, electronic nose, and dynamic headspace analysis/gas chromatography-mass spectrometry. *Journal of Food Science* 69: SNQ99–SNQ106.

Alexandratos, N. and Bruinsma, J. (2012). World agriculture towards 2030/2050: The 2012 revision; Agricultural Development Economics Division, Food and Agriculture Organization of the United Nations (FAO): Rome, Italy, June 2012; ESA Working Paper 12-03.

Andrewes, P., Busch, J.L.H.C., De Joode, T. et al. (2003). Sensory properties of virgin olive oil polyphenols: identification of deacetoxy-ligstroside aglycon as a key contributor to pungency. *Journal of Agricultural and Food Chemistry* 51: 1415–1420.

Angerosa, F. (2000). Sensory quality of olive oils. In: *Handbook of Olive Oil: Analysis and Properties* (eds. J. Harwood and R. Aparicio), 523–560. Gaithersburg, MD: Aspen Publications, Inc.

Angerosa, F. (2002). Influence of volatile compounds on virgin olive oil quality evaluated by analytical approaches and sensor panels. *European Journal of Lipid Science and Technology* 104 (9–10): 639–660.

Angerosa, F., Di Giancinto, L., Vito, R. et al. (1996). Sensory evaluation of virgin olive oils by artificial neural network processing of dynamic head-space gas chromatographic data. *Journal of the Science of Food and Agriculture* 72 (3): 323–328.

Angerosa, F., Dalessandro, N., Basti, C. et al. (1998). Biogeneration of volatile compounds in virgin olive oil: their evolution in relation to malaxation time. *Journal of Agricultural and Food Chemistry* 46 (8): 2940–2944.

Angerosa, F., Basti, C., and Vito, R. (1999). Virgin olive oil volatile compounds from lipoxygenase pathway and characterization of some Italian cultivars. *Journal of Agricultural and Food Chemistry* 47 (3): 836–839.

Angerosa, F., Mostallino, R., Basti, C. et al. (2000). Virgin olive oil odour notes: their relationship with volatile compounds from the lipoxygenase pathway and secoiridoid compounds. *Food Chemistry* 68: 283–287.

Aparicio, R. and Aparicio-Rúiz, R. (2000). Authentication of vegetable oils by chromatographic techniques. *Journal of Chromatography A* 881: 93–104.

Aparicio, R. and Morales, M.T. (1994). Optimization of a dynamic headspace technique for quantifying virgin olive oil volatiles. Relationships between sensory attributes and volatile peaks. *Food Quality and Preference* 5 (1–2): 109–114.

Aparicio, R. and Morales, M.T. (1998). Characterization of olive ripeness by green aroma compounds of virgin olive oil. *Journal of Agricultural and Food Chemistry* 46 (3): 1116–1122.

Aparicio, R., Roda, L., Albi, M.A. et al. (1999). Effect of various compounds on virgin olive oil stability measured by rancimat. *Journal of Agricultural and Food Chemistry* 47 (10): 4150–4155.

Ashton, O.B.O., Wong, M., Mcghie, T.K. et al. (2006). Pigments in avocado tissue and oil. *Journal of Agricultural and Food Chemistry* 54: 10151–10158.

Aydeniz Güneşer, B., Yılmaz, E., and Ok, S. (2017). Cold pressed versus refined winterized corn oils: quality, composition and aroma. *Grasas y Aceites* 68 (2) ISSN-L: 0017-3495.

Baccouri, O., Bendini, A., Cerretani, L. et al. (2008). Comparative study on volatile compounds from Tunisian and Sicilian monovarietal virgin olive oils. *Food Chemistry* 111: 322–328.

Barden, L. and Decker, E.A. (2016). Lipid oxidation in low-moisture food: a review. *Critical Reviews in Food Science* 63 (15): 1812–1818.

Bendini, A., Cerretani, L., Salvador, M.D. et al. (2009). Stability of the sensory quality of virgin olive oil during storage: an overview. *Italia, Journal of Food Science* 21: 389–406.

Bendini, A., Barbieri, S., Valli, E. et al. (2011). Quality evaluation of cold pressed sunflower oils by sensory and chemical analysis. *European Journal of Lipid Science and Technology* 113: 1375–1384.

Benincasa, C., De Nino, A., Lombardo, N. et al. (2003). Assay of aroma active components of virgin olive oils from southern Italian regions by SPME-GC/ion trap mass spectrometry. *Journal of Agricultural and Food Chemistry* 51 (3): 733–741.

Berasategi, I., Barriuso, B., Ansorena, D. et al. (2012). Stability of avocado oil during heating: comparative study to olive oil. *Food Chemistry* 132: 439–446.

Bernard, J.K. (2016). Oilseed and oilseed meals. In: *Reference Module in Food Science* (ed. G.W. Smithers), 349–355. Amsterdam: Elsevier.

Bocci, F. and Frega, N. (1996). Analysis of the volatile fraction from sunflower oil extracted under pressure. *Journal of the American Oil Chemists' Society* 73: 713–716.

Bolling, B.W. (2017). Almond polyphenols: methods of analysis, contribution to food quality, and health promotion. *Comprehensive Reviews in Food Science and Food Safety* 16: 346–368.

Brühl, L. and Matthäus, B. (2008). Sensory assessment of virgin rapeseed oils. *European Journal of Lipid Science and Technology* 110: 608–610.

Brühl, L., Matthäus, B., Scheipers, A. et al. (2008). Bitter off-taste in stored cold-pressed linseed oil obtained from different varieties. *European Journal of Lipid Science and Technology* 110: 625–631.

Burdack-Freitag, A. and Schieberle, P. (2010). Changes in the key odorants of Italian hazelnuts (*Coryllus avellana* L. Var. Tonda Romana) induced by roasting. *Journal of Agricultural and Food Chemistry* 58: 6351–6359.

Buttery, R.G. and Takeoka, G.R. (2004). Some unusual minor volatile components of tomato. *Journal of Agricultural and Food Chemistry* 52 (20): 6264–6266.

Byrareddy, K., Uppar, D.S., Vyakaranahal, B.S. et al. (2008). Effect of integrated nutrient management on sunflower hybrid (KBSH-1) seed production. *Karnataka Journal of Agricultural Sciences* 21 (2): 171–175.

del Castillo, M.L.R., Caballero, E.G., Blanch, G.P. et al. (2002). Enantiomeric composition of filbertone in hazelnuts and hazelnut oils from different geographical origins. *Journal of the American Oil Chemists' Society* 79: 589–592.

Cert, A., Moreda, W., and Pérez, C.M.C. (2000). Chromatographic analysis of minor constituents in vegetable oils – review. *Journal of Chromatography A* 881: 131–148.

Chang, C., Wu, G., Zhang, H. et al. (2019). Deep-fried flavor: characteristics, formation mechanisms, and influencing factors. *Critical Reviews in Food Science and Nutrition* 60: 1496–1514.

Choe, E. and Min, D.B. (2006). Chemistry and reactions of reactive oxygen species in foods. *Critical Reviews in Food Science and Nutrition* 46: 1–22.

Choe, E. and Min, D.B. (2007). Chemistry of deep-fat frying oils. *Journal of Food Science* 72: 77–86.

Choo, W.S., Birch, J., and Dufour, J.P. (2007). Physicochemical and quality characteristics of cold-pressed flaxseed oils. *Journal of Food Composition and Analysis* 20: 202–211.

Codex standards for named vegetable oils, (2005). CODEX-STAN 210 (amended 2003, 2005).

Costagli, G. and Betti, M. (2015). Avocado oil extraction processes: method for cold-pressed high-quality edible oil production versus traditional production. *Journal of Agricultural Engineering* 46: 115–122.

Davis, J.P., Dean, L.O., Faircloth, W.H. et al. (2008). Physical and chemical characterizations of normal and high-oleic oils from nine commercial cultivars of peanut. *Journal of the American Oil Chemists' Society* 85: 235–243.

Davis, J.P., Price, K.M., Dean, L.L. et al. (2016). Peanut oil stability and physical properties across a range of industrially relevant oleic acid/linoleic acid ratios. *Peanut Science* 43: 1–11.

Dayrit, F.M., Dimzon, I.K.D., Valde, M.F. et al. (2011). Quality characteristics of virgin coconut oil: comparisons with refined coconut oil. *Pure and Applies Chemistry* 83: 1789–1799.

De Leonardis, A., Macciola, V., and De Felice, M. (2001). Chemical and commercial characteristics of pressed sunflower seed oils. *Italian Food and Beverage Technology* 25: 46–54.

Di Giovacchino, L., Sestili, S., and Di Vincenzo, D. (2002). Influence of olive processing on virgin olive oil quality. *European Journal of Lipid Science and Technology* 104: 587–601.

Diamante, L.M. and Lan, T. (2014). Absolute viscosities of vegetable oils at different temperatures and shear rate range of 64.5 to 4835 s^{-1}. *Journal of Food Processing* 2014: 1–6.

Donno, D., Beccaro, G.L., Mellano, G.M. et al. (2013). Setting a protocol for hazelnut roasting using sensory and colorimetric analysis: influence of the roasting temperature on the quality of Tonda gentile delle Langhe Cv. Hazlenut. *Czech Journal of Food Sciences* 31: 390–401.

El-Mageed, M.A.A. (2007). Development of volatile compounds of avocado and casimiroa during fruit maturation. *Arab Universities Journal of Agricultural Sciences* 15: 89–100.

Elmore, J.S., Nisyrios, I., and Mottram, D.S. (2005). Analysis of the headspace aroma compounds of walnuts (*Juglans regia* L.). *Flavour and Fragrance Journal* 20: 501–506.

Faiza, S. and Debnath, S. (2016). Molecular finger printing of nutra-coconut oil with improved health protective phytoceuticals and its efficacy as frying medium. *Journal of the American Oil Chemists' Society* 93: 1301–1310.

Frandsen, H.B., Wæhrens, S.S., Petersen, M.A. et al. (2017). Supercritical fluid extracted rapeseed oil, its chemical and sensory profile. Conference abstract in the 14th Euro Fed Lipid Congress.

Franklin, L.M. and Mitchell, A.E. (2019). Review of the sensory and chemical characteristics of almond (*Prunus dulcis*) flavor. *Journal of Agricultural and Food Chemistry* 67: 2743–2753.

Franklin, L.M., King, E.S., Chapman, D. et al. (2018). Flavor and acceptance of roasted California almonds during accelerated storage. *Journal of Agricultural and Food Chemistry* 66: 1222–1232.

Gharibzahedi, S.M.T., Mousavi, S.M., Hamedi, M. et al. (2014). Determination and characterization of kernel biochemical composition and functional compounds of Persian walnut oil. *Journal of Food Science and Technology* 51: 34–42.

Giuffrè, A.M., Capocasale, M., Macrì, R. et al. (2020). Volatile profiles of extra virgin olive oil, olive pomace oil, soybean oil and palm oil in different heating conditions. *LWT* 117: 108631.

Goicoechea, E. and Guillén, M.D. (2014). Volatile compounds generated in corn oil stored at room temperature. Presence of toxic compounds. *European Journal of Lipid Science and Technology* 116 (4): 395–406.

Grob, K., Biedermann, M., Bronz, M. et al. (1994). Recognition of mild deodorization of edible oils by the loss of volatile components. *European Food Research and Technology* 199: 191–194.

Guichard, E. (2002). Interactions between flavor compounds and food ingredients and their influence on flavor perception. *Food Reviews International* 18: 49–70.

Gunstone, F. (2011). *Vegetable Oils in Food Technology: Composition, Properties and Uses*. Chichester: Wiley.

Gutierrez-Rosales, F., Rios, J.J., and Gomez-Rey, M.L. (2003). Main polyphenols in the bitter taste of virgin olive oil: structural confirmation by on-line high-performance liquid chromatography electrospray ionization mass spectrometry. *Journal of Agricultural and Food Chemistry* 51: 6021–6025.

Guzmán-Gerónimo, R.I., López, M.G., and Dorantes-Alvarez, L. (2008). Microwave processing of avocado: volatile flavor profiling and olfactometry. *Innovative Food Science & Emerging Technologies* 9: 501–506.

Ha, T.J., Nihei, K., and Kubo, I. (2004). Lipoxygenase inhibitory activity of Octyl Gallate. *Journal of Agricultural and Food Chemistry* 52 (10): 3177–3181.

Haiyan, Z., Bedgood, D.R., Bishop, A.G. et al. (2007). Endogenous biophenol, fatty acid and volatile profiles of selected oils. *Food Chemistry* 100: 1544–1551.

Ho, C.T. and Shahidi, F. (2005). Flavor Components of Fats and Oils. In: *Bailey's Industrial Oil and Fat Products* (ed. F. Shahidi). Hoboken, NJ: Wiley https://doi.org/10.1002/047167849X.bio009.

Hornero-Mendez, D., Gandul-Rojas, B., and Minguez-Mosquera, M. (2005). Routine and sensitive SPE-HPLC method for quantitative determination of pheophytin a and pyropheophytin a in olive oil. *Food Research International* 38 (8–9): 1067–1072.

Hou, J.C., Jiang, L.Z., and Zhang, C.W. (2012). Effects of frying on the trans-fatty acid formation in soybean oils. *European Journal of Lipid Science and Technology* 114 (3): 287–293.

http://www.thegoodscentscompany.com

https://apps.fas.usda.gov/psdonline/circulars/oilseeds.pdf.

Hu, W., Zhang, L., Li, P. et al. (2014). Characterization of volatile components in four vegetable oils by headspace two-dimensional comprehensive chromatography time-of-flight mass spectrometry. *Talanta* 129: 629–635.

Huang, C.F., Lin, Y.S., Chiang, Z.C. et al. (2014). Oxidized frying oil and its polar fraction fed to pregnant mice are teratogenic and alter mRNA expressions of vitamin a metabolism genes in the liver of dams and their fetuses. *Journal of Nutritional Biochemistry* 25 (5): 549–556.

Isleib, T.G., Pattee, H.E., Sanders, T.H. et al. (2006). Compositional and sensory comparisons between normal- and high-oleic peanuts. *Journal of Agricultural and Food Chemistry* 54: 1759–1763.

Ivanova-Petropulos, V., Mitrev, S., Stafilov, T. et al. (2015). Characterisation of traditional Macedonian edible oils by their fatty acid composition and their volatile compounds. *Food Research International* 77: 506–514.

Jia, X., Zhou, Q., Wang, J. et al. (2019). Identification of key aroma-active compounds in sesame oil from microwaved seeds using E-nose and HS-SPME-GC×GC-TOF/MS. *Journal of Food Biochemistry* 43: e12786.

Juita, J., Dlugogorski, B.Z., Kennedy, E.M. et al. (2012). Low temperature oxidation of linseed oil: a review. *Fire Science Reviews* 1: 3.

Jung, M.Y. and Min, D.B. (1990). Effects of α-, γ-, and δ-tocopherols on oxidative stability of soybean oil. *Journal of Food Science* 55 (5): 1464–1465.

Kalua, C.M., Allen, M.S., Bedgood, D.R. et al. (2007). Olive oil volatile compounds, flavour development and quality: a critical review. *Food Chemistry* 100: 273–286.

Kamisah, Y., Adam, A., Wan Ngah, W.Z. et al. (2005). Chronic intake of red palm olein and palm olein produce beneficial effects on plasma lipid profile in rats. *Pakistan Journal of Nutrition* 4: 89–96.

Kendirci, P. and Onoğur, T.A. (2011). Investigation of volatile compounds and characterization of flavor profiles of fresh pistachio nuts (*Pistacia vera* L.). *International Journal of Food Properties* 14: 319–330.

Kiefl, J. and Schieberle, P. (2013). Evaluation of process parameters governing the aroma generation in three hazelnut cultivars (*Corylus avellana* L.) by correlating quantitative key odorant profiling with sensory evaluation. *Journal of Agricultural and Food Chemistry* 61: 5236–5244.

Kinderlerer, J.L. (1994). Degradation of the Lauric acid oils. *International Biodeterioration & Biodegradation* 33 (4): 345–354.

King, E.S., Chapman, D.M., Luo, K. et al. (2019). Defining the sensory profiles of raw almond (*Prunus dulcis*) varieties and the contribution of key chemical compounds and physical properties. *Journal of Agricultural and Food Chemistry* 67: 3229–3241.

Kiritsakis, A.K. (1998). Flavor components of olive oil – a review. *Journal of the American Oil Chemists' Society* 75 (6): 673–681.

Koprivnjak, O., Procida, G., and Zelinotti, T. (2000). Changes in the volatile components of virgin olive oil during fruit storage in aqueous media. *Food Chemistry* 70 (3): 377–384.

Kraljić, K., Stjepanović, T., Obranović, M. et al. (2018). Influence of conditioning temperature on the quality, nutritional properties and volatile profile of virgin rapeseed oil. *Food Technology and Biotechnology* 56: 562–572.

Krist, S., Stuebiger, G., Bail, S. et al. (2006). Detection of adulteration of poppy seed oil with sunflower oil based on volatiles and triacylglycerol composition. *Journal of Agricultural and Food Chemistry* 54: 6385–6389.

Krumreich, F.D., Borges, C.D., Mendonça, C.R.B. et al. (2018). Bioactive compounds and quality parameters of avocado oil obtained by different processes. *Food Chemistry* 257: 376–381.

Lee, J., Vázquez-Araújo, L., Adhikari, K. et al. (2011). Volatile compounds in light, medium, and dark black walnut and their influence on the sensory aromatic profile. *Journal of Food Science* 76: C199–C204.

Liao, C.H., Shaw, H.M., and Chao, P.M. (2008). Impairment of glucose metabolism in mice induced by dietary oxidized frying oil is different from that induced by conjugated linoleic acid. *Nutrition* 24 (7–8): 744–752.

Ling, B., Yang, X., Li, R. et al. (2016). Physicochemical properties, volatile compounds, and oxidative stability of cold pressed kernel oils from raw and roasted pistachio (*Pistacia vera* L. Var Kerman). *European Journal of Lipid Science and Technology* 118: 1368–1379.

Liu, X., Jin, Q., Liu, Y. et al. (2011). Changes in volatile compounds of peanut oil during the roasting process for production of aromatic roasted peanut oil. *Journal of Food Science* 76 (3): C404–C412.

Liu, Y., Li, J., Cheng, Y. et al. (2020). Volatile components of deep-fried soybean oil as indicator indices of lipid oxidation and quality degradation. *European Food Research and Technology* 246: 1183–1192.

Luna, G., Morales, M.T., and Aparicio, R. (2006). Characterisation of 39 varietal virgin olive oils by their volatile compositions. *Food Chemistry* 98: 243–252.

Majchrzak, T., Lubinska, M., Różańska, A. et al. (2017). Thermal degradation assessment of canola and olive oil using ultra-fast gas chromatography coupled with chemometrics. *Monatshefte fur ciehem* 148): 1625–1630.

Mansouri, F., Moumen, A.B., Richard, G. et al. (2017). Flavor profiles of monovarietal virgin olive oils produced in the oriental region of Morocco. *Oilseeds Fats Crops Lipids* 24 (5): A501.

Mariotti, M. (2014). Virgin olive oil; definition and standards. In: *The Extra Virgin Olive Oil Handbook* (ed. C. Peri), 11–19. Oxford, UK: Wiley.

Martínez, M.L., Mattea, M.A., and Maestri, D.M. (2008). Pressing and supercritical carbon dioxide extraction of walnut oil. *Journal of Food Engineering* 88: 399–404.

Martínez, M., Barrionuevo, G., Nepote, V. et al. (2011). Sensory characterisation and oxidative stability of walnut oil. *International Journal of Food Science and Technology* 46: 1276–1281.

Mateos, R., Cert, A., Pérez-Camino, M.C. et al. (2004). Evaluation of virgin olive oil bitterness by quantification of secoiridoid derivatives. *Journal of the American Oil Chemists' Society* 81: 71–75.

Matheis, K. and Granvogl, M. (2016). Characterization of key odorants causing a fusty/musty off-flavor in native cold-pressed rapeseed oil by means of the Sensomics approach. *Journal of Agricultural and Food Chemistry* 64: 8168–8178.

http://Statista.com, (accessed on May 9th, 2020). http://www.statista.com/statistics/263978/global-vegetable-oil-production-since-2000-2001 and http://www.statista.com/statistics/263937/vegetable-oils-global-consumption.

Mexis, S.F., Badeka, A.V., Riganakos, K.A. et al. (2009). Effect of packaging and storage conditions on quality of shelled walnuts. *Food Control* 20: 743–751.

Morales, M. and Tsimidou, M. (2000). *Handbook of Olive Oil* (eds. J. Harwood and R. Aparicio), 393–458. New York:: Springer.

Moreno, A.O., Dorantes, L., Galíndez, J. et al. (2003). Effect of different extraction methods on fatty acids, volatile compounds, and physical and chemical properties of avocado (*Persea americana* Mill.) oil. *Journal of Agricultural and Food Chemistry* 51: 2216–2221.

Namiki, M., Fukuda, Y., Takei, Y. et al. (2002). Changes in functional factors of sesame seed and oil during various types of processing. In: *Bioactive Compounds in Foods*, ACS Symposium Series (eds. T.-C. Lee and C.-T. Ho), 85–104. Washington, DC: American Chemical Society.

Naz, S., Sheikh, S.H., and Sayeed, S.A. (2005). Deterioration of olive, corn and soybean oils due to air, light, heat and deep-frying. *Food Research Internatioal* 38 (2): 127–134.

Neta, D.C. and Ramos, E.M.B. (2010). Application of sensory and instrumental analyses for the characterization of flavor in roasted peanuts and farmstead cheddar cheese. PhD thesis, North Carolina State University.

Nevin, K.G. and Rajamohan, T. (2004). Beneficial effects of virgin coconut oil on lipid parameters and in vitro LDL oxidation. *Clinical Biochemistry* 37: 830–835.

Noguera-Artiaga, L., Salvador, M.D., Fregapane, G. et al. (2019). Functional and sensory properties of pistachio nuts as affected by cultivar. *Journal of the Science of Food and Agriculture* 99: 6696–6705.

Nykter, M., Kymäläinen, H.R., and Gates, F. (2008). Quality characteristics of edible linseed oil. *Agricultural and Food Science* 15: 402.

Obenland, D., Collin, S., Sievert, J. et al. (2012). Influence of maturity and ripening on aroma volatiles and flavor in 'Hass' avocado. *Postharvest Biology and Technology* 71: 41–50.

Ojeda-Amador, R.M., Fregapane, G., and Salvador, M.D. (2018). Composition and properties of virgin pistachio oils and their by-products from different cultivars. *Food Chemistry* 240: 123–130.

Ojeda-Amador, R.M., Fregapane, G., and Salvador, M.D. (2019). Chemical characterization of virgin almond and hazelnut oils and their by-products. *European Journal of Lipid Science and Technology* 121: 1900114.

Ojeda-Amador, R.M., Fregapane, G., and Salvador, M.D. (2020). Influence of cultivar and technological conditions on the volatile profile of virgin pistachio oils. *Food Chemistry* 311: 125957.

O'Keefe, S.F., Wiley, V.A., and Knauft, D.A. (1993). Comparison of oxidative stability of high- and normal-oleic peanut oils. *Journal of the American Oil Chemists' Society* 70: 489–492.

Olías-Jiménez, J.M., Gutiérrez-Rosales, F., and Dobarganes, M.C. (1980). Volatile components in the aroma of virgin olive oil. IV. Their evaluation and influence on the aroma during the fruit ripening process in the "Picual" and "Hojiblanca" varieties. *Grasas y Aceites* 31: 391–402.

Ortner, E., Granvogl, M., and Schieberle, P. (2016). Elucidation of thermally induced changes in key odorants of white mustard seeds (*Sinapis alba* L.) and rapeseeds (*Brassica napus* L.) using molecular sensory science. *Journal of Agricultural and Food Chemistry* 64: 8179–8190.

Park, M.H., Jeong, M.K., Yeo, J. et al. (2011). Application of solid phase-microextraction (SPME) and electronic nose techniques to differentiate volatiles of sesame oils prepared with diverse roasting conditions. *Journal of Food Science* 76: C80–C88.

Peng, C.Y., Lan, C.H., Lin, P.C. et al. (2017). Effect of cooking method, cooking oil, and food type on aldehyde emission in cooking oil fumes. *Journal of Hazardous Materials* 324: 160–167.

Pollner, G. and Schieberle, P. (2016). Characterization of the key odorants in commercial cold-pressed oils from unpeeled and peeled rapeseeds by the Sensomics approach. *Journal of Agricultural and Food Chemistry* 64: 627–636.

Prenzler, P.D., Bedgood, D.R., Bishop, A.G. et al. (2002). Volatile profile of olive oils. *Advances in Horticultural Science* 16 (3–4): 246–252.

Raghavendra, S.N. and Raghavarao, K.S.M.S. (2010). Effect of different treatments for the destabilization of coconut milk emulsion. *Journal of Food Engineering* 97: 341–347.

Ranalli, A., Tombesi, A., Ferrante, M.L. et al. (1998). Respiratory rate of olive drupes during their ripening cycle and quality of oil extracted. *Journal of the Science of Food and Agriculture* 77 (3): 359–367.

Ranalli, A., Contento, S., Schiavone, C. et al. (2001). Malaxing temperature affects volatile and phenol composition as well as other analytical features of virgin olive oil. *European Journal of Lipid Science and Technology* 103: 228–238.

Ranalli, A., Pollastri, L., Contento, S. et al. (2003). Effect of olive paste kneading process time on the overall quality of virgin olive oil. *European Journal of Lipid Science and Technology* 105: 57–67.

Rangrej, V., Shah, V., Patel, J. et al. (2015). Effect of shortening replacement with flaxseed oil on physical, sensory, fatty acid and storage characteristics of cookies. *Journal of Food Science and Technology* 52: 3694–3700.

Rani, A.K.S., Reddy, S.Y., and Chetana, R. (2010). Quality changes in trans and trans free fats/oils and products during frying. *European Food Research and Technology* 230: 803–811.

Ridolfi, M., Terenziani, S., Patumi, M. et al. (2002). Characterization of the lipoxygenases in some olive cultivars and determination of their role in volatile compounds formation. *Journal of Agricultural and Food Chemistry* 50 (4): 835–839.

van Ruth, S.M., Roozen, J.P., and Jansen, F.J. (2000). Aroma profiles of vegetable oils varying in fatty acid composition vs. concentrations of primary and secondary lipid oxidation products. *Nahrung* 44: 318–322.

Sacchi, R., Mannina, L., Fiordiponti, P. et al. (1998). Characterization of Italian extra virgin olive oils using H-l-NMR spectroscopy. *Journal of Agricultural and Food Chemistry* 46 (10): 3947–3951.

Sano, T., Okabe, R., Iwahashi, M. et al. (2017). Effect of furan fatty acids and 3-methyl-2,4-nonanedione on light-induced off-odor in soybean oil. *Journal of Agricultural and Food Chemistry* 65 (10): 2136–2140.

Santos, J.E.R., Villarino, B.J., Zosa, A.R. et al. (2011). Analysis of volatile organic compounds in virgin coconut oil and their sensory attributes. *The Philippine Journal of Science* 140 (2): 161–171.

Schirack, A.V., Drake, M.A., Sanders, T.H. et al. (2006). Characterization of aroma-active compounds in microwave blanched peanuts. *Journal of Food Science* 71: C513–C520.

Sena-Moreno, E., Pardo, J.E., Catalán, L. et al. (2015). Drying temperature and extraction method influence physicochemical and sensory characteristics of pistachio oils. *European Journal of Lipid Science and Technology* 117: 684–691.

Servili, M., Selvaggini, R., Esposto, S. et al. (2004). Health and sensory properties of virgin olive oil hydrophilic phenols: agronomic and technological aspects of production that affect their occurrence in the oil. *Journal of Chromatography A* 1054: 113–127.

Shi, A., Wang, Q., Liu, H. et al. (2016). Peanut processing quality evaluation technology. In: *Peanuts: Processing Technology and Product Development* (ed. Q. Wang), 23–61. Amsterdam: Academic Press.

Shimoda, M., Nakada, Y., Nakashima, M. et al. (1997). Quantitative comparison of volatile flavor compounds in deep-roasted and light-roasted sesame seed oil. *Journal of Agricultural and Food Chemistry* 45: 3193–3196.

Smith, A.L., Perry, J.J., Marshall, J.A. et al. (2014). Oven, microwave, and combination roasting of peanuts: comparison of inactivation of salmonella surrogate *Enterococcus faecium*, color, volatiles, flavor, and lipid oxidation. *Journal of Food Science* 79: S1584–S1594.

Smouse, T.H. (1985). Flavor reversion of soybean oil. In: *Flavor Chemistry of Fats and Oils* (eds. D.B. Min and T.H. Smouse), 85–116. Urbana, IL: American Oil Chemists' Society.

Syed, A. (2015). Specialty/GM vegetable oils: properties and applications in food (rapeseed, sunflower, soybean). In: *Specialty Oils and Fats in Food and Nutrition Woodhead Publishing Series in Food Science, Technology and Nutrition* (ed. G. Talbot), 173–205. Amsterdam: Woodhead Publishing.

Tan, C.X., Chong, G.H., Hamzah, H. et al. (2018). Comparison of subcritical CO_2 and ultrasound-assisted aqueous methods with the conventional solvent method in the extraction of avocado oil. *Journal of Supercritical Fluids* 135: 45–51.

Tarmizi, A.H.A., Ismail, R., and Kuntom, A. (2016). Effect of frying on the palm oil quality attributes – a review. *Journal of Oil Palm Research* 28 (2): 143–153.

Temime, S.B., Campeol, E., Cioni, P.L. et al. (2006). Volatile compounds from Chétoui olive oil and variations induced by growing area. *Food Chemistry* 99: 315–325.

Temime, S.B., Manai, H., Methenni, K. et al. (2008). Sterolic composition of Chétoui virgin olive oil: influence of geographical origin. *Food Chemistry* 110 (2): 368–374.
Tuan HQ. 2011. Characterization of natural edible oils regarding their quality and safety related constituents. Master thesis, Division of Food Chemistry, Department of Food Sciences and Technology, University of Natural Resources and Life Sciences, Vienna, Austria.
Vaidya, B. and Eun, J.B. (2013). Effect of roasting on oxidative and tocopherol stability of walnut oil during storage in the dark. *European Journal of Lipid Science and Technology* 115: 348–355.
Vázquez-Araújo, L., Enguix, L., Verdú, A. et al. (2008). Investigation of aromatic compounds in toasted almonds used for the manufacture of turrón. *European Food Research and Technology* 227: 243–254.
Veena, N., Surendra Nath, B., Srinivas, B. et al. (2017). Quality attributes of dahi prepared from milk fortified with omega-3 fatty acids, phytosterols and polydetxrose. *Journal of Food Science and Technology* 54: 1765–1775.
Vichi, S., Romero, A., Gallardo, J. et al. (2009). Volatile phenols in virgin olive oils: influence of olive variety on their formation during fruits storage. *Food Chemistry* 116: 651–656.
Vichi, S., Cortés, N., Romero, A. et al. (2014). Determination of volatile thiols in virgin olive oil by derivatisation and LC–HRMS, and relation with sensory attributes. *Food Chemistry* 149: 313–318.
Villarino, B.J., Dy, L.M., and Lizada, M.C.C. (2007). Descriptive sensory evaluation of virgin coconut oil and refined, bleached and deodorized coconut oil. *LWT – Food Science and Technology* 40: 193–199.
Warmund, M.R., Elmore, J., Drake, M. et al. (2009). Descriptive analysis of kernels of selected black and Persian walnut cultivars. *Jornal of the Science of Food and Agriculture* 89: 117–121.
Wei, C.Q., Liu, W.Y., Xi, W.P. et al. (2015). Comparison of volatile compounds of hot-pressed, cold-pressed and solvent-extracted flaxseed oils analyzed by SPME-GC/MS combined with electronic nose: major volatiles can be used as markers to distinguish differently processed oils. *European Journal of Lipid Science and Technology* 117: 320–330.
Whitfield, F.B. (1992). Volatiles from interactions of Maillard reactions and lipids. *Critical Reviews in Food Science and Nutrition* 31: 1–58.
Xu-Yan, D., Ping-Ping, L., Fang, W. et al. (2012). The impact of processing on the profile of volatile compounds in sesame oil. *European Journal of Lipid Science and Technology* 114: 277–286.
Yalcin, H., Ozturk, I., Hayta, M. et al. (2011). Effect of gamma-irradiation on some chemical characteristics and volatile content of linseed. *Journal of Medicinal Food* 14: 1223–1228.
Yeboah, S.O., Mitei, Y.C., Ngila, J.C. et al. (2012). Compositional and structural studies of the oils from two edible seeds: *Tiger nut, Cyperus esculentum*, and asiato, *Pachira insignis*, from Ghana. *Food Research International* 47: 259–266.
Zhang, Z.-S., Wang, L.-J., Li, D. et al. (2011). Characteristics of flaxseed oil from two different flax plants. *International Journal of Food Properties* 14: 1286–1296.
Zhang, Q., Qin, W., Lin, D. et al. (2015). The changes in the volatile aldehydes formed during the deep-fat frying process. *Journal of Food Science and Technology* 52 (12): 7683–7696.
Zhao, F., Liu, J., Wang, X. et al. (2013). Detection of adulteration of sesame and peanut oils via volatiles by GC×GC-TOF/MS coupled with principal components analysis and cluster analysis. *European Journal of Lipid Science and Technology* 115: 337–347.
Zhou, Q., Tang, H., Jia, X. et al. (2018). Distribution of glucosinolate and pungent odors in rapeseed oils from raw and microwaved seeds. *International Journal of Food Properties* 21: 2296–2308.

Index

Oil and Oilseed Processing: Opportunities and Challenges, First Edition.
Edited by Tomás Lafarga, Gloria Bobo, and Ingrid Aguiló-Aguayo.

www.ingramcontent.com/pod-product-compliance
Lightning Source LLC
Chambersburg PA
CBHW081415220926
51613CB00010B/225
* 9 7 8 1 1 1 9 5 7 5 2 7 6 *